15th International Conference on ICT, Society, and Human Beings (ICT2022); 19th International Conference on Web Based Communities and Social Media (WBC2022); and 14th International Conference on e-Health (EH2022)

Held at the 16th Multi-Conference on Computer Science and Information Systems (MCCSIS 2022)

Lisbon, Portugal
19-22 July 2022

Editors:

Piet Kommers
Mário Macedo
Pedro Isaías
Luís Rodrigues

ISBN: 978-1-7138-6089-1

Printed from e-media with permission by:

Curran Associates, Inc.
57 Morehouse Lane
Red Hook, NY 12571

Some format issues inherent in the e-media version may also appear in this print version.

International Association for Development of the Information Society (IADIS)
Rua Sao Sebastiao Da Pedreira
No. 100, 30, 1050-209
Lisbon, Portugal

Phone: 351 21 3151373
Fax: 351 21 3151244

secretariat@iadis.org

Additional copies of this publication are available from:

Curran Associates, Inc.
57 Morehouse Lane
Red Hook, NY 12571 USA
Phone: 845-758-0400
Fax: 845-758-2633
Email: curran@proceedings.com
Web: www.proceedings.com

INTERNATIONAL CONFERENCES ON

ICT, SOCIETY AND HUMAN BEINGS 2022

WEB BASED COMMUNITIES AND SOCIAL MEDIA 2022

and

E-HEALTH 2022

part of the

MULTI CONFERENCE ON COMPUTER SCIENCE AND INFORMATION SYSTEMS 2022

PROCEEDINGS OF THE INTERNATIONAL CONFERENCES ON

ICT, SOCIETY AND HUMAN BEINGS 2022

WEB BASED COMMUNITIES AND SOCIAL MEDIA 2022

and

E-HEALTH 2022

JULY 19 - 21, 2022

Organised by

iadis

international association for development of the information society

As a member of Crossref (a non-profit membership organization for scholarly publishing working with the purpose to make content easy to find, link, cite and assess) each published paper in this book of proceedings will be allocated a DOI (Digital Object Identifier) number for its fast and easy citation and indexation.

Volume Editors:
Piet Kommers and Mário Macedo

Computer Science and Information Systems Series Editors:
Piet Kommers and Pedro Isaías

Associate Editor: Luís Rodrigues

ISBN: 978-989-8704-40-5

TABLE OF CONTENTS

SHORT PAPERS

ICT, SOCIETY, AND HUMAN BEINGS

WEB BASED COMMUNITIES AND SOCIAL MEDIA

FOREWORD

These proceedings contain the papers of the 15th International Conference on ICT, Society and Human Beings (ICT 2022), the 19th International Conference Web Based Communities and Social Media (WBCSM 2022) and of the 14th International Conference on e-Health (EH 2022), which were organised by the International Association for Development of the Information Society, from 19 - 21 July, 2022. These conferences are part of the Multi Conference on Computer Science and Information Systems 2022, 19-22 July, which had a total of 608 submissions.

The Network period in the evolution of computer technology is very much based on the convergence and integration of three main technologies; computer technology, tele technology and media technology. Telecommunication technology is playing a more and more dominant role in this convergence, especially internet and web technology. Embedded (ubiquitous) computer technology is making the process invisible, and media technologies converge within itself (multimedia and cross media). The convergence process is enforced all the time by smaller, cheaper, and more powerful components.

ICT and its applications are interacting with environments, roles, and processes which can also be modelled by converging circles. The process of social and psychosocial change and ICT from a global perspective is described graphically in the convergence model in figure 1 (Bradley 2006 Routledge) with concepts and their interrelations. Both "convergence" and "interactions" are important features in the model. Read from the left hand side in the model for the titles of some main tracks of the conference:

- Globalisation and ICT: When technology, economy, norms/values and labour market are converging on a global level, what are the hard questions? When the geographical space in the future will be both global and beyond – including virtual reality (VR) what is the state of art in research? (see the list of key words under ′globalisation′)
- Information and Communication Technology (ICT), next cluster of circles to the left in the figure, what applications contribute to desirable goals in the society?
- When Work Environment, Home Environment, and Public Environment are converging and the work and public issues tend to merge into the private sphere of our homes – what main changes in peoples Life Environment occur?.
- If the Professional Role (Work Life), Private Role (Private Life) and Citizen's Role (Public Life) converge forming a Life Role, what are the main social-psychologial changes?
- Four circles representing Virtual Reality (VR) are marked with dotted lines and are surrounding the set of converging circles. These circles reflect our participation in cyberspace on various levels. To the left part in figure we could talk about Virtual Worlds on the global level. Within the concept of ICT, the steps taken by applied Embedded and ubiquitous technology make technology more hidden to the individual and society as a whole.

  Virtual Environments are already a common concept. Finally, we could talk about Virtual Human Roles, which could in more extreme forms be another personality that you play e g avatars. The converging circles are forming a Life Role and new life styles are being shaped.
- Effects on humans become more multi faceted and complex. Research focusing upon the individual is crucial i. e. research on how the use of ICT interacts with and impacts identity, social competence, creativity, integrity, trust, dependency etc.

A compass rose (card) for "Effects on Humans" (to the right) is used as a metaphor reminding us of the importance to keep the direction towards desirable human and societal goals and qualities at the development and use of ICT.

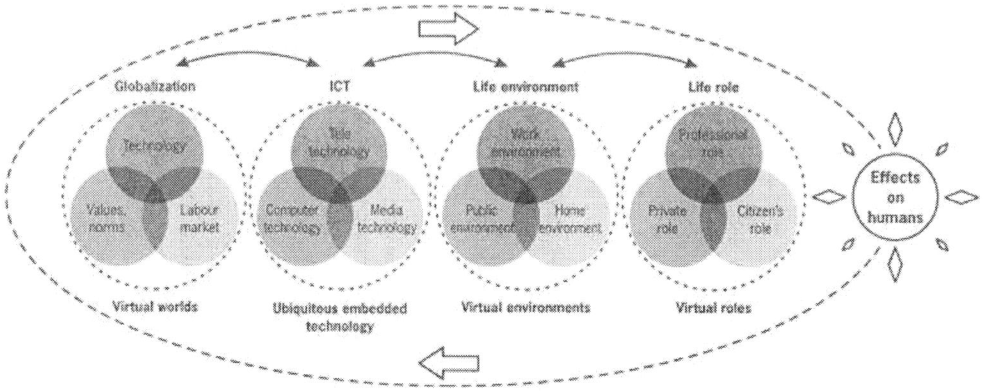

Figure 1. Convergence Model on ICT and Psychosocial Life Environment (Source: Bradley, 2005, 2006)

ICT can provide tools for promoting sustainability (environmental, economic, and social sustainability) but can also be a threat for sustainability. Sustainability as a guiding principle involves system perspective, holism, human aspects, bottom up approach, common good, and equality. A change in focus regarding research and development is taking place. Analysis and design increasingly address both the work process and management connected to the sphere of production life and people's life environment. Analysis and design of ICT and societal systems both at local level and globally become important. What research in the field exists or is needed?

Community research in a broad sense comes to the fore – both physical and virtual communities. There is also a requirement to involve new and additional actors at the deeper and broader integration of ICT in the society (children, elderly, and consumer organisations). Educational programs on Community Informatics and Social Informatics are appearing in many academic institutions. Can a new infrastructure of the society be identified?

The effects of ICT on human beings as well as the interaction between ICT, individuals, and society are all within the focus of this conference. Both analyses of interactions and effects are important. Changes in behaviour, perspectives, values, competencies, human and psychological aspects and feelings are all of interest. Reflections on past, present, and future challenges – especially planning for handling the latter – are encouraged.

Today, computer science and ICT-related disciplines are working more and more together with various behavioural and social sciences including child psychology and developmental psychology. For this reason, the conference pays attention to societal changes, global and more local organisational and institutional changes, changes in values and in lifestyles, as well as individual cognitive effects and changes, motivational and emotional changes. It also appeals to solution-building in terms of desirable goals and actions for reaching a Good Information Society.

In general, all types of research strategies are encouraged, and especially cross-disciplinary and multi-disciplinary studies. Case studies, broader empirical field studies, theoretical analyses, cross-cultural studies, scenarios, ethnographic studies, epistemological analyses may all be presented.

The ICT, Society and Human Beings 2022 conference addresses in detail seven main aspects:
- Globalization and ICT
- Life environment and ICT
- Life role and ICT
- ICT and effects on humans
- Perspectives on ICT
- Desirable goals and ICT
- Actions for reaching the Good Information Society

Significant societal challenges in the form of demographics, urbanisation, climate change, resource crises and global competition are driving profound changes within our cities. In order to cope, governments and businesses are turning to 'smart city' concepts with the aim to enhance the efficiency of key infrastructure, utilities and services to create a sustainable urban environment that improves the quality of life for its citizens and enhances economic development. Smart cities are essentially built by utilising a set of latest information and communication technologies (ICT), including Wi-Fi and mobile networks, wireless sensors, the Internet of things, big data analytic tools, cloud services, mobile devices, and mobile apps. In this context, ICT as an enabling Smart City technology will generate radically new "smart" services and facilities. The UK's Department of Business, Innovation and Skills values the smart city industry at more than $400 billion globally by 2020.

The World Wide Web has migrated from information space into opportunities for social communication. Social Media are growing rapidly and play an increasingly important role in the development of Online Communities. They are all about identity, reputation, presence and relationships. Web based communities announce themselves both in your professional and private life through several new media such as LinkedIn, Twitter, Plaxo, etc. In order to keep you up to date with the pace of these new technological developments this Conference offers a dedicated overview and informative discussion on today's most relevant issues in new media for social life on the web.

Social Media are growing rapidly and play an increasingly important role in the development of Online Communities. Social Network Sites and Web-based communities announce themselves both in your professional and private life through new media such as Facebook, LinkedIn, Twitter, Plaxo, etc. Social media allow more dynamic roles in participation, virtual presence and online communities. These new ways to communicate via online social media have great societal effects and are motivating the creation of best practices to help individuals, corporations and authorities to make the best of it. It raises the awareness of the growing impact of social media and the influence of web based communities in today's users / consumers behavior; many organizations spend an increasing share of their budget in online social marketing strategies.

The mission of the Web Based Communities and Social Media 2022 conference is to publish and integrate scientific results and act catalytically to the fast developing culture of web communities, while helping to disseminate and understand the latest developments social media and their impact.

Submissions were accepted under the following main topics:
- The History, Architecture and Future of Virtual Communities
- Cyborgs, Teleworking, Telemedicine, Art Games and Learning Communities
- Virtual Communities for People with Special Needs
- Group Processes and Self-Organization
- Expanding Markets through Virtual Communities
- Collaborative Technologies
- Social Media

The use of ICTs (Information and Communication Technologies) in Healthcare Services is the main mechanism to improve efficiency and effectiveness. Nowadays ICTs are being developed to achieve the following objectives:
– To integrate heterogeneous systems;
– To develop frameworks to make all data meaningful, accessible and available everywhere and permanently;
– To develop AIDC (Automatic Identification and Data Collectors) systems;
– To develop intelligent systems to support clinical and management decisions;
The use of these technologies also improves the quality of patient care and reduces clinical risk. At the same time, the patient will be part of the healthcare process, having more information about diseases and access to his/her electronic health record.

The e-Health (EH) 2022 conference aims to draw together information systems, practitioners and management experts from all quadrants involved in developing computer technology to improve healthcare quality.

Submissions were accepted under the following 3 main areas in the field of e-Health within specific topics:

-Research Issues
-Management Issues
-Applications

These conferences received 188 submissions from more than 25 countries. Each submission has been anonymously reviewed by an average of four independent reviewers, to ensure that accepted submissions were of a high standard. Consequently, only 28 full papers were approved which means an acceptance rate of 15%. A few more papers were accepted as short papers and reflection papers. An extended version of the best papers may be published in the IADIS International Journal on Computer Science and Information Systems (ISSN: 1646-3692), IADIS International Journal on WWW/Internet (ISSN: 1645-7641), and also in other selected journals.

Besides the presentation of full, short and reflection papers, these conferences also included one keynote presentation from an internationally distinguished researcher. We would therefore like to express our gratitude to Professor Piet Kommers, UNESCO Professor of Learning Technologies, The Netherlands, for being our keynote speaker.

This volume has taken shape as a result of the contributions from a number of individuals. We are grateful to all authors who have submitted their papers to enrich the conference proceedings. We wish to thank all members of the organizing committee, delegates, invitees and guests whose contribution and involvement are crucial for the success of the conference.

Last but not least, we hope that everybody enjoyed the presentations, and we invite all participants for next year's edition of these conferences.

Piet Kommers, University of Twente, The Netherlands
ICT 2022 & WBCSM 2022 Program Chair

Mário Macedo, Universidade Atlântica, Portugal
EH 2022 Program Chair

Piet Kommers, University of Twente, The Netherlands
Pedro Isaias, The University of New South Wales (UNSW – Sydney), Australia
MCCSIS 2022 General Conference Co-Chairs

July 2022

PROGRAM COMMITTEE

ICT, SOCIETY AND HUMAN BEINGS 2022
&
WEB BASED COMMUNITIES AND SOCIAL MEDIA 2022
PROGRAM CHAIR

Piet Kommers, University of Twente, The Netherlands

E-HEALTH 2022
PROGRAM CHAIR

Mário Macedo, Universidade Atlântica, Portugal

MCCSIS 2022 GENERAL CONFERENCE CO-CHAIRS

Piet Kommers, University of Twente, The Netherlands
Pedro Isaias, The University of New South Wales (UNSW – Sydney), Australia

ICT, SOCIETY AND HUMAN BEINGS 2022

COMMITTEE MEMBERS

Ana Isabel Paraguay, Officio e Ambiente, Brazil
Anastasija Nikiforova, University of Latvia, Latvia
Antonio Marturano, University of Rome Tor Vergata, Italy
Arianit Kurti, Department of Informatics, Linnaeus University, Sweden
Bertil P. Marques, Instituto Superior de Engenharia do Porto (ISEP), Portugal
Caroline Parker, Glasgow Caledonian University, United Kingdom
Christophe Trefois, Université du Luxembourg, Luxembourg
Daniel Pimienta, Observatory of Linguistic and Cultural Diversity in the Internet, Dominican Republic
Eila Jarvenpaa, Aalto University, Finland
Elvis Mazzoni, University of Bologna, Italy
Fabio Cassano, Università Degli Studi di Bari, Italy
Farshad Badie, Aalborg University, Denmark
Fatma Abdelkefi, Ecole Supérieure des Communications de Tunis, Tunisia
Federica Baroni, University of Bergamo, Italy
Francisco Garcia-Sanchez, University of Murcia, Spain
Jan Kubicek, Technical University of Ostrava, Czech Republic
Jan Meyer (Pr.M), Deputy Director NWU Business School, South Africa
Jari Multisilta, Satakunta University of Applied Sciences, Finland
John Sören Pettersson, Karlstad University, Sweden
Jorge Franco, Universidade Presbiteriana Mackenzie, Brazil
Kevin Caramancion, University at Albany, State University of New York, USA
Marco Lazzari, University of Bergamo, Italy
Margaret Tan, A*STAR IHPC, Singapore
Mikael Collan, Lappeenranta University of Technology, Finland
Nicola Doering, Ilmenau University of Technology, Germany
Przemyslaw Falkowski-Gilski, Gdansk University of Technology, Poland
Reima Suomi, University of Turku, Finland
Robert Pintér, Corvinus University of Budapest, Hungary

WEB BASED COMMUNITIES AND SOCIAL MEDIA 2022

COMMITTEE MEMBERS

E-HEALTH 2022

COMMITTEE MEMBERS

KEYNOTE LECTURE

NEEDS AND TOOLS FOR ARTIFICIAL INTELLIGENCE IN 21ST CENTURY SOCIETY

Professor Piet Kommers, UNESCO Professor of Learning Technologies, The Netherlands

ABSTRACT

The Covid-19 era unexpectedly made all sectors dependent from remote communication, virtual- and vicarious learning.

This lecture is based upon the new book: "Sources for a Better Education: Lessons from Research and Best Practices".

It signals parallels in society, technology, and demonstrates the risk for biased information; not just lacking knowledge or naïve misconceptions. Starting from abundant information access we now see tempting options for learners to restructure and even reconceive existing information. From the perspective of cognitive growth, the last four decades let learners 're-construct meaning' to stimulate highly individualized understanding: Simulations, modelling, concept mapping, and lately the cultivation of storytelling; they have been promoted as an extra to just absorbing new knowledge. So far, education still underestimated the flip side of constructivist learning practices: Critical thinking seemed to be a good candidate for a more active learning attitude; It may create more authentic students who build upon existential drive: "What do I need to 'make a difference' in life. Problem- and challenge-based learning are the keywords. The book appetizer "Sources for a better Education" exposes the landscape of learning theories and how teachers can benefit from the larger spectrum of A.I. tools: big data, data mining, deep learning, machine learning, learning analytics and multi-variate inductive reasoning? This lecture will guide you to the main questions: What didactic measures allow teachers to make students resilient to fake news? What scenarios for thematic- rather than mono-disciplinary courses need to be developed? For instance, in the attempts to implement and disseminate STEAM (Science, Technology, Engineering, Arts and Mathematics)? What social media mechanisms lead to web-based communities? And: What are valid ways to assess the quality of learning outcomes? I hope to meet you and your counter questions imaginary Lisbon.

FULL PAPERS

UA INFORMA CONTRIBUTION TO ATTRACT PROSPECTIVE STUDENTS: AN EXPLORATORY STUDY

Margarida M. Marques and Lúcia Pombo
Research Centre on Didactics and Technology in the Education of Trainers (CIDTFF)
Department of Education and Psychology, University of Aveiro, Campus Universitário de Santiago, 3810-193 Aveiro, Portugal

ABSTRACT

Student recruitment rates are essential for Higher Education institutions' sustainability. Universities may try to attract prospective students by providing information in their institutional website, advertising or offering campus visits, among other initiatives. In this line, the "UA Informa" is a project towards the promotion of extension activities for the community, to promote the image of the University of Aveiro (UA) and enhance education for sustainability. The project is relevant to UA students, prospective students, and other visitors. In this context, a set of open educational resources was developed to be accessed through QR codes spread across the campus. This exploratory study analyses the contribution of a non-formal game-based university campus visit into two dimensions: a) promotion of the institution's image; and b) students' satisfaction with the proposed activity. The game prompts the players to find nine points of interest with specific QR codes, resulting on a path through the campus. A total of 23 students attending grade 10, from a school out of the UA influence zone, participated in the campus visit. At the end, students filled in an individual and anonymous questionnaire exposing their opinion on the experience. The students revealed an overall favorable perception on the university and game-based campus visit, as they classified the activity as interesting and with good value for learning about the university. Nineteen students considered they would like to attend a UA course in the future (after grade 12), although many presented a neutral position regarding this possibility. This study indicates that the UA Informa may enhance the university image to capture prospective students, but its utility does not end here, as it may also facilitate the integration of students who attend the UA for the first time, and opens the university to the overall community. Furthermore, the QR codes are a visible and practical way to provide outreach and promote involvement of the community with sustainability issues, so it might have impact in the society sustainable habits as well.

KEYWORDS

Higher Education, Institution Promotion, Open Educational Resources, Outdoor Games, Campus Visit

1. INTRODUCTION

Student recruitment rates are essential for Higher Education institutions' sustainability, as their mission is usually focused on knowledge creation and teaching (Brock & Zhong, 2021). Universities may resort to a broad range of initiatives to stimulate and motivate students to enroll in their course offer, such as making information available through their institutional website, advertising or campus visits (Han, 2014). Campus visits are pointed as highly influential for students' choice of a postsecondary course and institution (Birch & Rosenman, 2019; Johnston, 2010).

One underexplored approach in campus visits is outdoor gaming, which can be supported by mobile technologies (Groff et al., 2015). For example, a literature review on mobile apps supporting campus visits retrieved only one work presenting an outdoor game approach (Andri et al., 2018). However, outdoor games are pointed as important for individuals' self-development and self-awareness and, when combined with collaborative approaches, they may strengthen their social relations as well (Baysal et al., 2022). Moreover, when game's wining conditions require working with other players, collaborative dynamics can also be promoted (Marques & Pombo, 2021; Robson et al., 2015). On the other hand, the competition between different groups created by games may increase students' engagement in challenging learning situations and improve their overall sense of enjoyment (Hwang et al., 2016).

As the access to mobile devices, such as laptops, tablets, smartphones and game consoles, increases in many educational contexts, the debate around concepts, such as Bring Your Own device – BYOD (Song, 2014) and Mobile Learning (Clarke & Svanaes, 2015), and their educational potential, become more acute. The potential of the use of the mentioned devices in educational contexts includes the development of digital competences by students. For example, a simple technology, such as the Quick Response codes (QR codes), allows students to develop meaningful and contextualized learning on curricular topics, while simultaneously gaining experience in the use of digital technologies (Uçak, 2019). Hence, and related with the pervasive access of student population to mobile devices, QR codes are becoming widespread in educational contexts as well. Their use can be directed at giving access to specific apps (either educational or generalist), so students may explore them for learning. QR codes can also be used to access desired information and contents, thus, preventing students from wasting time on search engines and find irrelevant or unreliable information. Other advantages include giving access to animated or interactive content, besides the prevention of paper wastage (Uçak, 2019).

When the digital content accessed through QR codes is freely available for all to explore, we face a truly democratization of education. In this line of thought, open educational resources (OER) gain relevance. These are educational materials available for the community under an open license that permits their use and re-use by anyone. By motivating students to take initiative in their learning, OER can support autonomous and ubiquitous learning, outside the classroom (Kim et al., 2020).

In this line, the "UA Informa" is a project grounded on a plural approach, articulating Education, Training and Research towards the promotion of extension activities for the community, including prospective students (Pombo et al., n.d.). The Education dimension is reflected on the aim of the project focused on the enhancement of education for sustainability for all; the Training dimension is translated into the involvement of Higher Education students in research projects in the area of Education, towards their scientific initiation; the Research dimension is based on the scientific investigation conducted in this project, under a social responsibility umbrella. The UA Informa project is integrated in the "Smart Knowledge Garden" and "Open Educational Smart Campus", which are programmatic projects of the Research Centre on Didactics and Technology in Education of Trainers (CIDTFF; https://www.ua.pt/en/cidtff/) of the University of Aveiro (UA), whose mission is anchored on the responsibility of research in Education to produce knowledge contributing to educate qualified and critical citizens, and to the creation of a better world.

Grounded on the social responsibility commitment, concerning namely knowledge-transfer practices and tools, the main aims of the UA Informa project are to promote the image of the UA and to enhance education for sustainability. The project endorses the participatory contribution of students, in initial and advanced training, so, at this stage, a student of High Degree in Basic Education was integrated into the research team of the UA Informa, for two years, under the PIC-Edu program. This is a program of initiation into research in the area of Education promoted by CIDTFF. The UA Informa team developed, tested and has been evaluating a set of multimedia OER, accessed through QR codes spread across the UA campus. The resources concern sustainability topics and are integrated in the UA Informa subweb (https://www.ua.pt/pt/uainforma), within the institutional Portal. The project opens the university to the broader community, but it has special interest to prospective students, UA students, particularly those attending the institution for the first time, and other campus visitors.

To promote the exploration of the resources, the team developed and implemented one non-formal university campus visit, entitled "UA Informa on Campus Sustainability", targeting secondary students. It is based on a quiz-game that prompts the exploration of QR codes installed in strategic points of interest in the UA campus, giving access to the UA Informa multimedia OER, towards the promotion of scientific literacy on sustainability issues (Pombo et al., n.d.). This work presents an exploratory study that analyses the contribution of the above described campus visit into the promotion of the institution's image, and the satisfaction of participating students concerning the proposed activity.

This document proceeds with the methodological options of the study, which integrates a mixed methods approach; the presentation and discussion of the main results; and finally, some conclusions are put forward.

2. METHODOLOGY

This exploratory study attempts to answer the research question: What is the contribution of a non-formal campus visit for secondary students, based on a quiz-game with QR codes giving access to multimedia OER towards sustainability learning, into two dimensions: a) promotion of the institution's image; and b) students' satisfaction with the proposed activity. For that, quantitative and qualitative data were obtained from a questionnaire applied after the campus visit.

This section comprises three subsections: i) an introduction contextualizing the campus visit; ii) data collection methods and data analysis; iii) participants brief description.

2.1 The Campus Visit "UA Informa on Campus Sustainability"

The UA Informa campus visit comprises three main stages of development: i) the creation of multimedia OER integrated in the subweb UA Informa; ii) the development of an outdoor game, with QR codes that give access to the above-mentioned educational resources; and iii) the implementation of the UA Informa resources with secondary students, that is, students' exploration of the OER in a game format campus visit.

In the first stage of development, the UA Informa team created a set of multimedia OER covering sustainability topics presented on a manual published by the UA group for sustainability: Energy, Water, Paper and Plastic, Waste, Mobility, Food and Green Events. Each section presents relevant impacting information, based on credible sources, illustrated with appealing images and small videos, which were conceived and produced by Higher Education students, under the supervision of experienced researchers. Moreover, the videos exhibit also sustainability actions and strategies that can be adopted on the campus (Figure 1).

Figure 1. Open educational resources on the UA Informa subweb (https://www.ua.pt/pt/uainforma)
A: texts and images regarding food waste; B: video featuring a cooking oil recycling bin at UA

UA's Communications, Image and Public Relations Services mediated the Rectory's authorization to publish the resources in a subweb of the university's portal that was created for this purpose: the UA Informa subweb (https://www.ua.pt/pt/uainforma). Furthermore, authorization was also given by the institution Rectory for the installation of a set of QR codes giving facilitated access to UA Informa OER, accordingly to their relevance to specific points of interest on the institution main campus. The installation of signals with the QR codes throughout the campus draws the attention of the passersby to sustainability issues, related to each visited point of interest. This instills curiosity about the information contained therein, which is quickly accessed by reading it on a mobile smartphone with a simple QR code reader. These contents are not static nor merely informative; they can provide moments of interaction between users and sustainability actions on campus.

In a second stage of development, the team created a peddy-paper game presented in a flyer. It starts with an informative section with the goals and instructions to play the game, and also a map with all the points of interest that comprise a path on the campus. Hence, the game prompts teams of players to find nine points of interest in the campus: 1- Rectory, 2- UA Informa, 3- Paper and plastic, 4- Waste, 5- Energy, 6- Green events, 7- Water, 8- Food, and 9- Mobility. In each point of interest, players must find a specific QR code to access a specific UA Informa OER comprising useful information in textual, image and video formats. For each point, three multiple choice questions with four answer options are presented, summing up to 27 questions in the game. The first question in each point of interest requires players to read, interpret, and select textual information. The second question involves the access to information presented in video, which usually includes actions and strategies people can take to reduce their environmental footprint in the campus. In the third question, players must observe their surroundings. Each correct answer is valued with one point and a score is kept, to find the winner team.

In a third stage, in order to test and evaluate the OER and outdoor game, a campus visit activity was offered to the community under an annual event targeting basic and secondary students, promoted by UA. The event, the XPERiMENTA, is the largest annual extension event of UA designed to demonstrate its skills, such as the training offer, as well as to present its conditions of study, research, personal and social development. Students are invited to work on hands-on activities, interactive projects, science shows and guided tours in the campuses.

Twenty-three secondary students (15 to 17 years-old), from a school outside of the UA influence region, participated in the UA Informa game-based campus visit under XPERiMENTA event, in teams of two or three, during 45 to 60 minutes. At the end, students filled in an evaluation questionnaire and certificates were distributed to all participants, as well as small prizes for the teams with the best performance.

2.2 Data Collection and Analysis Options

Data collection involved inquiry through a questionnaire collecting participants' perceptions regarding the campus visit activity. The questionnaire was developed to sustain a discussion about the informative and educational contribution of the UA Informa. Thus, the following dimensions where considered: i) sustainability learning value of the game; ii) informative value to know the UA infrastructures and conditions for the academy; and iii) activity appraisement. Hence, the questionnaire included three sections devoted to these dimensions and one additional section to briefly characterize the participants socio-demographics (age, gender, school year, and the area of the high degree course the respondent would like to attend in the future). However, as the section "i)" was analyzed in previous work (Pombo et al., n.d.), in this study the focus is on the remaining dimensions: "ii)" and "iii)".

The part of the questionnaire regarding the informative value to know the UA infrastructures and conditions for the academy included 6 closed-ended questions in 5-points Likert scale, from "totally disagree" to "totally agree". This part is followed by one section regarding the activity appraisement. This last part included 6 closed-ended questions in a similar scale, and contained as well one closed-ended question in a 5-points Likert scale, from "very uninteresting" to "very interesting". This part also included an open-ended question, where students should complete three sentences: "In this activity, I liked ....", "In this activity, I did not like..." and "I think this UA Informa campus visit is ..."

Quantitative data was analyzed through descriptive statistics with graph creation. Qualitative data from the open-ended questions were analyzed through descriptive analysis and presented in a table format.

All data was collected anonymously and students participated in this study voluntarily.

2.3 Participants Brief Description

The group of 23 secondary students that participated in the campus visit attended grade 10. From these, 11 students were 15 years-old, 10 students were 16 years-old and 2 students were 17 years-old. Regarding gender, 14 were female and 9 were male. About one quarter of the students (6) mentioned they did not know what course they plan to apply for post-secondary education. Other 6 students mentioned they desired sciences courses with no specification, 5 preferred medicine courses and the remaining mentioned a diverse set of areas: education, economy, multimedia, sports, engineering and no course desired (1 each).

3. RESULTS AND DISCUSSION

This section presents and briefly discusses the results of this study that focuses on the analysis of the contribution of a non-formal game-based campus visit into two dimensions: a) promotion of the institution's image; and b) students' satisfaction with the proposed activity.

3.1 Promotion of the Institution's Image

Graph 1 presents the students' opinion about the UA Informa campus visit value for the UA promotion. It reveals an overall favorable perception about the university. The first question (3.1.) showed that the campus visit supported by a mobile device was one unexpected experience, as 14 students (totally) agreed that, before this activity, they did not imagine they could use cell phones to learn more about UA. Moreover, most students considered that this activity allowed them not only to know the UA campus a little better, but also to know important places (questions 3.2. and 3.3.). Among the visited places are the Rectory building, the canteen, the book shop, bicycle park, recycle station, and several departments in the main campus.

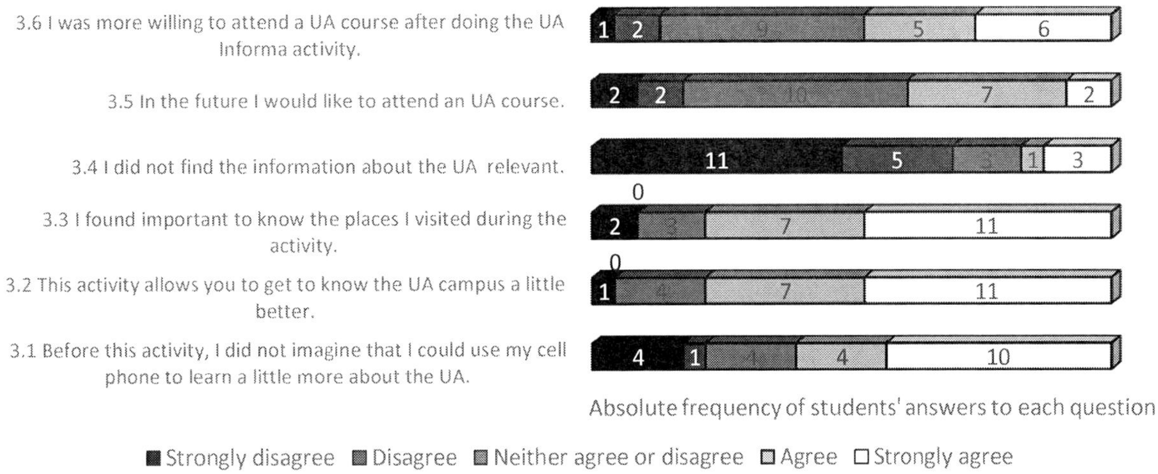

Graph 1. Secondary students' opinion about the UA Informa campus visit value for the UA promotion

In the question formulated in a negative way (3.4), most students disagreed, indicating that students considered the information about the UA relevant. Finally, the questions concerning their will to attend the UA institution in the future (3.5 and 3.6.), gathered the highest number of neutral answers (neither agree or disagree). This seems to be in accordance with the students' answers to the course desired. Maybe related with the grade students are attending (10th degree), there seems to be some undefinition regarding their post-secondary education, as half the students either did not know or mentioned a very broad area, as Sciences. Nevertheless, 11 students showed that the activity increased their willingness to attend UA in the future. This result supports the literature that reports campus visits as highly influential for students' choice of a post-secondary course and institution (Birch & Rosenman, 2019; Johnston, 2010).

3.2 Students' Satisfaction with the Proposed Activity

The participant students' global satisfaction with the campus visit is presented in Graph 2, which reveals an overall good perception. For instance, most students appreciated the game-based format, as 15 strongly agreed and 5 agreed with the sentence "4.1. I liked the UA Informa activity because it involved a game." Moreover, the appreciation of the activity does not seem to be linked with the small prizes offering at the end, as half of the students (strongly) disagreed with the related sentence (4.2).

The outdoor learning feature seems to be appreciated by students, as many acknowledged they can study curricular topics outside the school, which is revealed by 18 (strong) agreement answers in the sentence 4.3. Students also mentioned they prefer to learn in the outdoors, rather than in school, as 13 respondents (strongly) agreed with the sentence 4.4.

The mobile technology also seems to be appreciated, as students mostly mentioned liking to learn with cell phones and QR codes (16 respondents) and preferring to use mobile devices, rather than textbooks to learn (16 respondents), as revealed by their answers to 4.5 and 4.6. questions.

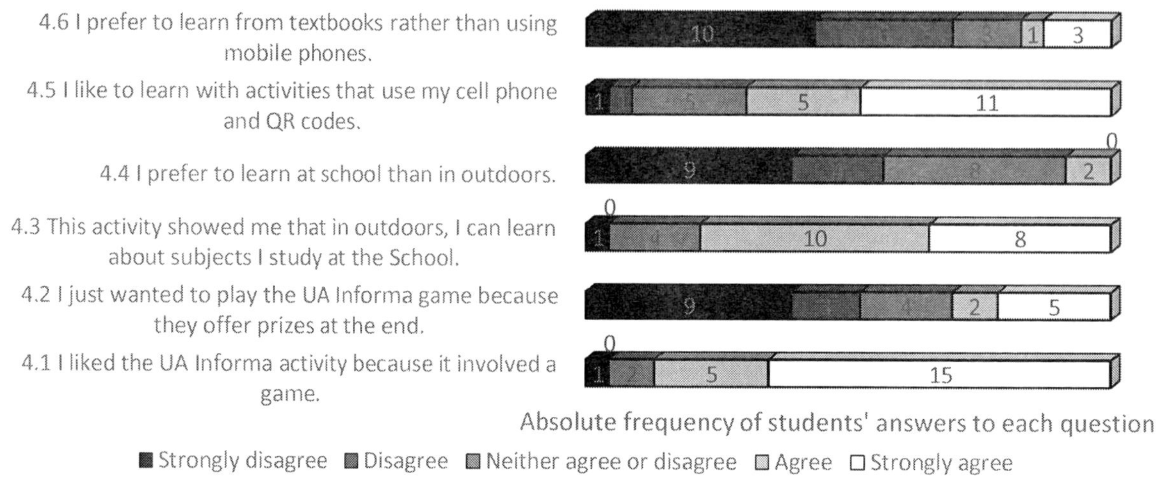

Graph 2. Secondary students' global satisfaction with the activity

Students' acceptability of this outdoor game-based format supported by mobile technologies is not yet reflected on the literature, as campus visits combining these features are not frequently reported (Andri et al., 2018; Groff et al., 2015).

Considering students opinions regarding the UA Informa activity interest, the majority classified it as very interesting (16 respondents) or interesting (4 respondents), as revealed in Graph 3. No student selected a (very) uninteresting answer.

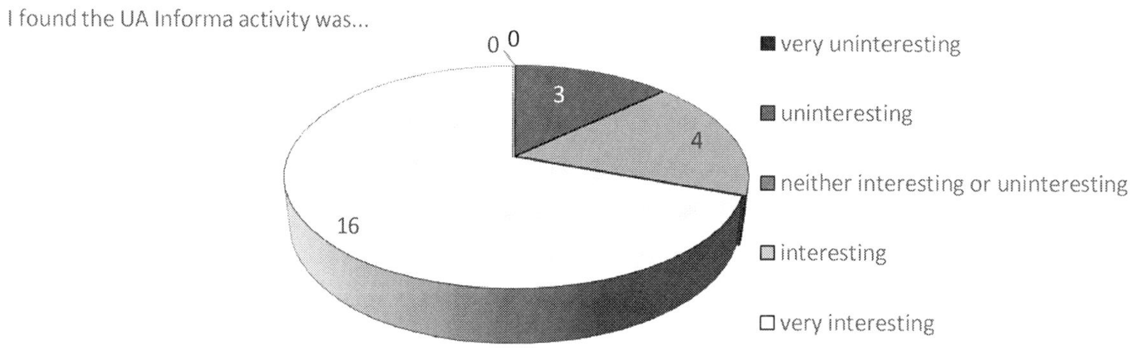

Graph 3. Secondary students' appraisal of the activity interest

Regarding the open-ended question, students overall revealed a very positive perception about the UA Informa campus visit (Table 1). For instance, they mentioned having enjoyed the exploration of the buildings and to get to know the university campus (11 answers), while having a fun experience (5 answers), in spite of the too sunny weather conditions (14) and the prompt to walk around the campus (4 answers). In general, students reinforced their evaluation of the activity as interesting (6 answers) and fun (3 answers), and acknowledged the activity learning value and importance (3 answers each). It is worth mentioning that the activity ran in the Spring, after 11 a.m., in a hot day.

Table 1. Students' answers to the open-ended question about the UA Informa campus visit

Category	Frequency	Citation
I liked...		
...the outdoors and to know the UA	11	"...to explore all the buildings and to know the university"
...having fun	5	"...to do very fun activities"
...the experience	1	"...of the experience, a lot of sympathy and it was innovative"
Unspecific answer	5	"...everything"
I did not like...		

Category	Frequency	Citation
...the weather	14	"...being very hot"
...to walk	4	"...having to walk"
Unspecific answer	5	"I have nothing to point out"
I think this UA Informa campus visit is...		
...interesting	6	"...very interesting"
...learning promoter	3	"...innovative and deepens our knowledge"
...fun	3	"...very fun"
...important	3	"...important and necessary"
Unspecific answer	8	"...different"

4. CONCLUSION

The exploratory study analyses the contribution of a non-formal game-based campus visit into two dimensions: a) promotion of the institution's image; and b) students' satisfaction with the proposed activity. The game is supported by mobile devices that are used to access OER on sustainability issues accessible through specific QR codes in nine points of interest in the campus. The activity was implemented in an annual event targeting basic and secondary students, promoted by UA.

A total of 23 grade 10 students participated in the study and revealed an overall favorable perception on the university. Students considered the campus visit allowed them to know important locations of the campus (such as the Rectory building) and to learn relevant information about the university. From the 23 students, from a school out of the influence zone of the institution, 19 considered they would like to attend a UA course in the future, although many (10) presented a neutral position regarding this possibility. Considering students' school year and undefinition about their desired post-secondary course, the high frequency of the neutral position is not surprising. It is also worth noting that these students have several other higher institutions geographically closer to their home city. Other issue to highlight is that almost half of the students acknowledged that the campus visit increased their willingness to attend UA in the future.

In what concerns students' satisfaction with the activity, they revealed also an overall positive perception and they classified the activity as interesting (4 answers) or very interesting (16 answers). Students mentioned having appreciated the: i) outdoor game format, making this a fun activity, ii) learning sustainability issues in the outdoors, involving topics they study at school, and iii) use of their own mobile devices in this type of activity. Facing these results on student acceptability of outdoor game-based campus visits supported by mobile devices, which are unusual features in this type of institution promotion (Andri et al., 2018; Groff et al., 2015), higher education institutions should consider to explore this approach. This recommendation becomes more relevant when considering the benefits of outdoor games for players pointed in the literature, namely personal self-development, increased digital competence, and engagement in challenging learning situations (Baysal et al., 2022; Hwang et al., 2016; Marques & Pombo, 2021; Robson et al., 2015; Uçak, 2019). Institutions can also consider that students may not appreciate too sunny weather conditions, and program these visits preferably outside the hotter hours of the day.

From the results, this exploratory study indicates that the UA Informa project may enhance the university image to capture prospective students. This is supported by the fact that the participant students, from a distant geographical region, mentioned they are open to attend the university in the future (after year 12). These results are in line with the literature, where campus visits are documented as highly influential for institution choice by prospective students (Birch & Rosenman, 2019; Johnston, 2010).

In addition, the project may also facilitate the integration of students who attend the UA for the first time, as they are supported in getting to know the campus infrastructure and functioning. However, as the QR codes are permanently available on the campus, any passerby with a smartphone can explore the OER on the UA Informa subweb, thus opening the university to the overall community. Furthermore, the QR codes are a visible and practical way to provide outreach and promote involvement of the community with sustainability issues, so it might have impact in the society sustainable habits.

This was the first experience of implementing this campus visit. Hence, the number of participants is small and does not allow the generalization of results, which was not the aim of this study. The intention was to enhance the discussion about the contribution of the game format articulating OER accessed through QR codes to the university image promotion, whilst making sure this activity is satisfactory for the participating students.

Further work involves the exploration of the UA Informa resources in new activities, for an extended target public, involving adults. Moreover, it is previewed to expand the UA Informa subweb with new themes and new games, and also to disseminate results.

ACKNOWLEDGEMENT

The UA Informa is funded by National Funds through FCT – Fundação para a Ciência e a Tecnologia, I.P., under the UIDB/00194/2020 project. The work of the first author is funded by national funds (OE), through University of Aveiro, in the scope of the framework contract foreseen in the numbers 4, 5 and 6 of the article 23, of the Decree-Law 57/2016, of August 29, changed by Law 57/2017, of July 19.

REFERENCES

Andri, C., Alkawaz, M. H., & Sallow, A. B. (2018). Adoption of Mobile Augmented Reality as a Campus Tour Application. *International Journal of Engineering & Technology*, *7*(4.11), 64–69. https://doi.org/10.14419/ijet.v7i4.11.20689

Baysal, E. A., Ocak, İ., & Öztürk, K. (2022). Attitudes of secondary school students towards outdoor games: A scale development study. *Pegem Journal of Education and Instruction*, *12*(1), 115–130. https://doi.org/10.47750/PEGEGOG.12.01.11

Birch, M., & Rosenman, R. (2019). Is it the visit or the scholarship? An analysis of a special campus visitation program. *Https://Doi.Org/10.1080/09645292.2019.1696750*, *28*(2), 179–195. https://doi.org/10.1080/09645292.2019.1696750

Brock, C., & Zhong, Z. (2021). The Many Contexts of the Social Responsibilities of Universities. *Journal of International and Comparative Education (JICE)*, *10*(2), 133–141. https://doi.org/10.14425/JICE.2021.10.2.0612

Clarke, B., & Svanaes, S. (2015). *Updated review of the global use of mobile technology in education.* http://www.kidsandyouth.com/pdf/T4S FK%26Y Literature Review 11.12.15.pdf

Groff, J., Clarke-Midura, J., Owen, V. E., Rosenheck, L., & Beall, M. (2015). *Better Learning in Games: A Balanced Design Lens for a New Generation of Learn-ing Games.* http://education.mit.edu/wp-content/uploads/2015/07/BalancedDesignGuide2015.pdf

Han, P. (2014). A Literature Review on College Choice and Marketing Strategies for Recruitment. *Family and Consumer Sciences Research Journal*, *43*(2), 120–130. https://doi.org/10.1111/FCSR.12091

Hwang, G.-J., Wu, P.-H., Chen, C.-C., & Tu, N.-T. (2016). Effects of an augmented reality-based educational game on students' learning achievements and attitudes in real-world observations. *Interactive Learning Environments*, *24*(8), 1895–1906. https://doi.org/10.1080/10494820.2015.1057747

Johnston, T. C. (2010). Who And What Influences Choice Of University? Student And University Perceptions. *American Journal of Business Education*, *3*(10), 15–24. https://www.clutejournals.com/index.php/AJBE/article/view/484/471

Kim, D., Lee, Y., Leite, W. L., & Huggins-Manley, A. C. (2020). Exploring student and teacher usage patterns associated with student attrition in an open educational resource-supported online learning platform. *Computers & Education*, *156*, 103961. https://doi.org/10.1016/J.COMPEDU.2020.103961

Marques, M. M., & Pombo, L. (2021). Teachers' experiences and perceptions regarding mobile augmented reality games: A case study of a teacher training. In L. G. Chova, A. L. Martínez, & I. C. Torres (Eds.), *Proceedings of INTED2021 Conference* (pp. 8938–8947). IATED.

Pombo, L., Marques, M. M., & Guimarães, F. (n.d.). UA Informa: Education for sustainability, from the academia to the community. *EDULEARN22 Proceedings: 14th International Conference on Education and New Learning Technologies, July 4th-6th, 2022.*

Robson, K., Plangger, K., Kietzmann, J. H., McCarthy, I., & Pitt, L. (2015). Is it all a game? Understanding the principles of gamification. *Business Horizons*, *58*(4), 411–420. https://doi.org/10.1016/j.bushor.2015.03.006

Song, Y. (2014). "bring Your Own Device (BYOD)" for seamless science inquiry in a primary school. *Computers and Education*, *74*, 50–60. https://doi.org/10.1016/j.compedu.2014.01.005

Uçak, E. (2019). Teaching Materials Developed Using QR Code Technology in Science Classes. *International Journal of Progressive Education*, *15*(4), 215–228. https://doi.org/10.29329/IJPE.2019.203.16

TECHNOLOGICAL, ORGANIZATIONAL AND PERSONAL FACTORS OF REMOTE WORK: AN EXPLORATORY STUDY

Ina Kayser and Martin Lange
IST University of Applied Sciences
Erkrather Straße 220 a-c, 40233 Düsseldorf, Germany

ABSTRACT

This article presents an exploratory approach to identify different clusters of employees' perceived stress, work anxiety and self-efficacy according to their individual technological, organizational, and personal parameters. In more detail, our research focuses on the context of remote work during the Covid 19 pandemic to analyze which combination of these parameters co-occur with particularly high or low levels of resilience-related variables such as self-efficacy, work stress, and work anxiety. We conduct a two-step cluster analysis to present findings to resolve the ambiguity of previous research on the role and understanding of resilience in the context of remote work and in light of the pandemic. As a result, this study shows that study participants with different levels of experienced stress and anxiety have different technological, personal, and organizational contexts. Moreover, we were able to identify technological factors associated with work resources that are linked to less stress and work anxiety. A deeper understanding of the factors underlying lower levels of stress in remote work through this study can help identify potential areas of improvement for individuals and organizations and provides a basis for further research in this area.

KEYWORDS

Cluster Analysis, Exploratory Research, Information Systems, Remote Work, Health

1. INTRODUCTION

The Covid 19 pandemic has led to many workers working remotely from home. Thus, the pandemic accelerated the trend to work remotely and posed new challenges for employers and employees. To address this, we examine work-related self-efficacy, job stress, and job anxiety as related variables to the resilience of remotely working employees. Previous research presented these variables in different contexts without a clear causal structure (Lloyd et al. 2017; Sadri and Robertson 1993; Zompa and Bompiedi 2021). There is evidence that all three variables are associated with resilience, but there is disagreement on what exactly the direction of the cause-and-effect relationship is, or whether it is moderation or mediation, if any, or "merely" correlation (García-León et al. 2019; Lehrer et al. 2020; Shi et al. 2015). Recent research approaches on resilience are shifting the focus from a vulnerability-oriented view to a protective and coping-based perspective identifying factors that buffer health effects. One of these protective factors that can be assigned to the concept of resilience is self-efficacy. Self-efficacy is the belief that one can perform novel or challenging tasks and achieve desired outcomes, as outlined in Social Cognitive Theory (Bandura 1992). More narrowly, remote work self-efficacy is the subjective certainty of being able to cope with challenging situations based on one's own competencies. This is important to study because self-efficacy is a critical antecedent to resilience. Previous research found that general self-efficacy beliefs have positive effects on coping with various stressors and on proactive preparation for potential stressors (Schwarzer and Warner 2013). Hence, another important factor for resilience is perceived stress. Job-related perceived stress can be defined as a subject's response to physical or mental job demands in relation to a subject's resources (Havnen et al. 2020). However, there are divergent findings from previous research on the presumed relationship between perceived stress and resilience. For example, García-León et al. (2019) found that individuals with low resilience showed higher scores in perceived stress levels. Moreover, Lehrer et al. (2020) disclosed that resilience moderated the indirect association of perceived stress with health-related dependent variables.

Contrastingly, Norris et al. (2008) conceptualized resilience not as an antecedent or psychological trait that determines certain levels of stress, but as a potential outcome after exposure to stress. Hence, there is evidence that perceived stress and resilience are at least correlated, however the causal relationship seems to be uncertain. Further literature supports this ambiguity by reporting that resilience is mediating the effects of stress on job anxiety (Shi et al. 2015). According to Havnen et al. (2020), job anxiety describes a result of exposure to job-related stress having various negative effects on health. In contrast to Shi et al. (2015), other results underline the moderating effects between job stress and job anxiety symptoms (Havnen et al. 2020).

As the pandemic has created new conditions for remote work, and now employees are working from home who did not do so before, the question arises how employees cope with the situation. One way of studying this matter would be to consider the cause-and-effect relationships between resilience-related variables to analyze which factors can positively contribute to resilience. Such a causal analysis would require that a path model can be derived from previous research. However, due to the divergent findings in previous research and the novelty of the pandemic and its dynamics, we pursue an exploratory approach. The pandemic has decisively changed the lives of many people. It may follow that established causal relationships have to be re-evaluated from different perspectives and include factors that had less relevance before. Possible influences are personal circumstances such as an emerging conflict within the family, triggered by closures of schools and kindergartens or also by the situation that people now spend time with their partner and children on a daily basis - even during work hours. From another perspective, the mingling of work and private life can lead to conflicts because work remains visible in private households even outside of working hours. In addition to the personal environment, technological and organizational conditions must also be considered. Therefore, another possible factor is that many employees were sent home without appropriate information systems (IS) for remote work being in place or appropriate training. Out of an organizational perspective, an employee's work autonomy is considered as an important factor giving employees a certain degree of freedom to deal with job-demands, work-family conflicts and other work associated aspects. Previous research shows that autonomy has a major positive impact on self-efficacy (Sousa et al. 2012) and plays a mediating role in work performance and work-family facilitation (Wattoo et al. 2020). In summary, it can be stated that the relationship between work-related stress and symptoms of anxiety as well as the role of self-efficacy and other environmental resilience factors are not adequately investigated.

Against this background, we are pursuing two research questions (RQ) in this study: First, we empirically investigate the question how employees cope with the situation and what kind of specific resilience groups can be formed based on the three variables self-efficacy, perceived stress, and job anxiety (RQ1). Second, we exploratively investigate the question which patterns of further possible influencing variables form to these groups (RQ2).

The goal is to identify which employees are particularly resilient through the pandemic situation. This can help to identify technological and/or organizational factors that can be investigated in further research for their causal effect and can be used in practice to provide assistance to those who have so far exhibited high levels of stress and anxiety as well as low self-efficacy.

2. DATA COLLECTION, SAMPLE CHARACTERISTICS AND METHOD

We conducted a wide-ranging survey among employees in Germany in 2021. Thus, the survey took place during the period in which the obligation to work remotely, as far as possible, was adopted by the legislator and thus also affected employees for whom remote work was not an option previously. In more detail, this law means that employers in Germany are legally obliged to allow their employees to work from home to reduce contacts and decrease mobility wherever and whenever possible. Participants were randomly selected by a panel service provider. Only employees who work predominantly at a desk were surveyed; participants who stated that they did not work remotely at all were excluded. With the assistance of the panel provider, we attempted to stratify respondents by demographics to be approximately representative of the labor force in terms of regional distribution, age, gender, and educational level.

To determine relevant variables to collect, we conducted an in depth-review of relevant publications from the Scopus database, along with a Google Scholar search. As presented in the first paragraph, remote work self-efficacy, perceived stress and job anxiety are the variables used to determine certain levels of resilience-associated behavior.

Table 1. Constructs

**. correlation sig. at .01; square root of the average variance extracted (AVE) in bold on the diagonal

Constructs	Reference	CR	1	2	3	4	5	6	7	8	9	10	11	12
1. Remote Work Self-Efficacy	Staples et al. (1999)	.902	**.732**											
2. Job Stress	Staples et al. (1999)	.906	-.386**	**.811**										
3. Job Anxiety	Lee and Keil (2018)	.914	-.311**	.566**	**.833**									
4. Remote Work Experience and Training	Staples et al. (1999)	.834	.330**	-.091**	.109**	**.761**								
5. IS Adequacy	Davison et al. (2019)	.841	.444**	-.271**	-.181**	.350**	**.801**							
6. IS Restrictiveness	Curreri and Lyytinen (2017)	.742	-.223**	.243**	.280**	.054**	-.107**	**.649**						
7. Work-Scheduling Autonomy	Morgeson and Humphrey (2006)	.893	.377**	-.237**	-.200**	.088**	.195**	-.335**	**.858**					
8. Decision-Making Autonomy	Morgeson and Humphrey (2006)	.912	.419**	-.195**	-.193**	.172**	.214**	-.343**	.649**	**.881**				
9. (work-related) Method Autonomy	Morgeson and Humphrey (2006)	.876	.413**	-.201**	-.185**	.160**	.210**	-.391**	.688**	.773**	**.838**			
10. Work-Family Conflict	Netemeyer et al. (1996)	.955	-.355**	.693**	.525**	-.039**	-.255**	.233**	-.216**	-.153**	-.163**	**.899**		
11. Resistance to Change	Laumer et al. (2016); Oreg et al. (2008)	.748	-.157**	.382**	.329**	.040**	-.081**	.199**	-.065**	-.084**	-.070**	.297**	**.709**	
12. Overall Job Satisfaction	Brayfield and Rothe (1951)	.908	.482**	-.424**	-.263**	.234**	.231**	-.202**	.308**	.406**	.364**	-.308**	-.213**	**.877**

Table 1 provides an overview of the constructs employed in this study. The exact wording of the items has been carefully adapted to the context of remote work during the COVID-19 pandemic. Items were measured on a 5-point Likert scale with anchors ranging from "strongly disagree" to "strongly agree". As presented in table 1, we surveyed different facets of workplace autonomy as suggested by Morgeson and Humphrey (2006). In accordance with Morgeson and Humphrey (2006), the surveyed items for work-scheduling autonomy, decision autonomy, and work-method autonomy do not load sufficiently on a common autonomy-construct are hence treated as individual variables of autonomy in the analysis.

We conducted a confirmatory factor analysis to test validity of the constructs. Each indicator loads higher on its respective construct than on other latent variables (Chin 1998). We evaluated internal consistency by assessing the composite reliability (CR) of the indicator items representing each construct. To determine discriminant validity, we analyzed the square roots of the AVE values. Table 1 shows that values for CR and AVE are satisfactory. CR values are all above the .7 threshold. The square roots of the AVE values are greater than the respective correlations for each construct (Fornell and Larcker 1981).

Our sample comprises n=5,161 cases. On average, participants worked 26.65 hours remotely per week with an SD of 12.32 hours. The mean age was 43 years with an SD of 11 years. 53.1 percent of the participants were female and 46.9 percent male.

Participants included in the survey had different educational backgrounds. Table 2 provides an overview of the highest level of education.

Table 2. Educational Background

	Secondary school leaving certificate (or equal, 9 or 10 years of education)	High school diploma (or equal, 12 or 13 years of education)	Studies without a degree (dropouts)	Bachelor's degree	Master's degree	PhD
Number of participants	1,227 (23.8%)	1,320 (25.6%)	200 (3.9%)	737 (14.3%)	1,566 (30.3%)	111 (2.2%)

To answer the two research questions, we analyzed the large dataset using two-step cluster analysis (Tkaczynski 2017). This is an explorative procedure developed by IBM for SPSS software (IBM 2021): Two-step cluster analysis is an empirical technique for identifying natural clusters within a dataset that would otherwise be undetectable. It is based on the BIRCH (Balanced Iterative Reducing and Clustering using Hierarchies) algorithm (Zhang et al. 1996). The main advantage of the two-step procedure is that it is suitable for very large sample sizes (in contrast to hierarchical clustering) and detects the number of clusters automatically (in contrast to k-means clustering) (Tkaczynski 2017; Zhang et al. 1996). The variables remote-work self-efficacy, job stress and job anxiety were used to compute the cluster solution (RQ1), while all other variables were used as evaluation fields (RQ2).

3. RESULTS OF THE CLUSTER ANALYSIS

The algorithm concluded with a three-cluster solution. Cluster 1 contains 1,973 observations (38.2%), cluster 2 contains 1,574 cases (30.5%), and 1,614 individuals are grouped in cluster 3 (31.3%). Table 3 summarizes the demographic data for each cluster.

Table 3. Cluster Demographics

		Cluster 1	Cluster 2	Cluster 3
	Size	n_1=1,973 (38.2%)	n_2=1,574 (30.5%)	n_3=1,614 (31.2%)
	Gender	52% female; 48% male	54.6% female; 45.4% male	52.9% female; 47.1% male
	Age (SD)	41.9 years (10.9)	45.6 years (11.1)	43.8 years (10.7)
Highest Degree of Education	Secondary school leaving certificate or intermediate school leaving certificate (9 or 10 years of education)	22.8%	26.9%	21.9%
	High school diploma or equivalent (12 or 13 years of education)	25.3%	24.6%	26.8%
	Studies without a degree (dropouts)	4.4%	4.3%	2.9%
	Bachelor's degree	15.6%	12.4%	14.5%
	Master's degree	29.6%	29.7%	31.9%
	PhD	2.3%	2.1%	2%
	Average number of years with current employer (SD)	10.4 years (9.4)	13.2 years (10.6)	11.1 years (9.4)

To visualize the results of the cluster analysis concisely, figure 1 provides an overview of the clusters and the distribution of variables (input and evaluation variables) using boxplots. In this way, the three identified clusters can be considered not only for the cluster-forming variables, but with respect to all variables surveyed. The boxplots also not only display the means, which would be susceptible to any outliers in the data set, but also the interquartile ranges and ranges for each cluster with respect to the variables examined.

Cluster 1 shows the highest values for job anxiety and job stress, and the lowest levels of remote work self-efficacy. Cluster 2 has the highest levels of remote work self-efficacy and the lowest levels of job stress and job anxiety. Cluster 3 still shows a high level of remote work self-efficacy, albeit lower than in cluster 2. Moreover, cluster 3 includes employees with a low level of both anxiety and stress, albeit higher than in cluster 2. As for the evaluation variables, remote work experience and training does not differ substantially between the three clusters. Adequacy of IS at home is highest in cluster 2, followed by cluster 3 and lowest in cluster 1.

The opposite is true for IS restrictiveness: Cluster 1 shows the highest level of restrictiveness, followed by cluster 3 and cluster 2. The three autonomy-related variables show a consistent pattern: All three levels of autonomy are highest in cluster 2, followed by cluster 3 and cluster 1 shows the lowest levels of autonomy. Work-family conflict differs greatly between the clusters. Cluster 2 has the lowest level, and cluster 3 is also on a low level, while cluster 1 shows a considerably higher degree. The resistance to change shows higher levels in cluster 1 as opposed to the other two clusters. All three clusters show an above-average degree of job satisfaction, whereas overall job satisfaction is highest in cluster 2 and lowest in cluster 1.

4. DISCUSSION

Overall, the results reveal versatile insights and well differentiated clusters regarding self-efficacy, job stress and job anxiety. Cluster 1 can be characterized by a younger age than the other clusters, the overall least professional experience, and a high degree of restrictiveness. Cluster 1 shows the lowest values in self-efficacy and the highest results on job-anxiety and job-stress. Cluster 2 seems to show opposite characteristics of cluster 1, with high self-efficacy and low values in the areas of job-stress and job-anxiety. Also, cluster 3 aligns with clusters 1 and 2 in a moderate way.

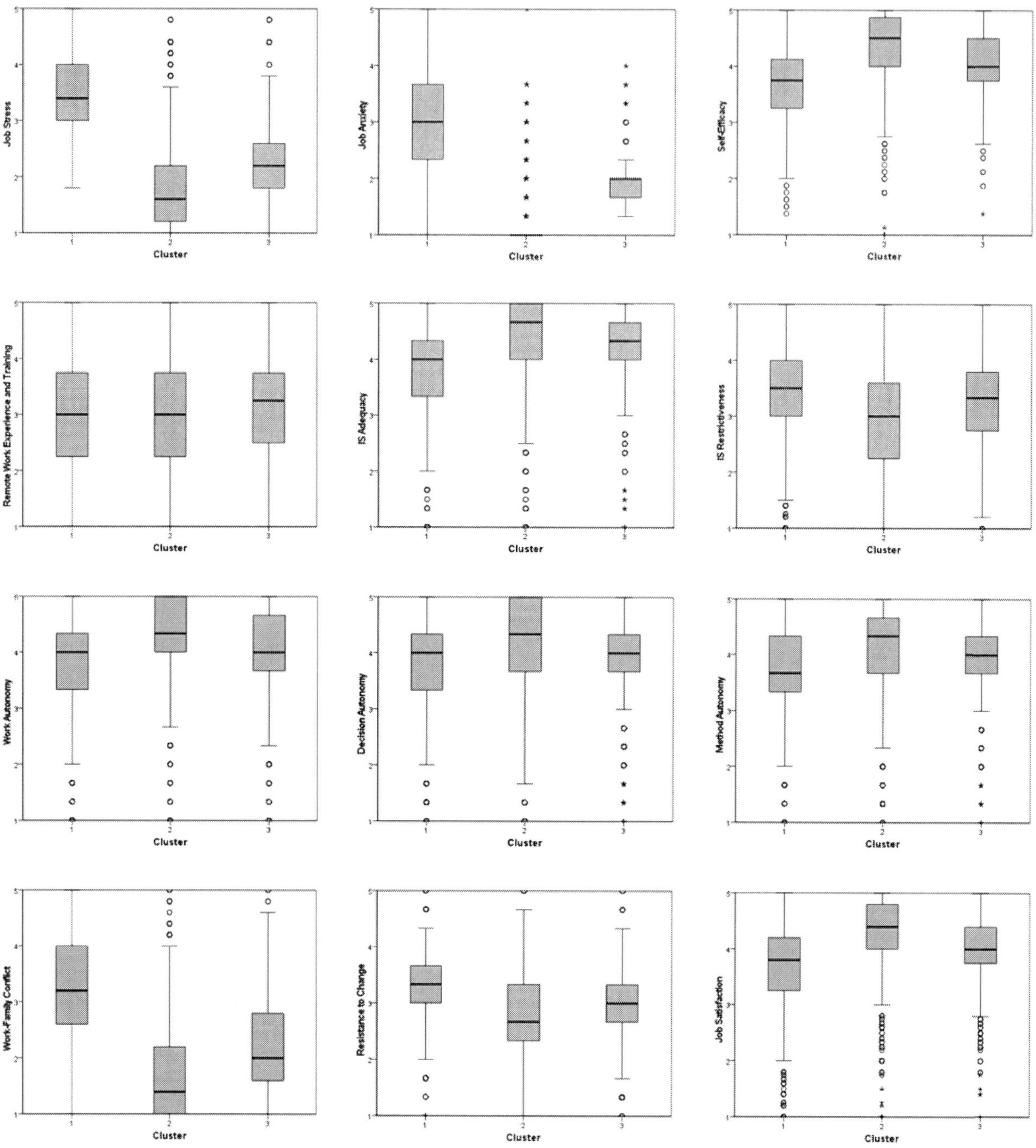

Figure 1. Cluster Overview

Especially the identified constellation of variables in cluster 1 and cluster 2 points in the direction of the results of Havnen et al. (2020), who postulated moderating effects of resilience-associated factors between job stress and job anxiety. Another supporting argument is the variable resistance to change indicating a resilience associated mindset (DeVerteuil and Golubchikov 2016). While it reaches higher values in cluster 1, the opposite applies to cluster 2. Even though the concept of a resilient mindset is still controversial, there are many variables associated with resilience in general such as self-efficacy, job stress, resistance to change, work autonomy and remote-work experience.

In response to the first research question, we reveal some initial insights. Apparently, there are different patterns of self-efficacy, job stress and job anxiety that can be investigated further. We aim to provide a richer understanding on how these three variables relate to resilience. We need more research on the direction of the effects and the issue whether resilience needs to be treated as independent or dependent variable in this context.

As for RQ2, the technological factors we investigated are the roles of IS adequacy and restrictiveness. While IS adequacy is high in cluster 2 and 3, resilience-linked variables show a higher degree of self-efficacy and a lower degree of stress and anxiety; the resistance to change is comparably lower at the same time. We aim to further investigate whether technological factors take over a supporting role (independent variables) or whether IS adequacy and restrictiveness are being perceived more positively (dependent variables): Studies suggest that resilient people are more open-minded to new content and technology (Blayone et al. 2020). Hence, this is an important contribution to an ongoing discussion on the causal structure of resilience and technology.

Moreover, personal factors such as work-family conflict shows a clear pattern that is consistent with the pattern of self-efficacy between the clusters. Organizational factors such as the autonomy-related variables show slight variations between clusters and there seems to be a tendency for higher levels of autonomy to be associated with lower stress, lower anxiety, and higher efficacy. Surprisingly, remote work experience and training does not differ considerably between the clusters. An important further development in this project will be a deeper analysis of this matter. However, there is also a limitation that we acknowledge. While we may be able to draw broad conclusions, we cannot consider specific industries and cultural contexts. This limitation will require further study to demonstrate broader validity. Future research should also look into the causal relationships of the variables investigated to derive managerial implication for the clusters identified in this present study.

5. CONCLUSION

Our study shows that different patterns of the resilience-related variables self-efficacy, job stress, and job anxiety occur amongst individuals who work remotely during the Covid 19 pandemic. This finding is a starting point to resolve the ambiguity of previous research on the role and understanding of resilience in the context of remote work. Furthermore, the present study shows that study participants with different resilient mindsets have divergent personal and organizational frameworks. We moreover identify technological factors that occurred together with a more resilient mindset. These are important findings that provide us with the next step for our ongoing research project. The next step is to derive a causal model based on these exploratory findings and previous research that accounts for interdependencies between variables and to disclose possible unobserved heterogeneity.

REFERENCES

Bandura, A. (1992) 'Self-efficacy mechanism in psychobiologic functioning', in *Self-efficacy: Thought control of action*, Washington, DC, US, Hemisphere Publishing Corp, pp. 355–394.

Blayone, T. J., Mykhailenko, O., Usca, S., Abuze, A., Romanets, I. and Oleksiiv, M. (2020) 'Exploring technology attitudes and personal–cultural orientations as student readiness factors for digitalised work', *Higher Education, Skills and Work-Based Learning*, ahead-of-print, ahead-of-print.

Brayfield, A. H. and Rothe, H. F. (1951) 'An index of job satisfaction', *The Journal of applied psychology*, vol. 35, no. 5, pp. 307–311.

Chin, W. W. (1998) 'The partial least squares approach for structural equation modeling', in *Modern methods for business research*, Mahwah, NJ, US, Lawrence Erlbaum Associates Publishers, pp. 295–336.

Curreri, A. and Lyytinen, K. (2017) 'Mindfulness, Information Technology Use, and Physicians' Performance in Emergency Rooms', *Academy of Management Proceedings*, vol. 2017, no. 1, p. 13828.

Davison, R. M., Wong, L., Alter, S. and Ou, C. (2019) 'Adopted globally but unusable locally: What workarounds reveal about adoption, resistance, compliance and non-compliance', *Proceedings of the 27th European Conference on Information Systems (ECIS)*. United States, Association for Information Systems.

DeVerteuil, G. and Golubchikov, O. (2016) 'Can resilience be redeemed?', *City*, vol. 20, no. 1, pp. 143–151.

Fornell, C. and Larcker, D. F. (1981) 'Evaluating Structural Equation Models with Unobservable Variables and Measurement Error', *Journal of Marketing Research*, vol. 18, no. 1, p. 39.

García-León, M. Á., Pérez-Mármol, J. M., Gonzalez-Pérez, R., Del García-Ríos, M. C. and Peralta-Ramírez, M. I. (2019) 'Relationship between resilience and stress: Perceived stress, stressful life events, HPA axis response during a stressful task and hair cortisol', *Physiology & behavior*, vol. 202, pp. 87–93.

Havnen, A., Anyan, F., Hjemdal, O., Solem, S., Gurigard Riksfjord, M. and Hagen, K. (2020) 'Resilience Moderates Negative Outcome from Stress during the COVID-19 Pandemic: A Moderated-Mediation Approach', *International journal of environmental research and public health*, vol. 17, no. 18.

IBM (2021): SPSS 28. https://www.ibm.com/analytics/spss-statistics-software.

Laumer, S., Maier, C., Eckhardt, A. and Weitzel, T. (2016) 'User Personality and Resistance to Mandatory Information Systems in Organizations: A Theoretical Model and Empirical Test of Dispositional Resistance to Change', *Journal of Information Technology*, vol. 31, no. 1, pp. 67–82.

Lee, J. S. and Keil, M. (2018) 'The effects of relative and criticism-based performance appraisals on task-level escalation in an IT project: a laboratory experiment', *European Journal of Information Systems*, vol. 27, no. 5, pp. 551–569.

Lehrer, H. M., Steinhardt, M. A., Dubois, S. K. and Laudenslager, M. L. (2020) 'Perceived stress, psychological resilience, hair cortisol concentration, and metabolic syndrome severity: A moderated mediation model', *Psychoneuroendocrinology*, vol. 113, p. 104510.

Lloyd, J., Bond, F. W. and Flaxman, P. E. (2017) 'Work-related self-efficacy as a moderator of the impact of a worksite stress management training intervention: Intrinsic work motivation as a higher order condition of effect', *Journal of occupational health psychology*, vol. 22, no. 1, pp. 115–127.

Morgeson, F. P. and Humphrey, S. E. (2006) 'The Work Design Questionnaire (WDQ): developing and validating a comprehensive measure for assessing job design and the nature of work', *The Journal of applied psychology*, vol. 91, no. 6, pp. 1321–1339.

Netemeyer, R. G., Boles, J. S. and McMurrian, R. (1996) 'Development and validation of work–family conflict and family–work conflict scales', *The Journal of applied psychology*, vol. 81, no. 4, pp. 400–410.

Norris, F. H., Stevens, S. P., Pfefferbaum, B., Wyche, K. F. and Pfefferbaum, R. L. (2008) 'Community resilience as a metaphor, theory, set of capacities, and strategy for disaster readiness', *American journal of community psychology*, vol. 41, 1-2, pp. 127–150.

Oreg, S., Bayazit, M., Vakola, M., Arciniega, L., Armenakis, A., Barkauskiene, R., Bozionelos, N., Fujimoto, Y., González, L., Han, J., Hrebícková, M., Jimmieson, N., Kordacová, J., Mitsuhashi, H., Mlacic, B., Feric, I., Topic, M. K., Ohly, S., Saksvik, P. O., Hetland, H., Saksvik, I. and van Dam, K. (2008) 'Dispositional resistance to change: measurement equivalence and the link to personal values across 17 nations', *The Journal of applied psychology*, vol. 93, no. 4, pp. 935–944.

Sadri, G. and Robertson, I. T. (1993) 'Self-efficacy and Work-related Behaviour: A Review and Meta-analysis', *Applied Psychology*, vol. 42, no. 2, pp. 139–152.

Schwarzer, R. and Warner, L. M. (2013) 'Perceived Self-Efficacy and its Relationship to Resilience', in Prince-Embury, S. and Saklofske, D. H. (eds) *Resilience in Children, Adolescents, and Adults*, New York, NY, Springer New York, pp. 139–150.

Shi, M., Liu, L., Wang, Z. Y. and Wang, L. (2015) 'The mediating role of resilience in the relationship between big five personality and anxiety among Chinese medical students: a cross-sectional study', *PloS one*, vol. 10, no. 3, e0119916.

Sousa, C. M. P., Coelho, F. and Guillamon-Saorin, E. (2012) 'Personal Values, Autonomy, and Self-efficacy: Evidence from frontline service employees', *International Journal of Selection and Assessment*, vol. 20, no. 2, pp. 159–170.

Staples, D. S., Hulland, J. S. and Higgins, C. A. (1999) 'A Self-Efficacy Theory Explanation for the Management of Remote Workers in Virtual Organizations', *Organization Science*, vol. 10, no. 6, pp. 758–776.

Tkaczynski, A. (2017): Segmentation Using Two-Step Cluster Analysis. In: Segmentation in Social Marketing: Springer, Singapore, pp. 109–125.

Wattoo, M. A., Zhao, S. and Xi, M. (2020) 'High-performance work systems and work–family interface: job autonomy and self-efficacy as mediators', *Asia Pacific Journal of Human Resources*, vol. 58, no. 1, pp. 128–148.

Zhang, T.; Ramakrishnan, R.; Livny, M. (1996): BIRCH. *SIGMOD Rec.*, vol. 25, no. 2, p. 103–114.

Zompa, A. and Bompiedi, R. (2021) 'Resilience During COVID-19 – A Look at How Employers and Their Employees Adapted', *SSRN Electronic Journal*.

PROMOTING THE ROAD SAFETY THROUGH THE AUGMENTED REALITY: AN ITALIAN EXPERIENCE IN OCCUPATIONAL SAFETY AND HEALTH

Emma Pietrafesa, Nunzia Bellantonio and Agnese Martini

Department of Occupational and Environmental Medicine, Epidemiology and Hygiene INAIL, Via Stefano Gradi, 55 – 00143 Rome RM, Italy

ABSTRACT

In light of the technological changes taking place in the world today, it is important that new tools, devices and technologies are used to improve and support also education and teaching, especially in scientific context. Positive technology, an emerging field based on theoretical and applied research, aimed to investigate how information and communication technologies can be used to enhance the quality of personal experience at three different levels: hedonic well-being, eudaimonic well-being and social well-being. This study highlights how the use of augmented reality learning experiences, could be considered an important tool to invite teachers and students to use virtual approaches to facilitate the processing of a new and different information transfer system. The aim of the project is to promote a road safety education and to raise awareness, stimulating the implementation of risk education initiatives. In this Italian experience, an info-training tool (multimedia augmented poster) has been realized; in particular, two possible immersive experiences were included: the first linked to the infographic multimedia elements, and the second to the videos to create a different visual and auditory stimulation. The study shows that innovation tools and methods are appreciated in occupational safety and health sector and confirms that they can have positive impacts for information, education and training.

KEYWORDS

Road Safety, Augmented Reality, Virtual Reality, Training and Education, OSH

1. INTRODUCTION

In recent years "digital games" with an explicit and well-defined educational purpose, not primarily designed for entertainment, but not excluding it (Abt, 1987), are emerging as innovative tools to promote opportunities for psychological growth and well-being and could be introduced as positive technologies. Positive technology is an emerging field based on theoretical and applied research, whose aim is to investigate how ICT can be used to enhance the quality of personal experience at three different levels: hedonic well-being, eudaimonic well-being and social well-being (Argenton, 2014).

Road traffic injuries are a major global public health issue: current trend in fact suggest that this will continue to be the case in the foreseeable future, and requiring concerted efforts for effective and sustainable prevention (WHO, 2013). Road transport is the most complex of all the systems that people have to deal with on a daily basis, and the most dangerous too with very serious consequences ranging from injury, disability and death. Since 2015 road safety is considered as a prerequisite for ensuring healthy lives, promoting well-being and making cities inclusive, safe, resilient and sustainable by the United Nations. Global data published by the World Health Organization (WHO) show that 1 million 350 thousand people die in the world every year due to road accidents. Road crashes represent the leading cause of death for children and young adults aged 5–29 years and the eighth leading cause of death for all age groups surpassing HIV/AIDS, tuberculosis and diarrhoeal diseases. More than 50% of all road traffic deaths are among vulnerable road users, still too often neglected in the design of road systems in many countries: pedestrians, cyclists and motorcyclists (WHO, 2018). In 2019, an estimated 22,800 deaths from road accidents have reported in the EU Member States; this represents nearly 7,000 fewer deaths than in 2010, a decrease of 23% compared to 2018, the number decreased by 2%. While the underlying trend remains downward, progress has slowed in most countries since 2013 and the EU target of halving the number of road fatalities by 2020 (compared to the 2010 baseline) it will not be met (Eurostat, 2019). The exact number of road work-related accidents is unknown but it estimated that 6 out of 10 work-related accidents resulting in death in Europe are road crashes, including both accidents while driving for work and road accidents with commuters. Furthermore, according to the European Commission, road collisions can account for up to 40% of occupational accidents resulting in deaths. However, many successful initiatives have put in place to fight road accidents: the awareness campaigns carried out by international, European and national institutions and the information campaigns on occupational road risk, aimed at spreading a new sensitivity towards this issue. In order to make people develop good safety habits on the streets, it is very important to educate them on this subject since early age.

2. AIMS

The project focuses on the extent and severity of data relating to road accidents at various levels (living environment and workplace). It started from the need to raise awareness on the issue of road safety with new and effective communication tools, in line with the most recent technological evolution and with the most effective communication strategies analysed so far. The aim of this study is to promote a RS education and to raise awareness, stimulating the implementation of risk education initiatives.

3. MATERIALS AND METHODS

The study started with a comparative analysis of the best innovative practices about road safety (RS) to identify tools and alternative methods of communication, education, information to promote RS culture, using new ideas and innovative technologies. The technological advancements allow the creation of applications for training assistance and support: Virtual Reality (VR) and Augmented Reality (AR) are some of the best suitable scientific domains for successful training applications.

The project led to the creation of an info-training tool in poster format entitled: "Safe road. You can choose!" containing simple, direct graphics recalling essential elements of the road recognisable to any user implemented with augmented elements as real scenes (videos), overlapping virtual and multimedia information levels (infographics and texts).

A questionnaire was drawn up to evaluate and validate the prototype. The items included in the evaluation tool were related to: a) personal data as anonymous user (year of birth, profession, region of domicile, nationality); b) clarity of communication; c) comprehensibility of the message; d) completeness of the information; e) pleasantness of the graphics; f) readability of the information; g) effectiveness of the tool as a vehicle for information and training on road safety; h) innovativeness of the product and i) functionality. The questionnaire was prepared using Google-modules application and administered to participant during a national public event on innovation. For all questions the scale of values was between 1 and 10 where 1 corresponds to "complete disagreement" and 10 to "complete agreement". The last item concerns the overall assessment of the training product with an overall scale between 1 and 5 (through a graphic representation in the form of an emoticon) where 1 represents "not at all satisfied" and 5 "completely satisfied".

4. RESULTS

4.1 Identification and Choice of the Better Digital Tool

An accurate analysis of the literature, showed studies revealing that in meta analysis studies, AR/VR improves post intervention knowledge and skills outcomes of health professionals when compared with traditional education or other types of digital education, such as online or offline digital education, even though further investigation has still to be done to evaluate other outcomes as attitude, satisfaction or behaviour changes. Randomised trials shows that it is still uncertain whether e-learning improves or reduces health professionals' skills and it may make little or no difference in health professionals' knowledge. It seems in fact that, due to the paucity of studies and data, when compared to traditional learning, e-learning may make little or no difference in patient outcomes or health professionals' behaviours, skills or knowledge even though, competency-based education in health care requires rigorous standards to ensure professional proficiency. Nevertheless, demonstrating competency in hands-on laboratories calls for effective preparation, knowledge, and experience, all of which can be difficult to achieve using traditional teaching methods. Various national and international studies have highlighted significant aspects related to the application of augmented reality in educational contexts, with particular reference to the ability to develop autonomy in cognitive processes; to promote self-learning by ensuring compliance with the times and rhythms of each student; to apply a methodology based on discovery, exploration and research; to enhance collaborative and cooperative learning among students; to highlight the complex structure of knowledge, favoring the integration of different disciplinary fields (Panciroli 2018).

In order to allow the development of an informative/training product that would allow a greater involvement and interaction with the user, it was inserted in the project an innovative element compared to the traditional communication campaigns on road risk (video campaigns, information obtained through the creation of posters or corporate awareness campaigns calibrated on strictly working targets) mixing the real world with advanced digital information. This objective was achieved through the use of new emerging communication technologies, such as the augmented reality, in this way users are able to interact with the real environment through a mobile display (smartphone, notebook, tablet etc.), which shows, once framed, the physical world 'augmented', i.e. enriched with certain visual information. In AR, in fact, the subject can maintain contact with reality, because this technology, in most of all cases, allows use even in the absence of specific devices (i.e helmets, viewers, and oculus) which are instead required in the case of virtual reality (Filomia, 2019). AR promotes a high customization of the educational experience, because it lets students to take an active role in the educational process too. In this context, the student has the opportunity to integrate the theory with the experience, deciding from time to time (independently, at the instigation of the teacher or in groups) which topics need to be deepened and then, supported by the informative feedback that can be provided by this type of technology. The aim is not only the quantitative enrichment of information. Increasing reality also means providing places of cognitive growth in which forms of collaborative construction of knowledge and skills are implemented as they get implemented (Panciroli, 2018). AR has a large use especially in teaching very specialized field as surgery, pediatrics, diabetes, dental care and mostly in nursing. Nevertheless, the field called largely health, having a more sociological aspect, is less represented and needs a further development.

4.2 The Augmented Poster "Safe Road. You Can Choose!"

The prototype "Safe road. You can choose!" focused on the following contents: a) data on road accidents differentiated into three levels (world, European and national); b) data on accidents, especially fatal ones, in the workplace, through national accident data with means of transport involved (INAIL source), differentiating the data on "overall" events and "fatal" events; c) illustrated safety prescriptions. The safety prescriptions were oriented towards the use of seat belts, the use of helmets for drivers of motorbikes, the warning not to exceed speed limits, the prohibition of driving under the influence of alcohol, the alerting drivers to driving in suboptimal lighting conditions and fatigue (journeys at night). It is possible to make the journey experience scanning the QR-code (see Figure 1.

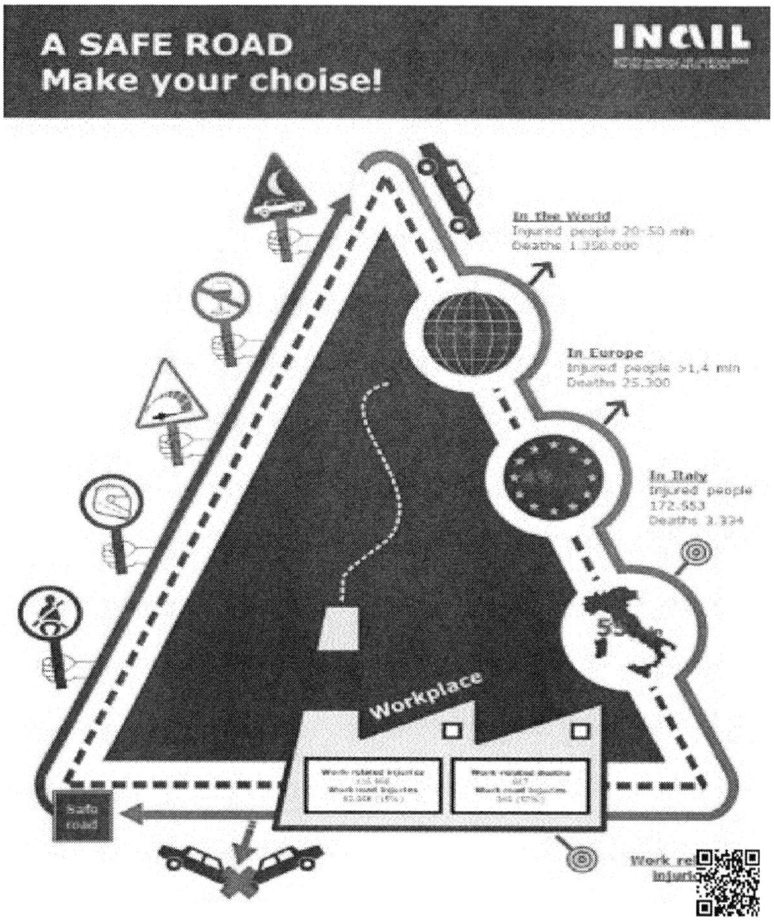

Figure 1. AR Poster "Safe road. You can choose!"

The poster used virtual elements within real scenes, overlapping virtual and multimedia information levels. Two possible type of experiences were included: the first linked to the infographics reproduced through the multimedia elements, and the second to the videos that create a different visual and auditory stimulation. The technology chosen for the realisation of the info-training product was AR with "marker recognition", a technology based on the use of markers, (such as AR tags, photos, images and drawings), which, shown to the webcam, are recognised by the device (iPad, iPhone, tablet or Android smartphone) and superimposed in real time on the multimedia content (video, audio, 3D objects, etc.). In particular, 7 elements of augmented reality were inserted into the poster: 2 images as infographics and 5 videos. The Poster starts from a simple graphic representation with a strong symbolic impact: the triangle, which in the symbolism of road signs indicates danger and metaphorically refers to the intrinsic danger of the road. The road is also recalled by the dotted line element that runs along all the sides of the triangle drawing an imaginary path that crosses "3 roundabouts", i.e. information points that contain numerical data (synthetic and easy to read) on the accident rate in comparison (world, European and national level) aimed at symbolising forms of danger or prohibition or attention. The Triangle shows, by means of a 'traffic light color model', two possible roads: the red road of insecurity and the green road of safety. The car ideally travels along the triangle from the vertex to the left side of the image. The color chosen was red to draw attention to the accident data. Data are represented in the 3 specific roundabouts and were taken from the Global Status Report On Road Safety (WHO, 2018), the 12th Road Safety Performance Index Report (European Transport Safety Council, 2018) and the ISTAT-ACI statistics, Road Accidents (ISTAT-ACI, 2019). The figure is represented as the number of fatalities per million inhabitants, in order to standardize the data and allow for comparison. Moreover, in order to better define and provide information, the data are directed, through an arrow, towards a detailed "window" that shows the numerical data (absolute value) of "injuries" and "deaths" caused by road accidents. AR elements (indicated by a line and a target) were inserted in correspondence with the 'roundabout' and show the national ones.

From the country image it is possible to analyse a first infographic by ISTAT, modified and reworked, in which is presented an urban street, visible from the dashboard of a car, enriched by the silhouettes of buildings which, from the left, show the number of road victims, with details of the type of vehicle used, followed by the number of deaths, with details of the vehicle (car, motorbike, pedestrian, cyclist and so on), then the number of injured (shown on the silhouette of a hospital). Infographic also shows the national accident figures differentiated by the months of the year in which the greatest number of accidents occurs (shown in central buildings), the circumstances of the accidents (shown on a billboard on the right), details of the types of road, details of the age and gender of the victims and details of the days of the week and the hours when the greatest number of accidents occur. The latter figure in particular has been included in order to highlight the importance of the home-work journey and its weight in the overall road accident rate. The symbolic journey of the car reaches the base of the triangle where another graphic element has been created and inserted, another 'information station' linked to the world of work and represented by the shape of a factory. Inside the shape of the factory, a new augmented reality element opens up in the form of an infographic in which the data on road accidents, overall and fatal, taken from the Ministry of Infrastructure-INAIL-CSA Report (2019) are shown. The infographic shows the data on accidents overall, i.e. "total", and that on road accidents (by means of transport) with details of accidents "in itinere" (travelling between home and work). The same logical pathway was also used to report the data on fatal accidents (total, by means of transport and 'en route').

In order to have a better detail of the accident phenomenon it has been decided to report further data, related to road accidents in the infographic that opens in AR at the line/target that is named: "Infographic INAIL". In this infographic a steering wheel is shown in the middle of the page indicating the topic that has been developed, i.e. "Accidents at work by means of transport" and around the central figure rotate the detailed data on the phenomenon (gender, age, economic activity, profession, month, day, time and territory of occurrence). The car's route continues after crossing the "factory" and, at the base of the triangular route, the dangerous red line leads to a deviation, a fork, a choice: downwards the red arrow is directed along a dangerous route (because safe driving behaviour is not taken into account) and there is a collision of two cars while to the left, i.e. towards the apex, the road continues along a route characterised by "safe" driving behaviour (safe route) the route then becomes green. The signposting included in the prototype, starting from bottom to top, represents: a) with the obligatory sign the use of seat belts; b) with the obligatory sign the use of helmets for drivers of motorbikes; c) with the danger sign the warning not to exceed the speed limit and to maintain an adequate speed, (represented by the odometer); (d) the prohibition sign and the symbol of a 'crossed-out' wine glass are used to warn against driving while under the influence of alcohol or alcohol, (e) the final caution/danger sign indicates a night-time context to alert drivers to the fact that they are driving in brightly lit and tired conditions (night-time driving). The choice of the safety prescriptions illustrated in the prototype refer to the main causes of road accidents/accidents recognized at world, European and national level; in fact, they were the subject of an awareness campaign video produced by the WHO, based precisely on the five simple rules identified as necessary to improve road safety. The WHO video was broken down and cut into five micro-parts corresponding to the five rules so that each graphic element corresponded to the video related to the specific rule and inserted as integrated Augmented Reality elements within the visual context of the information product. The objective is to draw attention to the messages of safe behaviour, following the indications of the green path to return safely to the starting point.

4.3 Questionnaire

Fifty subjects (mean age 41.1 years; SD 10.7) completed the questionnaire and assessed a mean value greater than 9 for each dimensions investigated (Likert scale 1 till 10). The mean value for the pleasantness dimension is 4.6 (SD 0.9; Likert scale 1 till 5). The study shows that innovation tools and methods are appreciated in occupational safety and health (OSH) sector and confirms that virtual and augmented reality applications can have positive impacts for information, education and training.

5. DISCUSSION AND CONCLUSION

In this research, we have been able to confirm that the use of AR in the road safety education, has been positive. The subjects who have participated in the study identify this technology as suitable for their use. They have presented as advantages that it is dynamic and fun, raises motivation and attention, and enhances interaction between people. The journey through road safety in this study was motivated by the need to communicate a risk, the road risk, which according to WHO estimates is the eighth leading cause of death worldwide. After reviewing the main communication tools, the augmented poster was considered the most interesting option for a communication that would go beyond the simple visual element and enrich the perception of our addressee with new information and knowledge about road safety. A different form of user involvement, more direct and interactive, which stimulates the user to feel part of the road safety issue, to make it his own by pointing the cursor of his smartphone and his brain at the message that the road can be safe for us and for others, first and foremost "by our own choice", through appropriate behaviour, well-considered choices supported by a full awareness of the risks. The project, given the presence of multimedia elements in AR, is currently the most appropriate choice as it is easily reproducible and replicable in electronic format for sharing as a possible "online" tool of information and communication, assuming the possibility of future insertion of additional information elements more descriptive and detailed, suitable for specific target audience. The prototype can, in fact, be easily reproduced in different sizes and formats (on paper or fabric, for example) so that it can be distributed to workers and made available in any sector or business sector, to inform and communicate in a direct, immediate and effective way the road risk, also in compliance with the law.

Augmented Reality seemed to be the most appropriate choice, as it is a recent and evolving technology, whose applications initially concerned the gaming sector (on which we were oriented) and which is now fully entitled to Industry 4.0 and Training 4.0 because of the various industrial applications that have now expanded the boundaries and diversified the functions. This research shows that the AR technology allows to explore, practice and interact digital contents and provides opportunities to experiment while working: to learn about insecurity and risk in a safe environment. Through the active behaviour of the learning subjects, the ability to make conscious choices develops, an attitude develops, a "mental habit", a social and emotional mastery.

REFERENCES

Abt, C. C. (1987). *Serious games*. University press of America.

Argenton, L., Triberti, S., Serino, S., Muzio, M., & Riva, G. (2014). Serious games as positive technologies for individual and group flourishing. In *Technologies of inclusive well-being* (pp. 221-244). Springer, Berlin, Heidelberg.

D'Amario, S., Tesei, C., Bucciarelli, A., Brusco, A., Salvati, A., Marcelloni, R., Fizzano, MR. (eds.): The trend of accidents at work and occupational diseases. In: Dati Inail, vol.6. Inail, Rome (June 2020). Available from: https://www.inail.it/cs/internet/docs/alg-dati-inail-2020-giugno-inglese.pdf?section=comunicazione

European Commission. (2019). Road Safety Facts & Figures. [Last update: 10/05/2021]. Available from: https://ec.europa.eu/transport/road_safety/road-safety-facts-figures-0_en

Filomia, M., (2019). Augmented reality and textbooks: systematic review. *Form@re - Open Journal Per La Formazione in Rete, 19*(1), 165-178. https://doi.org/10.13128/formare-24757

World Health Organization. (2016). Global status report on road safety 2018. Geneva: World Health Organization; 2018. Available from: https://www.who.int/publications/i/item/9789241565684

Incidenti stradali. Anno 2019. Automobile Club d'Italia, ISTAT; 2019. Available from: https://www.istat.it/it/files//2020/07/Incidenti-stradali-in-Italia-Anno-2019-aggiornamento27ottobre2020.pdf

Ministero delle Infrastrutture e dei Trasporti, Dipartimento per le Infrastrutture, i Sistemi Informativi e Statistici, Direzione generale per i Sistemi Informativi e Statistici (DGSIS) - Div. 3 - Ufficio di Statistica: Div. 3 - Ufficio di Statistica: Programmazione strategica 2019. Obiettivo strategico: miglioramento della sicurezza nelle costruzioni, nelle infrastrutture e nei cantieri. Obiettivo operativo: statistiche sull'incidentalità nei trasporti stradali, anche con riferimento alla tipologia di strada. Documento di presentazione del rapporto 2019. Available from: https://www.mit.gov.it/sites/default/files/media/pubblicazioni/2019-12/Presentazione%20Rapporto_1.pdf

Panciroli, C., Macauda, A., & Russo, V. (2018). Educating about art by augmented reality: New didactic mediation perspectives at school and in museums. *Multidisciplinary Digital Publishing Institute Proceedings*, *1*(9), 1107.

World Health Organization. (2013). *World health statistics 2013: a wealth of information on global public health* (No. WHO/HIS/HSI/13.1). World Health Organization. [cited 2018 Oct 29]. Available from: http://www.who.int/healthinfo/global_burden_disease/projections2015_2030/en/

EXPLORING CONSUMER ATTITUDE TOWARD SUSTAINABLE ENERGY-EFFICIENT APPLIANCE: PRELIMINARY FINDINGS FOR AUGMENTED REALITY APPLICATION

Gabriella Francesca Amalia Pernice[1], Valeria Orso[2] and Luciano Gamberini[3]
[1]*Departments of General Psychology, University of Padova, Padova, Italy*
[2]*Human Inspired Technology Research Center, University of Padova, Padova, Italy*
[3]*Departments of General Psychology, Human Inspired Technology Research Center University of Padova, Padova, Italy*

ABSTRACT

Increasing the energy efficiency (EE) of a household appliance is a step forward to reduce energy consumption. However, it will not be enough if consumers, during the purchase, are not fully aware of the sustainable characteristics of the household appliances and their environmental impact. Eco-labels contain a lot of information sustainable-related and are a powerful tool that is considered by 79% of consumers during purchases but it is unclear whether users really know this important information. However, the large number of different eco-labels on the market may decrease consumer confidence and increase misunderstanding. In our study we explored the eco-label knowledge of 101 respondents, highlighting the most relevant characteristics for them when purchasing a household appliance. Despite relying on eco-label, users are not aware of the sustainable information about electricity and water consumption and that could have a negative impact on the environment and on users' savings.

Results from this preliminary study could be translated into suggestions for the application of Augmented Reality (AR) in the sustainability market. This technology could be used to make energy labeling clearer for the user, showing relevant appliance information and providing the consumer with more eco-friendly options to encourage a more responsible choice of sustainable appliances. It can also be a powerful tool that can change non-ecological behaviors in the daily use of household appliances.

KEYWORDS

Purchase Intention, Sustainability, Household Appliance, Eco-Label; Refurbished, Augmented Reality

1. INTRODUCTION

Climate change is undoubtedly a major issue that forces both policy makers and individual citizens to take action. Manufacturing companies, industrial farming, and the usage of household appliances are sources of carbon emissions that contribute to global warming. Indeed, the European residential sector is on the top of the list for energy use, accounting for 26.3% of the entire European energy consumption (European Commission, 2019). End-users may play a key role in mitigating household energy consumption because they both purchase home appliances and manage their daily usage (Gamberini et al., 2012). However, if the end-user is not fully aware of the characteristics of the appliance at the time of purchase (or how to use it properly), and their impact on the environment, the efficiency of the appliance is jeopardized (Rosak-Szyrocka & Żywiołek, 2022). Previous studies underlined the importance of energy labeling to make sustainable purchase choices, despite this, the average users reported little understanding of labeled information and energy characteristics of household appliances.

In this exploratory study, we aimed to investigate users' general perceptions and knowledge about eco-labeling and their green trust to improve the sustainability behavior of users who purchase energy-efficient appliances (Waris & Hameed, 2020). Additionally, we addressed refurbishment, which is the process of replacing, repairing, or reprocessing parts of a used product to bring it to like-new condition, This process's benefits for the end user are already known in Industry 4.0, (e.g. in terms of economical savings) (Ijomah,

2009), but researchers investigated the perception of the purchaser and their expectations about the phenomenon in the household appliances sector are needed (Bressanelli, Saccani, Perona, & Baccanelli, 2020).

Our findings provide indications to assist consumers in the aware choice of energy-efficient appliances by leveraging mixed reality technology.

1.1 Related Work

In the European Union marketers must expose a label that provides information regarding the Energy Efficiency (EE) and environmental attributes of some specific product. Household appliances must be classified by efficiency, from A (most efficient) to G (least efficient). According to the Special Eurobarometer 492 (European Commission, 2019), eco-labels are considered by 79% of consumers during purchases. However, thanks to advances in energy efficiency, most appliances now have the A+ (or higher) label, and the difference between these high energy classes is not clearly defined or explained. To solve this problem, in March 2021 energy label categories were rescaled from A to G for product groups such as refrigerators, dishwashers, and washing machines, while for others (e.g., oven and tumble dryer) they have not been changed.

Eco-label is a powerful instrument because it helps consumers to evaluate the potential effects on the environment, improving their decision process (Thøgersen, Pedersen, Paternoga, Schwendel, & Aschemann-Witzel, 2017). However, this great amount of different active labels on the market can negatively affect the levels of trust perceived by the consumers in sustainable purchases (Nikolina, 2020).

Recent research underlined those green purchases are affected by several factors that decrease consumers' trust, such as high prices of green products, perceived lower product quality, ineffective marketing strategies, and lack of accurate information (Sheth, Sethia, & Srinivas, 2011). Current evidence emphasized that typically consumers are not always aware of the information shown by the eco-labels, and misunderstandings are possible (Brécard, 2014). Generally, the eco-label focuses specifically on water or energy consumption in a specific use phase and does not report clear information about other uses. In addition, important information about the Life Cycle of Appliances is completely missing in energy labels (Russo, Rossi, Germani, & Favi, 2018). Understanding the environmental impact of household appliances during their entire life span, from raw material extraction to final disposal, may be another factor affecting consumer decisions towards better environmentally-driven decisions, such as buying refurbished household appliances (Li, Wu, Jin, & Lai, 2017).

The role of the end user in label design is still limited, so future research should take a user-centered approach to eco-design to improve the labelling and make it clear and more understandable to consumers. Indeed, the literature points out that the knowledge of consumers about products' sustainability increases their purchase intention. Further studies should evaluate the content of these labels, not forgetting the importance of the medium in which the information is presented.

Augmented Reality (AR) technology aligns real and virtual objects with each other through different devices. AR enhances or overlays virtual content, including textual and visual content over the physical environment. (Álvarez Márquez & Ziegler, 2020), encouraging exploratory behavior in consumers, which will affect their intention to purchase. Augmented reality applications in retail have been shown to allow consumers to make more aware purchase decisions by bringing to the foreground product information that is poorly visible.

The capability of Augmented Reality to highlight relevant product-related information, makes it a powerful tool to assist consumers when choosing a new appliance.

In addition, several studies point out that this technology is associated with learning benefits, better understanding, and increased user motivation. However, knowledge alone is not always enough to change entrenched behavior. Exposing people to the consequences of their unsustainable choices, e.g., simulating them in AR, could encourage green behavioral change as shown in Virtual Reality studies (Ahn, 2011).

2. MATERIALS AND METHODS

2.1 Materials

The questionnaire comprised 50 items in total, grouped into three sections being, Demographics, Perceptions and knowledge of EE labels, and Perceptions and knowledge of refurbished electronic devices. The Demographic section investigated participants' background, namely gender, age, education, job position, monthly income, with how many people they live, and past experience in purchasing household appliances (8 items).

The second section explored respondents' perceptions and knowledge about EE labels. In addition, it explored the characteristics that participants consider at the purchase stage and their willingness to pay for energy-efficient appliances.

Perceived knowledge of energy-efficient label (PK), investigated consumers' evaluation and perception of the eco-labels (2 items for generic EE label adapted from Waris & Hameed, 2020; 7-point Scale).

Green trust (GT) assessed consumers' perceived credibility about EE labels (3 items adapted from Hameed & Waris, 2018; 7-point Scale).

Purchase intention (PI) explored intentions to purchase environmentally friendly products, such as high EE household appliances (1 item adapted from Nguyen et al., 2017; 7-point Scale).

Prioritization of purchase criteria; in order to understand which are the most important features of a household appliance that might influence the buying decisions, a rank with the product's features was created (12 features for each appliance). Respondents had to order the features from the most to the least important (Sonnenberg, Erasmus & Donoghue, 2011).

Knowledge of the new EE label; to deeply explore the knowledge of the new eco-label, an ad-hoc questionnaire was created. We asked participants to select all the statements that reported correct general information about the new label (6/11 were true). The 11 statements were based on the provided by the European Union. Finally, the last section assessed the respondent's knowledge and attitude toward refurbished electronic devices.

Perceived knowledge of refurbishment (PK) assessed the extent to which respondents considered themselves knowledgeable about refurbished devices (2 items adapted from Waris & Hameed, 2020; 7-point Scale).

Purchase intention (PI) assessed respondents' willingness to buy a refurbished device (3 items adapted from Hameed & Waris, 2018; 7-point Scale).

Environmental Concern (EC) explored participants' attitudes toward environmental issues (1 item adapted from Ward, Clark, Jensen & Yen, 2011, 7-point Scale).

Knowledge of refurbished electronic devices investigated participants' knowledge regarding the characteristics of refurbished technology (3/7 statements were true; statements were adapted from existing literature) (Ijomah, 2009).

2.2 Participants

The sample comprised 101 respondents, (F=49; M=50; Non-binary=1; Unspecified=1), with an average age of 37.8 years old (SD=15.6; age range 21- 81).

The majority of the respondents reported medium or high educational qualifications (master's degree=33.66%; high school degree=31.68%). Fifty-eight percent reported a medium-high income (while 30.7% had a medium-low income and 11% had a high income). The income consistently reflected participants' occupation: 49% was an employee (in the private sector 28%; in the public one 21%), students were the 15%; blue collar workers and the unemployed were both the 8% of the sample. Respondents reported living on average with other 2 people (SD=1,31). Finally, 62% of respondents had purchased a household appliance in the past, of them the 73% reported being directly involved in this kind of purchase.

2.3 Procedure

The questionnaire was pre-tested to reduce error through possible misinterpretation. The data were obtained through an online survey using the Qualtrics platform and was conducted in June 2022. Participants had accepted the informed consent to fill the questionnaire. To involve a diverse sample in terms of age, income, and education, snowball recruitment was employed. Each participant directly recruited through the researchers' social network send the questionnaire to others. Participation in the survey was voluntary, and no incentive was offered.

2.4 Analysis and Results

To explore respondents' answers, an overall score was computed for each of the dimensions explored by averaging the scores assigned to each item. It turned out that respondents' claimed to be familiar with the EE label (M= 5.52; SD= 1.3), reporting that they trust the information shown, and that it is useful for understanding the environmental impact of an appliance (M= 4.77; SD= 0.88), and that they also rely on it during purchase (M= 5.65; SD= 1.32). Finally, they reported that the meaning of the new energy label is clear (M= 5.52; SD= 1.30).

However, their tested knowledge about the new EE label (Table 1) reveals that it is still unclear which appliances feature the eco-label (58.42% of incorrect responses). The only device on which they know that the new eco-label is placed in the TV and monitor (only 11% of wrong answers). In addition, they seemed also confused about the energy-related information reported on the label (EI) (49% of incorrect responses). It is clear that the information reported by the new EE label is confusing for the majority of the sample. The only statement that seems not to be misunderstood is that the new energy label does not indicate that the same appliance is now less efficient (only 12% of wrong answers). However, it should be noted that some information reported such as "Specifically reports the daily electricity consumption, in kWh " (40% of incorrect responses) or "Indicates the energy efficiency of household appliances" (36.6% of incorrect responses) are contents also present in the old energy labels. A low match of correct answers for these items could also reflect a misunderstanding of the old energy label, which is still placed on other household appliances.

Table 1. Statements reported about the new Eco-Label True (T); False (F) and respondents' frequencies of the wrong answer in percentage (%)

Dimensions	Statement reported about the new Eco-Label True (T); False (F)	Frequencies of wrong answer (in %)	Mean % of the dimensions
WA			58, 42%
	It's placed on refrigerators, freezers, wine cellars (T)	48, 5%	
	It's placed on washing machines but not on tumble dryers (T)	97%	
	It's placed on air conditioners (T)	95%	
	It's placed on ovens (electric and gas) (F)	40, 6%	
	It's placed on TV and monitor* (F)	11%	
EI			49%
	It has a scale ranging from A+++ to G (F)	43, 6%	
	For some household appliances, it shows the stand-by consumption (T)	80%	
	For some household appliances, it shows the level of noisy (T)	80%	
	It Specifically reports the daily electricity consumption, in kWh (T)	40%	
	It Indicates the energy efficiency of household appliances (T)	36, 6%	
	If with the old label an appliance was rated A+++, now that is rated B, it means that is less efficient energy than before (F)	12%	

In Table 2 the most important characteristics chosen by the respondents when they have to buy a household appliance are reported (specifically the five most chosen options are shown, out of twelve alternatives proposed). In this questionnaire, we considered specifically three large household appliances (i.e., washing machine; refrigerator; dishwasher) because they feature the new energy label. After the price, the most important characteristics considered by the participants when buying a washing machine and a dishwasher are energy efficiency and general consumption. Considering the fridge, this choice seems to be related to the specific functionalities and physical characteristics of the appliance.

Table 2. Respondents ranking of the 5 most important characteristics (out of twelve alternatives proposed) considered during the purchase phase of household appliances

Rank	Characteristics	Frequencies in %
Washing Machine		
1st	Price	32
2nd	Electric Consumption	23
3rd	Water Consumption	20
4th	Energetic Class	13
5th	Ecological washing	14
Fridge		
1st	Fridge Capacity	35
2nd	Price	20
3rd	Water Consumption	22
4th	Noise	19
5th	No frost function	16
Dishwasher		
1st	Price	31
2nd	Electric Consumption	22
3rd	Water Consumption	19
4th	Energetic Class	17
5th	Ecological washing	17

Finally, respondents have to choose how much they would be willing to pay for a high energy class (A) household appliance with the most common features (Table 3). The prices range were six, based on the three major e-commerce of electronics, as follows: "Less than 400 €"; "400-599 €"; "559-749€"; "749-999€"; "1000-1249 €"; "More than 1249€". The majority of the participants underestimated the value of such appliances, compared to the average market price. Notably, regardless of the type of household appliance, participants would be willing to pay the same lower price range.

Table 3. Respondents willing to pay for high energy class appliances compared with the Italian market average price

High energy-efficient household appliance	Italian market average price	Most frequently chosen range of price (€)	% of participants who underestimated the market value
Washing Machine	952 €	400-599 €	82%
Fridge	2199€	400-599 €	100%
Dishwasher	886€	400-599 €	80%

The last section of the questionnaire aimed to understand respondents' past experiences with refurbished electronic devices. The 87% of the sample had never bought a refurbished electronic device. Despite this, exploring in-depth their actual knowledge about reconditioning, participants seem to clearly know who resells these devices and the features and quality standards these devices achieve (on average, only 14 % wrong answers were given). They claimed to know them (M= 4.85; SD= 1.32), understood the benefit to the environment (M= 5.6; SD= 1.5) and would be willing to buy them (M= 4.4; SD= 0.71). Although 68% of participants would be willing to purchase a small refurbished appliance (e.g., microwave, coffee maker, blender), the choice to purchase a large refurbished appliance (e.g., dishwasher, fridge, washing machine) does not appear to be as clear-cut, even if it is high (50 %).

3. DISCUSSION AND CONCLUSION

In this exploratory study, we analysed the importance of eco-label as a tool capable to influence the purchase phase of household appliances. Consistently with the Special Eurobarometer 492 (European Commission, 2019), participants reported a positive attitude towards the EE, highlighting a general medium-high level of perceived knowledge (both with regard to the old labels and the new ones) green trust, and purchase intention, that the literature links to consumers' environmental attitudes and green consumption behavior (Nguyen et al., 2017). However, some issues emerged while testing users' true knowledge about the new EE label, for example in which appliances they can find it or the reported information about electricity consumption related to the environmental impact of the household appliances. This finding may also be relevant considering that some statements that carried general information about sustainability are also true statements carried in the old eco-labels. The current presence of many different eco-labels on the market could be an element of misunderstanding. Consistently with the literature, participants emphasize that the price guides them in their purchase choice, followed by the features that may be most related to sustainability (such as energy and water consumption, or energy class) (Sonnenberg, Erasmus & Donoghue, 2011). Although the average monthly income of the sample is medium-high, they are willing to pay much less than the average price for a high-energy class household appliance. Energy label information should be easily understandable and report more obvious and detailed information about the sustainability of the appliance, such as water or electricity consumption.

This study also aimed to explore the participants' perceptions of refurbished electronic devices. Most studies have focused on the role of reconditioning in Industry 4.0 (Ijomah, 2009), but few paid attention to common electronic devices. Although most participants had never purchased a refurbished device, the preliminary results of the present study show a good understanding of the phenomenon and general confidence in the green potential of these devices. In addition, participants showed a propensity towards the purchase of small refurbished appliances rather than larger ones.

The above considerations are fertile ground for future research. If price drives participants' purchase decisions, refurbished and energy-efficient household appliances represent a valid eco-friendly alternative. AR represents an emerging digital market trend that is supporting users' decision-making process by offering functions for product comparison and recommendation. The potential of this tool in helping consumers make healthy and environmentally friendly choices is still under evaluation, but the benefits are clear. AR allows customers to have a real-time interactive experience examining product information, boosting exploration behavior, and improving buy intentions. (Lee, Kaipainen & Väänänen, 2020).

Since users generally rely on energy labels without really understanding them, the implementation of AR could improve user knowledge and prevent erroneous usage behaviors in daily life. Customers are known to be reluctant to invest time and cognitive effort during the purchase decision. AR technology may help users in gathering, analysing, and integrating new information, lowering the cognitive load and cost of accessing information (lvarez Márquez & Ziegler, 2020). Augmented reality makes it easier for consumers to visually examine product features by highlighting relevant product-specific information. This enhances product knowledge in favor of sustainable choices, boosting their willingness to pay for that product, and poses as a powerful tool in the retail of household appliances. As results from this preliminary study suggest, AR could be used to make energy labelling clearer for the user, and to clearly show relevant appliance information (e.g., consumption of specific wash cycles), warranty, and reparability index. In addition, it would be possible to provide the consumer with more eco-friendly options (e.g., buying a refurbished appliance) in line with the user's needs. In conclusion, AR is proposed as a powerful and promising tool for the sustainability market. This technology has the potential for creating positive behavioral changes, that could be exploited to promote long-term behaviors in the use of household appliances.

REFERENCES

Ahn, S. J. (2011). *Embodied experiences in immersive virtual environments: Effects on pro-environmental attitude and behavior*. Stanford University.

Álvarez Márquez, J. O., & Ziegler, J. (2020, September). In-store augmented reality-enabled product comparison and recommendation. *In Fourteenth ACM Conference on Recommender Systems* (pp. 180-189).

Brécard, D. (2014). Consumer confusion over the profusion of eco-labels: Lessons from a double differentiation model. *Resource and Energy Economics, 37*, 64–84

Bressanelli, G., Saccani, N., Perona, M., & Baccanelli, I. (2020). Towards circular economy in the household appliance industry: An overview of cases. *Resources, 9*(11), 128.

Commissione Europea (2019) Il Green Deal Europeo (No. COM/2019/640 final). https://eur-lex.europa.eu/legal-content/IT/TXT/?qid=1576150542719&uri=COM%3A2019%3A640%3AFIN

European Commission. (2019). Special Eurobarometer 492: Europeans' attitudes on EU energy policy. data.europa.eu. https://data.europa.eu/data/datasets/s2238_91_4_492_eng?locale=en

European Commission. (2022). Energy consumption in households - Statistics Explained. https://ec.europa.eu/eurostat/statistics-explained/index.php?title=Energy_consumption_in_households#Energy_consumption_in_households_by_type_of_end-use

Fujiwara, N., van Asselt, H., Bößner, S. et al. The practice of climate change policy evaluations in the European Union and its member states: results from a meta-analysis. *Sustain Earth 2*, 9 (2019). https://doi.org/10.1186/s42055-019-0015-8

Gamberini, L., Spagnolli, A., Corradi, N., Jacucci, G., Tusa, G., Mikkola, T., ... & Hoggan, E. (2012, June). Tailoring feedback to users' actions in a persuasive game for household electricity conservation. In *International conference on persuasive technology* (pp. 100-111). Springer, Berlin, Heidelberg.

Hameed, D., & Waris, I. (2018). Eco labels and eco conscious consumer behavior: The mediating effect of green trust and environmental concern. Hameed, Irfan and Waris, Idrees (2018): Eco Labels and Eco Conscious Consumer Behavior: The Mediating Effect of Green Trust and Environmental Concern. Published in: *Journal of Management Sciences, 5*(2), 86-105.

Ijomah, W. L. (2009). Addressing decision making for remanufacturing operations and design-for-remanufacture. *International Journal of Sustainable Engineering, 2*(2), 91-102.

Lee, J., Kaipainen, K., & Väänänen, K. (2020, January). Local foodie: Experience design of a mobile augmented reality application for tourists to encourage local food consumption. In *Proceedings of the 23rd International Conference on Academic Mindtrek* (pp. 110-119)

Li, W., Wu, H., Jin, M., & Lai, M. (2017). Two-stage remanufacturing decision makings considering product life cycle and consumer perception. *Journal of Cleaner Production, 161*, 581-590.

Nguyen, T. N., Lobo, A., & Greenland, S. (2017). Energy efficient household appliances in emerging markets: the influence of consumers' values and knowledge on their attitudes and purchase behaviour. *International journal of consumer studies, 41*(2), 167-177.

Nikolina, S. A. J. N. (2020). Sustainable consumption: Helping consumers make eco-friendly choices.

Rosak-Szyrocka, J., & Żywiołek, J. (2022). Qualitative Analysis of Household Energy Awareness in Poland. *Energies, 15*(6), 2279.

Russo, A. C., Rossi, M., Germani, M., & Favi, C. (2018). Energy Label Directive: current limitations and guidelines for the improvement. *procedia CIRP, 69*, 674-679.

Sheth, J. N., Sethia, N. K., & Srinivas, S. (2011). Mindful consumption: A customer-centric approach to sustainability. *Journal of the Academy of Marketing Science,39*(1), 21–39

Solà, M.d.M., de Ayala, A., Galarraga, I. et al. Promoting energy efficiency at household level: a literature review. *Energy Efficiency 14*, 6 (2021). https://doi.org/10.1007/s12053-020-09918-9

Sonnenberg, N. C., Erasmus, A. C., & Donoghue, S. (2011). Significance of environmental sustainability issues in consumers' choice of major household appliances in South Africa. *International Journal of Consumer Studies, 35*(2), 153-163.

Thøgersen, J., Pedersen, S., Paternoga, M., Schwendel, E., & Aschemann-Witzel, J. (2017). How important is country-of-origin for organic food consumers? A review of the literature and suggestions for future research. *British Food Journal, 119*(3), 542–557.

Ward, D. O., Clark, C. D., Jensen, K. L., & Yen, S. T. (2011). Consumer willingness to pay for appliances produced by Green Power Partners. *Energy Economics, 33*(6), 1095-1102.

Waris, I., & Hameed, I. (2020). Promoting environmentally sustainable consumption behavior: an empirical evaluation of purchase intention of energy-efficient appliances. *Energy Efficiency, 13*(8), 1653-1664.

THE HUMAN IN THE HOME: PRIVACY INVASION RISKS OF SMART HOME APPLIANCES AND DEVICES

Kalala T Nshima[1] and Roelien Goede[2]
[1]*Vaal University of Technology, Vanderbijlpark, South Africa*
[2]*North-West University, Potchefstroom, South Africa*

ABSTRACT

Homeowners are generally very eager to add the latest gadgets to their homes. They buy smart speakers, smart televisions, or any other smart appliance without regard to the security risks these devices might pose. Not only does a smart device collect data for its basic operation, but it also stores the information on the device and often on the cloud. Most often the device is managed from a mobile phone application. Unauthorized access to the devices can have a disastrous impact for the homeowner. Consequences include jeopardy of access controls when locks are opened, financial lost when fridges are compromised, loss of private data and many more risks when cloud data is compromised. The aim of this paper is to raise awareness to the education of the unsuspected homeowner on the privacy invasion risks involved in the modern smart home. Guidelines for homeowners are developed from combining results from a literature review and an empirical study. The empirical investigation was done as a study on data from documents. Data was collected from two main sources, manufacturers' documents (in the form of technical manuals) and from online reviews. Data from manufacturers are viewed as data coming from official documents, and data from reviewers are subjective viewpoints which can be seen as open-ended interviews without any detail interview questions. The two sources provide two main viewpoints on the data: firstly, the official view of the manufacturer, and secondly, the subjective view of various reviewers. Content analysis is used to analyze and code the data from an interpretive research paradigm perspective. The contribution of the paper is list of guidelines for homeowners of the risks of smart devices in general and guideline from specific smart devices such as televisions, fridges, speakers, and locks.

KEYWORDS

Smart Home Security; Security Threats, Security Awareness, Smart Appliance Security

1. INTRODUCTION

With the rapid advancement in smart technology, many users have adopted smart appliances into their homes. It is estimated that by the year 2030, up to 500 billion devices will be connected to the internet (Cisco, 2019). These devices will be equipped with sensors, allowing them to be able to collect environmental data or any other type of data. Due to their network connectivity, these devices will be able to transmit collected data to servers in the cloud for analysis, which in turn will help these devices to make intelligent decisions or take better actions.

Among the 500 billion devices which will connect to the internet by 2030, a portion of that number will comprise of household appliances, such as fridges, microwave ovens, stoves, coffee makers, air conditioners, electricity meters, televisions and many more. A number of these devices already have internet connectivity, such as smart televisions.

With such a high number of devices and home appliances connecting to the internet, security becomes a major concern for most homeowners.

In view of the advancement in IoT, data protection will become even more important. It is true that IoT generates a tremendous amount of data (Aazam et al., 2014:414) due to its many connected devices. It was estimated that by the end of 2019 the amount of data produced by humans, machines, and things was going to be in the range 500 zettabytes (Ni et al., 2018:601).

The introduction the various smart devices and appliances in a Home Area Network, put users' privacy at risk. For instance, some of these devices, such smart speakers and smart TVs may come with microphones and cameras that are at risk of being hacked by intruders. Often users are always not aware of the amount and type of data being collected and how this data is being used.

The purpose of this paper is to demonstrate how the introduction of smart devices and appliances in a HAN may compromise the data privacy of its users in a smart home.

The paper is structured as follow: Section 2 introduces the literature review upon which the research is based on, Section 3 covers the empirical study, in Section 4 we present our findings, and in Section 5 our conclusion is presented.

2. LITERATURE REVIEW: IOT SECURITY

The core function of every secure system, according to Fisch et al. (2017:2) is to provide confidentiality, integrity, and availability. Bhaskar (2008:14) goes even further at defining what security is, by dividing the concept into computer security and network security. He describes computer security as mechanisms and procedures taken to provide confidentiality, integrity, and availability of the data stored on a computer. The components of Figure 1 show that data is at the core of security in any type of network, including IoT networks.

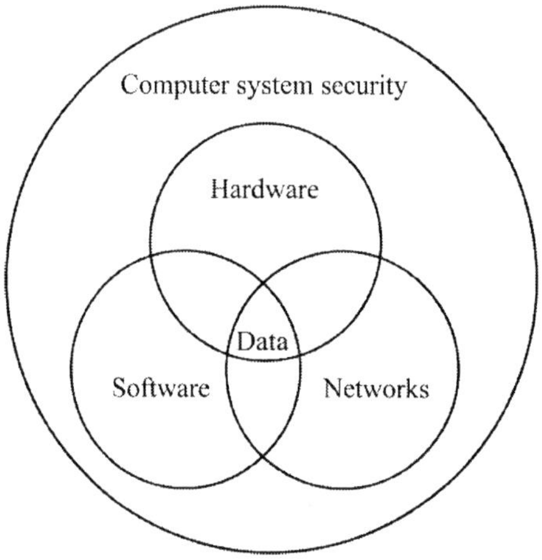

Figure 1. Aspects of computer security, adapted from Carroll, 2014:8

Experts believe that IoT networks are extremely vulnerable to attacks due to the following facts, (1) the unattended nature of most IoT devices; (2) most communications in IoT are wireless, making eavesdropping easy; (3) the low capabilities of IoT devices in both power and hardware, makes it difficult to implement strong security measures (Atzori et al., 2010:2801). According to Ali et al. (2019:1), resource constrained IoT devices makes it difficult to implement traditional security mechanisms such as cryptography, exposing these devices to data integrity and confidentiality issues. In addition, Zhang and Zhu (2011:511) describe hidden dangers to information security in IoT. Such as the wide application of RFID tags, and the way information is processed and stored. Taking into consideration the risks which IoT possess, it is necessary that measures be implemented to protect the privacy of users in HANs. Lack of privacy has been the root cause of most concerns expressed alongside IoT. Atzori et al. (2010) acknowledges that "IoT represents an environment in which the privacy of individuals is utterly jeopardized in various ways".

In this paper, the IoT network of focus is a Home Area Network (HAN) which part of a smart home. A typical home area network may consist of smart appliances, smart sockets, smart lights, security systems smart locks, smart TVs, smart wearable devices etc. (Ali et al., 2017). All these IoT objects are network capable, allowing them to be remotely accessed and operated, thus making them all susceptible to attacks. A HAN may consist of several smart devices and objects, all of which can be vulnerable in one form or another. Although some of these devices may have been designed with security in mind, but to have just one insecure device in a network may present significant vulnerabilities (Patton et al., 2014:232) and may just be the weakest point for the whole network.

Sathu and Shukla (2007:87) have also concluded that smart homes will have significant security challenges due to the lack of security knowledge by most of these households. For instance, most users operate in the administrator account on their PCs, password control is always feeble, computers are sometimes left on the whole night connected to the internet, and there are no backups done regularly or at all. These poor security practices can also be brought into IoT networks, resulting in real security challenges for smart homes where most smart appliances will have no ability to detect malicious infections.

In a survey done by Canonical (2017:5), it was discovered that several factors contribute to vulnerabilities in IoT devices today, these factors include:

- Hard-coded passwords: they argue that the majority of IoT devices are shipped with no password or the standard admin password which is known and exploitable.
- Fundamentally weak security at both the software and hardware levels: they argue that many IoT devices are not designed with security best-practice as a priority.
- Lack of software updates: the survey argues that when a vulnerability is found, manufacturers are not quick to roll out updates in time, or it might happen that no updates are released at all.
- The size of the opportunity: the billions of IoT devices which have been sold in the past few years mean the attack surface has also increased tremendously. This poses a high risk for attacks such as Distributed Denial of Service (DDoS). As these devices become more integrated into the industry and society at large, the survey argues that organized attacks will also increase.

It has been argued in this paper that connectivity in an IoT network contributes substantially to the risk in the smart home. Network security threats can be explained according to Bhaskar (2008:24), under the following four categories: interruption, interception, modification, and fabrication.

I. Interruption: in a network setup, most communication takes place between clients and servers. The communication between these clients and servers may be interrupted using various techniques resulting in the services being provided by the servers to be unavailable to clients (Bhaskar, 2008:24). Several attacks exist that may cause interruption to a system, such as denial of service (DoS).

II. Interception: in network communication, an attacker can make use of various tools, such as packet sniffers, to intercept traffic between the sender and the receiver, and be able to read the contents of the packets. This type of threat may result in a problem of confidentiality (Bhaskar, 2008:24). Interception is passive in nature, whereby the intention of the attacker is to gain access to information by using methods such as eavesdropping, no tampering of data takes place.

III. Modification: unlike interception, which is passive, modification is active (Bishop, 2005:5). With modification, packets are captured, and the contents of these packets are altered before the packets are forwarded to the targeted recipient; this causes a problem of data integrity (Bhaskar, 2008:25). Modification can also be used by an attacker to alter data in a database, or even change the working of hardware by changing certain settings.

IV. Fabrication: fabrication aims at corrupting the authenticity of the data (Bhaskar, 2008:25). With fabrication, an attacker will try to send a message to a recipient pretending to be who they are not, or the attacker might insert a record in a database without permission. (Bishop, 2005:5).

Data security is vital to any network, since protecting data is the main objective of computer and network security. Data security involves the protection of sensitive data to avoid theft and protect privacy (MIT, 2016). Not all data is sensitive, hence it is important to have data classified. Stewart et al. (2012:225) describes the primary objective of data classification is to provide security mechanisms for storing, processing, and transferring of data. Data can be classified based on its sensitivity, cost, risk, vulnerability and many more.

For instance, in a smart home, a large amount of data is generated from smart meters to smart thermostats, to the many sensors that may be connected in the home, and smart appliances. Most of this data are of no value if it is not sent to server applications in the cloud where the data can be analyzed to benefit the householder. Majority of IoT devices suffers from resource limitations such as memory, thus they cannot store a large amount of data. Therefore, IoT will greatly rely on cloud computing for the storage, processing, and interpretation of its data. The data generated in a HAN will face many challenges. First, there is a high risk of data being stolen while it is transmitted between the HAN and the internet. Secondly data collected by the utility companies and device manufacturers will be stored online making it a target for cybercriminals.

How secure is data stored online? The slow adoption of cloud computing, especially by organizations, has always been around the issue of data security and privacy protection (Chen & Zhao, 2012:647) due to the insecure nature of the internet. In essence, storing data in the cloud means that the user has no control over it. In the current setup of things, service providers have been providing users with the assurance of their data using a service level agreement (SLA) (Kandukuri et al., 2009:517), but it is proven through research that cloud computing is not as safe as many service providers might claim (Chen & Zhao, 2012:647).

According to Wang et al. (2010), cloud storage is not yet safe for sensitive data; this is mainly because cloud storage does not guarantee integrity and availability of data. With the establishment of bodies such as the cloud security alliance (CSA) and the use of security content automation protocol (SCAP) (Ali et al., 2019:460), cloud security can see a significant improvement in data security. It will take time before everyone follows any standards set, as it was discussed before, the lack of standardization in IoT is a significant problem.

Ali et al. (2019:460) argue that large cloud vendors, such as Microsoft, Amazon, Google, and others, use proprietary technologies. This heterogeneity may result in security problems. Data security does not only involve storage but also protection during transmission as well.

3. EMPIRICAL STUDY

The approach this study adopted is that of interpretive case study and the source of data is mainly from documents, thus document research in context of a case study, was used as the main method of data collection. Data was collected from two main sources, manufacturers' documents (in the form of technical manuals) and from online reviews. Data from manufacturers are viewed as data coming from official documents, and data from reviewers are subjective viewpoints which can be seen as open-ended interviews without any detail interview questions. The two sources provide two main viewpoints on the data: firstly, the official view of the manufacturer, and secondly, the subjective view of various reviewers.

A large amount of data was collected from these two sources. To be able to analyze this data, it was vital that a fitting analysis method be used. In this research a deductive qualitative content analysis approach was adopted to analyze manufacturers' manuals and reviewers' documents from a security perspective. Both categories of the data are in electronic format. In the instance of electronic documents, although the data are already captured, time is needed to prepare them for analysis.

Content analysis is defined by Vaismoradi et al. (2013:400) as "a systematic coding and categorizing approach used for exploring a large amount of textual information unobtrusively to determining trends and patterns of words used, their frequency, their relationship, and the structure and discourse of communication".

3.1 Data Collection

Before data could be analyzed, the first step is identifying the data of data you want to collect and where to collect it from. For instance, Table 1 gives an example of data sources on smart speakers. Data sources for smart speakers come from three of the largest manufacturers of smart speakers globally according to market share. These speakers in no way represent the whole spectrum of smart speakers out there, but the data obtained is enough for us to have an understanding on the security issues smart speakers are facing.

The data in Table 1 come from documents from manufacturers, these include device manuals, privacy documents of smart speakers, and manufacturers' websites. For each chosen device, data are presented in terms of both sources, namely the manufacturer's documents and reviewer information. Reviewer documentations were collected by following the following process:

- For each device a model was selected to gather reviews for. In many cases, slightly older models of devices were selected to ensure that enough reviews are available to collect valuable data.
- The process starts by searching for a particular term or statement, such as "Sony smart TVs data privacy".
- From the millions of results from the search, reviews are carefully read to see if they cover the main point required. This is done for several reviews until a satisfactory one is found.
- Once the review is selected, it is included in the reviewers' table (discussed below).
- Then a code is created in the first column and a footnote is generated for extra information.

Table 1. Smart speakers data source

Manufacturer	Model	Motivation for selection#	Manufacturer documents
Amazon	Echo	By 2018, the USA alone had almost 118 million smart speakers in users' homes (Sterling, 2019). Amazon alone claims to have sold around 100 million Echo smart speakers worldwide (Leary, 2019). The Echo model seems to be the most popular smart speaker from Amazon. Also, Amazon is currently the leader in the number of smart speakers sold to date (Owen, 2019).	Amazon (2019c)
Google	Home	Google Home is the second most sold smart speaker, according to published reports (Owen, 2019). By the end of 2018, almost 11 million google home smart speakers were sold.	Google (2019a)
Apple	Homepod	Apple Homepod is added to this list because Apple produces premium products. Although it might be 6th globally, it is 3rd in the USA, and the brand is well known also globally as compared to Chinese brands which are currently only popular locally.	Apple (2019)

3.2 Data Representation: Coding

Coding can be described as the process of "transforming raw data into theoretical constructions of social processes" (Kendall, 1999:746). Corbin and Strauss (1990:12) though, think of coding as data analysis process. According to Kendall (1999:746), through the process of emergence, codes and categories should be able to fit the data.

3.2.1. Coding from Manufacturers' Documents

The first source of data namely manufacturer's documents include technical manuals, user manuals, and privacy policy document on the device. Data for each category of device is presented and analysed in a table, similar to the excerpt provided in Table 2. The table has four columns, the first represents codes developed from the content analysis method chosen for this study, and the other three represent each manufacturer; under each manufacturer, there are three columns with the headings of "Yes, No, and No Comment (NC)". NC represents a situation where the aspect is neither confirmed nor denied in the data.

Table 2. Representation of manufacturers' documentations

Manufacturers	Amazon			Google			Apple		
Model	Echo Dot			Home			Homepod		
Codes	Yes	No	NC	Yes	No	NC	Yes	No	NC
VULNERABILITY									
Acknowledges speaker having a microphone.	X			X			X		
Acknowledges that the speaker is always listening.	X			X			X		
Specifies how much RAM the speaker has.		X			X				X
Manufacturer specifies the type of encryption the speakers use.		X		X					X
Specifies the CPU being used by the speaker.			X		X		X[1]		
THREAT									

[1] Apple acknowledges that the smart speaker is using an A8 processor

Manufacturers Model Codes	Amazon Echo Dot Yes	 No	 NC	Google Home Yes	 No	 NC	Apple Homepod Yes	 No	 NC
Acknowledges that the speaker is always listening.	X			X			X		
Acknowledges collecting information on the usability of the service by users.			X	X			X		
Acknowledges storing collected data from smart speakers on their servers.	X			X					X
Acknowledges storing a recording of voice commands to improve the functionality of the service.			X	X					X
Acknowledges encrypting conversations stored locally on the device.			X	X					X

3.2.2 Coding from Reviewers' Documents

The final data representation table for each device (in this case smart speakers) presents the coding of the reviewer documents as depicted in Table 3. The reviewers' table used as example here, consists of three major columns, the first column representing the codes, the next two columns represent manufacturers. Under each manufacturer's column, there are sub-columns for reviewers. There are between four to five reviewers per item; this done so that saturation can be achieved by collecting data about the same topic from different viewpoints. In this table, a code coming from a reviewer is represented with a Y#, where the Y represents the reviewer agreeing to what the code says, then # represents the footnote. Because codes should be short, footnotes are provided where more detail of what the reviewer said is given. For instance, the code "Claims that speakers can be hacked to access locally stored data" that appears under the vulnerability row in Table 3 represents statement made by reviewer R1 about Amazon Echo dot, denoted by Y[4]. Reviewer R2 concurs with the views of reviewer R1 hence the notation Y[5] appearing under the same row. By referring to the relevant footnote, more info is given about reviewer's statements. Once again, the table is followed by a discussion of the data to identify vulnerabilities and threats.

Table 3. Coding example of online reviews

Code Reviewer references	Amazon Echo Dot R1 (Panda, 2018)	 R2 (Symantec, 2019)	 R3 (Koch, 2019)	Google Home R1 (Panda, 2018)	 R2 (Symantec, 2019)	 R3 (Koch, 2019)
			THREAT			
Claims smart speakers always listening.	Y[2]	Y[3]	Y[4]	Y1	Y2	Y3
			VULNERABILITY			
Claims that speakers can be hacked to access locally stored data.	Y[5]	Y[6]		Y4	Y5	

[2] The reviewer claims that apple, google, and Amazon admit that their smart speakers are always listening to the conversation where they are unless you switch off that function.
[3] The reviewer claims that the always-listening function of smart speakers scares most people.
[4] The reviewer argues that smart speakers represent a new frontier of corporate espionage due to their ability to always listening.
[5] The reviewer argues that, although the locally stored data is deleted once the speaker realises that you have stopped talking, it takes only a few minutes for a hacker to break into the speaker and steal local data.
[6] The reviewer confirms that smart speakers can be hacked.

4. FINDINGS

The aim of the findings is to establish how the introduction of smart appliances and devices put the data and privacy of users in a smart home at risk. This was accomplished by identifying the security concerns of each category of smart devices. The term security "concern" represents both vulnerabilities and threats and possible attacks as this distinction is dependent on context which is sufficiently described in the documents under investigation.

In this section the security concerns discussed in terms of vulnerabilities and threats identified in the data are summarised and combined from the two data sources. The aim is to identify specific concerns that needs to be considered in the development of guidelines for homeowners regarding security concerns in smart homes.

Although in this research data from five categories (fridges, speakers, TVs, Locks, and Cameras) of smart devices was analysed, we will now present the findings from the smart speakers' data analysis.

4.1 Smart Speaker Findings

The following threats were identified from manufacturer's documents:

- Always on speaker recording data.
- Storage of collected data in the cloud.
- Replication of user voice commands.

The following security concerns were identified from reviewers' documents:

- Data stored permanently in cloud for long periods.
- Data is stored on the device itself.
- Built-in security lacking.
- Third party plugin software not under scrutiny of manufacturer.
- Intruders may use device to control gateway.
- Misused of stored information.

These lists may be combined into a single list of threats posed using smart speakers:

- Always on speaker.
- Data is stored on the device itself.
- Data stored in cloud for long periods.
- Built-in security lacking.
- Third party plugin software not under scrutiny of manufacturer.
- Replication of user voice commands.
- Intruders may use device to control gateway.
- Are the findings rooted in the data analysis?

The coding of other devices was done similarly, resulting in similar list from the two sets of documents analysed.

4.2 Resulting Guidelines

We were able to provide generic guidelines fitting of must types of devices in addition to guidelines for owners of specific devices as indicated in tables 4 and 5 respectively.

Table 4. Guidelines for homeowners using smart devices from literature and device documentation

General Guidelines
For each device in the smart home, be sure to understand the user interface, pay special attention to user access and authentication.
For each device in the smart home, be sure to investigate which security settings can be configured by the user.
When using a mobile app to configure devices, be aware of vulnerabilities caused by outdated software and lack of updates.
Restrict access to mobile devices used to control the gateway.
Always set the mobile app to update automatically.
Ensure that smart device firmware is up to date.
Be aware of the service level agreements in terms of data protection, encryption and sharing of data.

Be aware of the security impact of the specific type of network used to upload data from the device to the cloud.
Be aware of different types of networks used in smart homes and that the connection between devices may be insecure.
Investigate which security and protocol settings can be controlled on the gateway to improve secure communication.
Investigate which OS is installed on a gateway, the well-known OSs used in current mobile phones and tablets are typically more secure than an unknown one.
Be aware of electricity options in terms of uninterrupted power supply of the hub since power cuts may be a form of attack.
Be aware of the dangers concerning communication between the mobile app and the gateway for control. Investigate the encryption provided and ensure it is activated.
Be aware of which sensors are installed in each device and which data they collect and importantly over what range they collect data.
Investigate the effects when a specific sensor fails on the functionality and safety of the appliance.
Be aware of the time settings of data collection and the reach of the sensors to know when and where data is collected.
Investigate the OS used in appliance or host device to understand secure data storage and transfer from the specific device.
Be aware that individual devices have identification tags that may be copied to create insecure entry.

Table 5. Additional device specific guidelines from empirical study

Additional Device Specific Guideline
Television
Be aware that all additional apps loaded on the TV may pose additional security concerns.
Users should be aware that they need much more knowledge than anticipated to protect themselves.
Pay special attention to data collection and sharing in the policy documents of the specific manufacturer.
Investigate the specific OS of a smart TV model. Different models of TVs use different operating systems, each with implications of security.
Smart Speakers
Users should delete old data from the device.
Access to the device in a smart home should be considered as a risk and therefore users should know the range of their speakers.
Smart Locks
Users should consider everything that can go wrong if security attach is experienced.
Users should use a qualified technician to install devices and discuss security concerns with them,
Smart Fridges
User access is of greater importance since fridge can be used as the gateway. Users need to be aware that their IoT fridge maybe not be able to detect that it has been affected by malware, so extra precaution is needed to make sure they keep their appliances unaffected.
Users should be aware that the durable lifespan of the fridge itself is much longer than that of the technology. Support might not be available while the fridge is still working. This is especially a concern for security updates.
When buying a fridge, a user should make himself/herself aware of all the security concerns of gateways.
Smart gateways
User access is of great importance, since all other devices the home is connected to the gateway. The consequences of a possible attack concern every aspect of the network.
Users need to be aware that the gateway has always watching cameras and always listening microphones. This also implies that users should know the ranges of these sensors.
Users should take responsibility to know what is recorded and what is done with the data.
Users should understand that secure network communication depends on hardware and OS capabilities of the gateway, as well as the continued support thereof in terms of updates.

5. CONCLUSION

The aim of this paper is to identify data privacy threats posed by the introduction of smart appliances and devices to unsuspecting homeowners in relation to their privacy in a home area network. The following conclusions can be reached:

Poor security management: Many smart home appliances are shipped with minimal security settings. It rests upon the user to implement other security settings, such as changing the default administrator login details or changing the password from time to time. The lack of knowledge from users is amplified by the technical know-how of configuring such settings. This in turn open loopholes in the network security for intruders.

Risk to data: It was established that a smart home generates a large amount of data, the majority of which is uploaded to the cloud for analysis to help these smart appliances make better decisions or give the users better advice, whichever the case maybe. From the analysis of some of the privacy documents of these devices, it appears that sensitive data may be collected in the process. Although the manufacturer's intention is not to collect such data, the presence of sensitive microphones in smart speakers and some smart TVs, already put these devices at risk of being exploited by intruders.

Due to a lack of proper hardware to support good encryption, some smart appliances and devices may be at risk of transmitting data which is easy to be intercepted by intruders, such as in the case of man-in-the-middle attack.

The final risk to users' data concerns the location and use of the data. The storage of users' data in the cloud presents a major challenge to their privacy. Sometimes not all users know what is done with their stored data. For instance, smart speakers collect a tremendous amount of audio data which can be analyzed in various ways.

In conclusion, there is no doubt of the many benefits that smart appliances bring to users, such as convenience. The value of a smart appliance is not limited only to what the appliance can do, but more so to the services, the appliance can provide through the cloud (Uehara, 2015:457). Despite these benefits, many home users are not technical people and are interested only to have a device or an appliance that is plug and play (PnP). Unfortunately, not all these smart devices and appliances provide PnP security, thus it rests upon manufacturers to improve on the security of these devices and appliances from their designs. It is the responsibility of the home users to educate themselves on security management of their smart home or to hire an expert to setup the security for them. Through this research, security guidelines for homeowners have been devised to help them mitigate security threats in their smart homes.

REFERENCES

Aazam, M., Khan, I., Alsaffar, A. A. & Huh, E.-N. Cloud of Things: Integrating Internet of Things and cloud computing and the issues involved. Applied Sciences and Technology (IBCAST), 2014 11th International Bhurban Conference on, 2014. IEEE, 414-419.

Ali, I., Sabir, S. & Ullah, Z. 2019. Internet of things security, device authentication and access control: a review. *arXiv preprint arXiv:1901.07309.*

Ali, W., Dustgeer, G., Awais, M. & Shah, M. A. IoT based smart home: Security challenges, security requirements and solutions. 2017 23rd International Conference on Automation and Computing (ICAC), 7-8 Sept. 2017 2017. 1-6.

Amazon. 2019. *Alexa, Echo Devices, and Your Privacy* [Online]. Available: https://www.amazon.com/gp/help/customer/display.html/ref=hp_left_v4_sib?ie=UTF8&nodeId=GVP69FUJ48X9DK8V [Accessed 18/7/2019 2019].

Apple. 2019. *Apple homepod* [Online]. Available: https://www.apple.com/homepod/ [Accessed 8/8/2019 2019].

Atzori, L., Iera, A. & Morabito, G. 2010. The Internet of Things: A survey. *Computer Networks,* 54, 2787-2805.

Bhaskar, S. 2008. *Information Security: A Practical Approach*, Alpha Science International, Ltd.

Bishop, M. 2005. *Introduction to computer security*, Addison-Wesley Boston, MA.

Canonical. 2017. *Taking charge of the IoT's security vulnerabilities* [Online]. Available: https://pages.ubuntu.com/IoT-Security-whitepaper.html [Accessed 24/10/2019 2019].

Carroll, J. M. 2014. *Computer security*, Butterworth-Heinemann.

Chen, D. & Zhao, H. Data security and privacy protection issues in cloud computing. Computer Science and Electronics Engineering (ICCSEE), 2012 International Conference on, 2012. IEEE, 647-651.

Cisco. 2019. *Internet of Things* [Online]. Available: https://www.cisco.com/c/dam/en/us/products/collateral/se/internet-of-things/at-a-glance-c45-731471.pdf?dtid=osscdc000283 [Accessed 14/10/2019 2019].

Corbin, J. M. & STRAUSS, A. 1990. Grounded theory research: Procedures, canons, and evaluative criteria. *Qualitative sociology,* 13, 3-21.

Fisch, E. A., White, G. B. & Pooch, U. W. 2017. *Computer system and network security*, CRC press.

Google. 2019. *Data security & privacy on Google Home* [Online]. Available: https://support.google.com/googlenest/answer/7072285?hl=en [Accessed].

Kandukuri, B. R., Paturi, V. R. & Rakshit, A. Cloud security issues. Services Computing, 2009. SCC'09. IEEE International Conference on, 2009. IEEE, 517-520.

Kendall, J. 1999. Axial coding and the grounded theory controversy. *Western journal of nursing research,* 21, 743-757.

Kinsella, B. 2019. *Alibaba Dominates China Smart Speaker Sales with 41.2% Share* [Online]. Available: https://voicebot.ai/2019/02/21/alibaba-dominates-china-smart-speaker-sales-with-41-2-share/ [Accessed 05/06/2019].

Koch, R. 2019. *The privacy risks of smart speakers like Amazon Echo, Apple HomePod, and Google Home* [Online]. Available: https://securityboulevard.com/2019/01/the-privacy-risks-of-smart-speakers-like-amazon-echo-apple-homepod-and-google-home/ [Accessed 18/7/2019 2019].

Leary, B. 2019. *As Smart Speakers Rise in Popularity, Synthetic Speech and Voice Synthesis is Something We Need to be Ready For* [Online]. Available: https://smallbiztrends.com/2019/01/voice-assistant-security-concerns.html [Accessed 05/06/2019 2019].

MIT. 2016. *Protecting Data* [Online]. Available: https://ist.mit.edu/security/protecting_data [Accessed 7/6/2016].

Ni, J., Zhang, K., Lin, X. & Shen, X. S. 2018. Securing Fog Computing for Internet of Things Applications: Challenges and Solutions. *IEEE Communications Surveys & Tutorials,* 20, 601-628.

Owen, M. 2019. *HomePod sales up in fourth quarter, Amazon and Google extending lead* [Online]. Available: https://appleinsider.com/articles/19/02/19/homepod-sales-up-in-fourth-quarter-amazon-and-google-Extending-Lead [Accessed 05/06/2019].

Panda. 2018. *Which is the safest smart speaker?* [Online]. Available: https://www.pandasecurity.com/mediacenter/technology/which-is-the-safest-smart-speaker/ [Accessed 04/06/2019 2019].

Patton, M., Gross, E., Chinn, R., Forbis, S., Walker, L. & Chen, H. Uninvited Connections: A Study of Vulnerable Devices on the Internet of Things (IoT). 2014 IEEE Joint Intelligence and Security Informatics Conference, 24-26 Sept. 2014 2014. 232-235.

Sathu, H. & Shukla, R. Home area network: a security perspective. Proceedings of the 6th WSEAS International Conference on Information security and Privacy (ISP'07), 2007. 14-16.

Sterling, G. 2019. *Survey: 118 million smart speakers in US, but expectation is low for future demand* [Online]. Available: https://marketingland.com/survey-reports-118-million-smart-speakers-in-u-s-but-the-expectation-of-future-demand-is-way-down-254937 [Accessed 05/06/2019].

Stewart, J. M., Chapple, M. & Gibson, D. 2012. *CISSP: Certified Information Systems Security Professional Study Guide*, John Wiley & Sons.

Symantec. 2019. *Can smart speakers be hacked? 10 tips to help stay secure* [Online]. Available: https://us.norton.com/internetsecurity-iot-can-smart-speakers-be-hacked.html [Accessed 04/06/2019].

Thestaronline.Com. 2018. *Chinese smart speaker models prove a hit at home* [Online]. Available: https://www.thestar.com.my/tech/tech-news/2018/11/08/chinese-smart-speaker-models-prove-a-hit-at-home/ [Accessed 05/06/2019 2019].

Uehara, M. The Design of a Framework for Smart Appliances. Advanced Information Networking and Applications Workshops (WAINA), 2015 IEEE 29th International Conference on, 24-27 March 2015 2015. 457-462.

Vaismoradi, M., Turunen, H. & Bondas, T. 2013. Content analysis and thematic analysis: Implications for conducting a qualitative descriptive study. *Nursing & health sciences,* 15, 398-405.

Wang, C., Ren, K., Lou, W. & Li, J. 2010. Toward publicly auditable secure cloud data storage services. *IEEE network,* 24, 19-24.

Zhang, H. & Zhu, L. Internet of Things: Key technology, architecture and challenging problems. Computer Science and Automation Engineering (CSAE), 2011 IEEE International Conference on, 2011. IEEE, 507-512.

VIRTUAL REALITY APPLICATIONS IN AUTISM SPECTRUM DISORDER: A SYSTEMATIC REVIEW

Mohd Amran Md Ali[1], Mohammad Nazir Ahmad[1], Wan Salwina Wan Ismail[2] and Nur Saadah Mohamad Aun[3]

[1]*Institute IR4.0, Universiti Kebangsaan Malaysia, Bangi, Selangor, Malaysia*
[2]*Department of Psychiatry, Faculty of Medicine, Universiti Kebangsaan Malaysia, Bangi, Selangor, Malaysia*
[3]*Centre for Research in Psychology and Human Well-Being, Faculty of Social Sciences and Humanities, Universiti Kebangsaan Malaysia, Bangi, Selangor, Malaysia*

ABSTRACT

Autism Spectrum Disorder (ASD) is a neurodevelopmental disorder that can cause significant social, communication, and behavioral challenges. Recently, Virtual Reality (VR) has emerged to play a significant role in enhancing some of the treatments of autism spectrum disorder. Studies have shown that VR has been used to enhance various aspects of communication, social, and cognitive skills among autism spectrum disorder patients. However, experts of the ASD community still lack a comprehensive understanding of how VR can be used in the autism spectrum disorder treatment process. This systematic review has the objective of exploring how VR has been applied in the ASD domain to date. The identification process produced 353 papers from 11 different databases. After applying the exclusion criteria, the set was reduced to 34 papers, which clearly fitted the criteria defined for the accomplishment of the systematic review, which were then further analyzed and classified. As a result, we highlight several key elements or factors that are imperative to understanding how VR is applied in the ASD domain. From the studies analyzed, there is evidence that indicates that VR based treatments can help children with ASD. Nevertheless, the promising results and the advantages of virtual reality (especially considering ASD symptomatology) should encourage the scientific community to further develop new or more advanced VR based treatments by further exploring the key factors that may hinder or challenge the development of VR for the ASD domain.

KEYWORDS

Virtual Reality, Autism Spectrum Disorder, Application, Treatment

1. INTRODUCTION

Autism Spectrum Disorder (ASD) results in impairments in three important domains: social skills, communication, and behaviour. Individuals suffering from ASD need proper, strategized help (American Psychiatric Association, 2019). Studies and development in VR for these individuals have been ongoing for over two decades (Didehbani et al., 2016). Mesa-Gresa et al., (2018) focused on VR and autism by conducting evidence-based systematic studies on the effectiveness of VR-based interventions in ASD. However, no systematic approach exists that helps in understanding how VR is used in ASD. In 2017, Brok and Sterkenburg contributed a systematic review that examines studies that use self-control technology to support skills achievement. While this review included other studies that focused on people with intellectual disabilities, it did not consider some interesting studies on VR and ASD. Although there have been past studies that have explored the relationship between VR and ASD, no studies have detailed a systematic approach to understand how VR is used in ASD.

The contributions of this study also have a significant added value as they include several other elements such as the type of VR technology and VR methodology used in the systematic search. With the addition of these elements, the contributions of this study cover a wider range of publications related to the use of VR in ASD. Therefore, this paper seeks to answer the following research questions:

RQ1: For which specific target group in ASD is VR applied?
RQ2: What skills in ASD are being trained using VR?

RQ3: What are the techniques used in interventions for patients with ASD?
RQ4: What are the technologies used in VR-based Training (VRT) for ASD?

This paper conducts a systematic literature review of current state-of-the-art research into VR in ASD. The paper is organized as follows; Section 2 describes the literature review with a focus on VR in ASD. Section 3 describes the method used for our systematic review. Section 4 presents the results of our synthesis of the literature and discussion of the findings. This is followed Section 5, which concludes the study.

2. LITERATURE REVIEW

VR is defined as a computer-generated simulation, such as a set of images and sounds that represent a real place or situation that can be interacted with in a seemingly real or physical way by a person using special electronic equipment. It can transmit visual, auditory, and other various sensations to users through a headset to make them feel as if they are in a virtual or imagined environment. The idea of VR was first presented in the 1950s, and it has now developed to a point where it may be used for entertainment (Grossard, 2018). More than 230 businesses, including major corporations like Samsung Electronics, Apple, Facebook, Amazon, and Microsoft, are now working on various VR-related products and doing research and development. VR systems are made up of a computer, a video, and VR headgear. Recently added items include seats, gloves, and sensors.

According to a study by Bellani et al. (2011), a VR setup imitates a real environment (which it overlays), offering a very accurate environment for VR validation because it gives the participant the same exact visual input regardless of position and orientation within the environment, real or virtual. Haptic feedback systems are virtual reality (VR) devices that send vibrations and other sensations to the user through a game controller, gloves, or chairs (Bekelis et al., 2017).

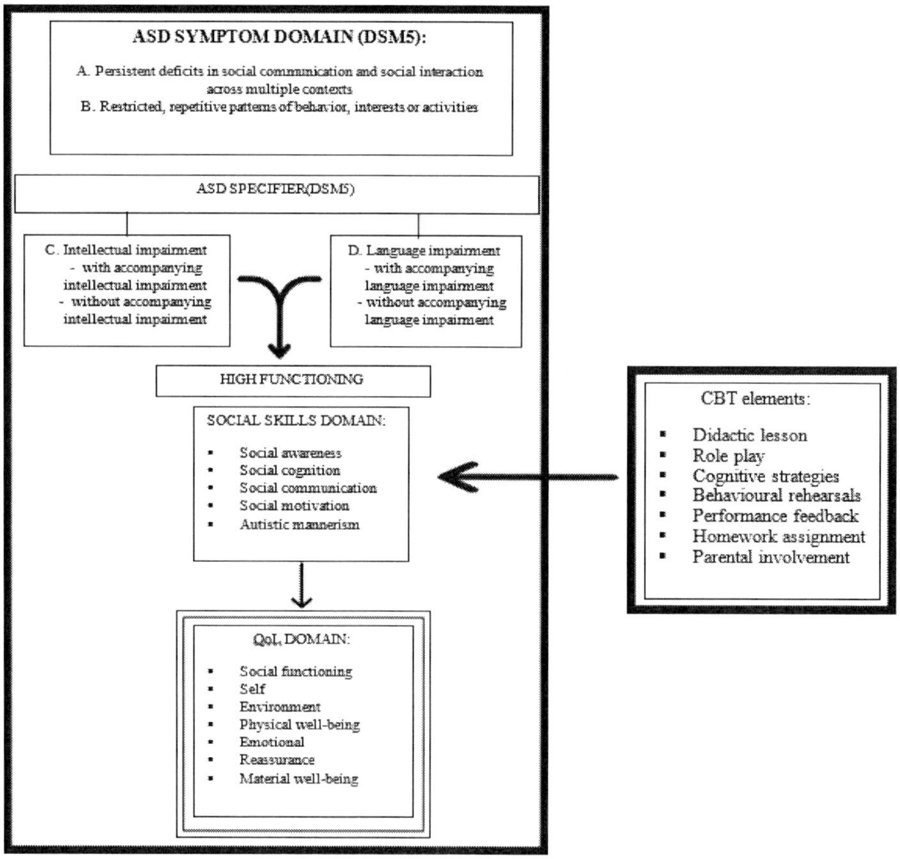

Figure 1. Body of Knowledge for ASD

ASD is a neurodevelopmental disorder characterized by persistent deficits in social communication and interaction, and restricted, repetitive patterns of behavior, interests, or activities. Figure 1 above depicts that, although the term High-Functioning Autism Spectrum Disorder (HFASD) is not included in the DSM, it is used to describe individuals on the spectrum that have IQs that are within the normal range (Blyth, 2018). Despite having a normal IQ, their inflexibility and ongoing social communication problems hinder their ability to do regular social tasks.

Elements of Cognitive behavioral therapy (CBT) for training social skills in ASD include didactic lessons, role play, cognitive strategies, behavioral training, performance feedback, homework assignments, and parental involvement (Brereton et al., 2007). Improvement in the functioning of social skills will potentially lead to improvements in the quality of life (QoL). Domains of QoL include social functioning, self, environment, physical well-being, emotional, reassurance, and material well-being. QoL among individuals with HFASD is influenced by social impairment, level of severity, intellectual, and language impairment (Botella et al., 2017). CBT intervention will improve social skills among individuals with high functioning ASD, optimize their potential, and consequently their QoL. In conclusion, it is evident that improvement in social skills using CBT will potentially improve QoL among individuals with HFASD. Figure 1 below combines all elements of CBT, social skills and QoL.

3. METHOD

The systematic literature review (SLR) is a step-by-step process that enables researchers to create their own search procedure. This review was carried out in accordance with the technique for conducting SLRs as proposed by (Bird et al., 2017). It is used in identifying the required information from selected articles. This method was chosen because it makes it easier to capture, summarize, synthesize, and critically comment on any of the topics reviewed. The SLR process consists of the following steps which are; 1) identify resources, 2) study selection, 3) inclusion and exclusion criteria, 4) data extraction and lastly 5) data analysis.

The stages of study selection for the systematic review in this paper using Burke et al., (2018) guidelines. The first stage involved searching for 7 keywords (virtual reality, VR, virtual reality treatment, autism spectrum disorder, ASD, high functioning autism spectrum disorder, HFASD) in 11 scientific databases. As a result, 353 primary studies were identified. Subsequently, Stages 2–4 were undertaken twice. In the first iteration, the papers identified in Stage 1 were examined, while in the second iteration, the references contained in those papers were examined. The systematic literature review was the methodology used to perform the review conducted in this study. In the first stage, 353 papers were identified from 11 different databases. The following databases were used to search keywords in the 'search terms' section: Science Direct; Business Source Premier; Inspec; Springer Link; AIS (Association for Information System) Electronic library; Scopus; ProQuest Science Journals; Google Scholar; ISI Web of Science; ACM Digital library; IEEE Explore.

The selection of material for inclusion in our systematic review was based on the following inclusion and exclusion criteria. The researcher included studies: (1) published from 2016 onwards (up to present day); (2) that studied the use of VR as intervention for any skills for ASD population; (3) published in related journals/conference proceedings, and (4) that directly answered our research questions. The researcher also excluded studies: (1) in languages other than English; (2) not focusing on the use of VR in ASD population; (3) focused on any other type of virtual reality such as augmented reality; (4) that were in the form of article summaries, news, or reviews, and (5) that were dated before year 2016. The data extraction and synthesis stage involved the extraction of some key details from the 34 studies included in our systematic review. For this study, the analysis was performed by combining multiple variables of interest which included: (1) year of publication and source name, (2) diagnosis spectrum (ASD/HFASD), (3) skills treated, (4) technologies involved, and (5) techniques used in treating ASD patients. Microsoft Excel was used for entering the data, undertaking descriptive analysis, and drawing diagrams.

4. RESULTS

This section explains the searching strategies of this literature survey to produce the results. The search is done manually through popular and familiar digital libraries and databases. 34 reports were identified relevant to the scope of the study, which is, the application of virtual reality in autism spectrum disorder. The

34 reports were re-screened to identify the factors underlying each study. 34 articles were identified similar to the factors addressed in this study. In other words, 11 reports were not focused on important factors of virtual reality for autism spectrum disorder. The data sources for all these articles are presented in Table 1.

In the previous years, many articles have been published relating to virtual reality applications in autism spectrum disorder. After selecting 34 articles about virtual reality in autism spectrum disorder, the author found that all 34 of the articles had all the elements that used important criteria. From the study, the expected findings were classified into percentages based on the results of the articles, where the participants' age, spectrum (ASD/HFASD), skills treated, technologies involved, and techniques used in VR to treat patients with ASD were evaluated. This section also discusses some elements related to VR development in ASD as well as HFASD, consisting of virtual reality effectiveness for ASD and HFASD, role of domain knowledge, general VR methodology in VR development in ASD, cultural elements in VR content, appropriate language used in the development of VR for Autism Spectrum Disorder, and the impact of VR used in ASD.

Table 1. Papers' classifications

No.	Year	Responden Age	ASD Spectrun	Skills Treated	Technique of Intervention	Tool / Technologies Applied
1	2020	Age not specified	ASD	Communication and collaboration skil - Verbal communication	ABA	CAVE
2	2020	9-16	ASD	Improvement in attention processes and spatial cognition skills	Physical therapy	HMD
3	2020	7-15	ASD	Showed improvements in their real-life targeted phobia	Sensory integration therapy	HMD
4	2019	6-16	ASD	Development of daily living skills (shopping skills)	ABA	HMD
5	2019	18-65	ASD	Emotion recognition in a social context	Not mentioned	HMD
6	2019	Age not specified	ASD	Social learning and imitation skills	Not mentioned	Mobile Devices
7	2019	26	ASD	Improve emotional skills; Specific emotional script	DIR	Avatar
8	2019	16	ASD	Social skills	Speech therapy	Avatar
9	2019	8-14	ASD	Treatment for fears/phobias	CBT	CVE (Blue Room)
10	2019	7-12	HFASD	Emotion recognition in a social context	DIR	HMD
11	2019	8–11	ASD	Therapy for Specific Phobias	CBT	CAVE
12	2018	4–6	ASD	Enhance social skills, emotion & attention	Occupational therapy	CAVE
13	2018	8–11	HFASD	Communication and interaction	Speech therapy	Avatar
14	2018	10–13	ASD	Identify the 6 core emotions	DIR	Avatar
15	2018	Age not specified	ASD	Improve communication ability	Speech therapy	HMD
16	2017	10–12	ASD	Improve social understanding and skills	Life skills classes	Haptic device
17	2017	8–13	HFASD	Examining approach and tendencies in the recognition of emotions	Physical therapy	Virtual reality (VR) and (CVE)
18	2017	4–7	ASD	Representation of pretense and promoting pretend play	Physical therapy	Haptic device
19	2017	12–15	ASD	Development of daily living skills	Life skills classes	HMD
20	2017	13–17	HFASD	Emotion recognition in a social context	Sensory therapy	Avatar
21	2017	11–13	ASD	Social learning and imitation skills	Occupational therapy	Virtual environment
22	2016	7–12	ASD	Improve emotional skills	Not mentioned	Mobile Devices
23	2016	6–15	ASD	Social skills	Life skills classes	Avatar
24	2016	8–15	ASD	Improve social skills	Physical therapy	Virtual environment
25	2016	10–12	ASD	Daily Skill	Speech therapy	VR and CVE
26	2016	Age not specified	ASD	Improvement in attention processes	Physical therapy	VR and CVE
27	2016	13 and above	HFASD	Showed improvements in their real-life targeted phobia	Not mentioned	Mobile Devices

28	2016	13–15	HFASD	Development of daily living skills	TEACCH	VR and CVE
29	2016	13–17	HFASD	Emotion recognition in a social context	DIR	VR and CVE
30	2016	12–15	ASD	Social learning and imitation skills	TEACCH	Virtual environment
31	2016	13–17	ASD	Improvement in attention processes and spatial cognition skills	Sensory integration therapy	HMD
32	2016	11–13	ASD	Showed improvements in their real life	Physical therapy	CVE
33	2016	7–12	ASD	Development of daily living skills	TEACCH	HMD
34	2016	7–12	ASD	Emotion recognition in a social context	Sensory integration therapy	VR and CVE

5. DISCUSSION

This section contains a detailed discussion which aims to answer the research questions that have been posed.

RQ1: For which specific target group in ASD is VR applied? In RQ1, researchers found that, among the specific target groups for which usage of VR is applied, its usage on the ASD and HFASD groups was more appropriate and effective. However, researchers also found that there is still a lack of studies on the effectiveness of the use of VR on the HFASD group.

RQ2: What skills in ASD are being trained using VR? For RQ2, the researchers found that the skills that need to be trained using VR are for the target group with conventional CBT treatment in ASD individuals, with the skills being social skills, communication, daily life development (shopping skills) and emotional control to improve quality of life.

RQ3: What are the techniques used in interventions for patients with ASD? For RQ3, it was found that among the techniques or treatments used for individuals with ASD is the conventional CBT technique. The technique has its own challenges with certain techniques that require skills not possessed by individuals with ASD, such as imagination and abstract thinking.

RQ4: What are the technologies used in VR-based Training (VRT) for ASD? In RQ4, the technologies used include hardware and software. In developing VR methodologies, elements such as language and culture are appropriate to be associated with VR contents.

In the following subsections, we reveal insightful thoughts on the factors that are crucial and need consideration when developing VR applications for ASD.

5.1 Effectiveness of Virtual Reality for ASD and HFASD

The effectiveness of the use of VR on the ASD spectrum is unknown, but there is potential in the usage as intervention in ASD (Mesa-Gresa et al. 2018). VR has emerged as an effective new treatment approach in different areas of the health field, such as rehabilitation (Bird et al. 2017), promotion of emotional wellbeing in inpatients (Bekelis et al. 2017), diagnosis, surgery training (Pulijala et al. 2017) and mental health treatment. VR is an evolving and feasible technology in the education of people with neurodevelopmental disorders including ASD individuals where they are exposed to stimuli in a 3D and interactive environment, almost resembling real and controlled situations (Mesa-Gresa et al. 2018). Exposure to the 3D environment, either immersive or non-immersive, can help these special individuals better understand the situations that need to be passed in real life.

However, studies have found that VR technology can improve the achievement of autistic individuals in social interactions (Gal et al., 2009). Parsons and Mitchell (2002) train social skills using cafés and virtual buses. This study found that ASD individuals can adapt to learning methods using VR and benefit from past experience. Therefore, autistic individuals who have poor cognitive ability and are less able to receive formal learning, can be actively involved in learning daily activities by using VR.

5.2 VR as a Tool to Increase Social Skills and Quality of Life

Most of the ASD children including HFASD children struggle to generate and sustain interaction. This impairment caused by poor social skills may become more prominent over time, and this reflects their quality of life. Thus, children with ASD lack social skills, which may restrict them from communicating effectively, and hence impact their quality of life. Quality of life is referring to a perception of an individual's general wellbeing, including emotional, social, and physical aspects of the individual's life (Lin & Huang, 2017). Therefore, any intervention to ASD children should focus on developing their social skills to improve their quality of life. Robots and games have been shown to be effective in facilitating social behavior between children with ASD and developing their social skills. Thus, recent VR developments may offer different benefits and advantages to people with ASD in many ways, especially to improve their quality of life.

5.3 Virtual Reality as a Tool for Cognitive Behavioral Therapy

Despite the promising results of conventional CBT treatment in ASD individuals, the application has its own challenges. For example, certain techniques of CBT require skills that ASD individuals lack, such as imagination and abstract thinking (Maskey et.al. 2019). Due to these challenges, researchers are exploring technology-based mediums into how they can help in simulating the CBT concept of imagination into an explicit visual and auditory output, hence reducing reliance on imagination and abstract thinking. CBT combined with immersive VRE has the potential to be developed, as there are many treatments for anxiety associated with certain fears and phobia in children with ASD. The addition of VRE apparently offers many advantages over CBT alone or CBT, with much more traditional exposure therapy for improving other skills such as social skills.

5.4 The Role of Domain Knowledge (Conceptual Model) for Developing Virtual Reality in ASD

One tool that is being embraced by therapists, counselors, teachers, parents, and their children to help those with autism to better communicate and connect with others and the world around them, is virtual reality (Mallari et al., 2019). The VR industry has a huge role in shifting how we use technology, to help support those on the autistic spectrum to connect, communicate, and navigate. Moreover, it can help those without the condition learn more about it. There are two papers that state the tendency of conceptual model development supporting VR in ASD (Reyes et al., 2021 & Polcar et al., 2016). But these do not describe the development measures or approach in detail. The role of domain knowledge in the development of VR for ASD needs to be emphasized in order to succeed in the concept of helping this group in an orderly manner and meet accuracy of domain ASD. In this light, by looking into the high potential use of ontology-based domain models such as a reviewed by Mohamad et. al, (2021), it would be able to further spark a new direction on ontological support for developing VR applications in the ASD context.

5.5 Using Common VR Methodology in the Development of VR in ASD

The traditional method of programming VR applications makes it all knowledge of a strictly coded product or process, effectively losing access to it from outside the programming software. Moreover, creating a new solution without any methodology makes the process longer and less effective (Gorski, 2017). In the rapid and advanced development in the field of medicine, the application of VR methodology still has gaps. For example, there is no standard VR development methodology used in the development of current VR applications for the orthopedic domain (Bajuri et al., 2021). This situation occurs because there is no gold standard VR application development methodology from the ICT domain. However, the use of this methodology is incomplete in its processing layout to guide VR development in ASD and HFASD. In this study, the use of VR requires a complete module in terms of the methodology and technology used.

5.6 Cultures in VR Content

Tzanavari et al. (2015) define culture as a collection of factors shared by people in social institutions, such as ideas, behaviors, and others. Besides, culture can moreover influence the determination of society in association with innovation. Furthermore, the learning medium (in this case, VR) moreover permits clients to make their claim culture in that medium. Understudies gotten to be portion of virtual learning encounters, they utilize modern social standards with which they connected. Other researchers have appeared how the socially imbued viewpoints of VR have driven to quickened picks up by learners. O'Brien and Require

(2008) note that in a dialect learning lesson, the affordances of VR to incorporate social stuff of the dialect that learners expected to memorize gives a conducive setting for outside dialect learners. In this study, none of the papers reviewed discuss culture in VR development. This proves that there are still gaps in the study regarding the development of VR through cultural elements.

5.7 Language in Development of VR for Autism Spectrum Disorder

One of the main features in diagnosing autism is a deficiency in language development, particularly the pragmatic aspect. The lack of language itself, however, is not sufficient to diagnose a child with ASD; there are other diagnostic criteria that need to be met. As a result, several studies have shown extreme variability in language development among autistic participants; the results show gross gaps in language development ranging from two standard deviations below average, to two standard deviations above average. The latter shows how language development delay cannot be relied upon solely as a diagnostic criterion for autism (Gernsbacher, Morson & Grace, 2015). This led to the importance of language in developing VR for ASD. In this study, there is a paper that describes the importance of language in the development of these ASD patients. There are still many literacy gaps in the description of the importance of language in the development of VR models for ASD.

5.8 Impact and Challenges of Using VR for ASD

VR is still undergoing a lot of research on application systems that show a variety of problems and challenges to overcome, to further minimize communication barriers between the user and the system. The impact of using different technology solutions for VR has been studied, and many benefits have been demonstrated and suggested. The language, culture, and interpretation of VR content are different potential challenges. VR apps are expensive to produce, which means that they are often only accessible in English as a lingua franca, perhaps even with subtitles, which can limit their use for younger children, and create misunderstanding or ambiguity for older students, because of translation problems and cultural differences (Simoes et al., 2020). However, learning a language with the use of VR can help if it contributes to creating interest and incentives to overcome language barriers to further explore VR applications.

6. CONCLUSION

This type of VR application (ASD, HFASD) has been used to work in areas affected by autism spectrum disorders, with an emphasis on developing apps that help autistic people communicate using pictures and speech. These systems are commonly accepted because they are straightforward to use and provide reasonably intuitive tools, and they work with commonplace commodities. As a result, it's vital to continue to improve these systems and conduct additional research in the field to address major challenges like communication and interaction

In theory, VR can help overcome this limitation by recasting the same dynamic skill-learning practice in different VR contexts, thus facilitating the generalization of skills learned in VR to everyday life interactions, since the same procedure is trained in several environments (Simoes et al., 2020). Furthermore, the potential of this technology to support the learning of children, young people, and adults on the autistic spectrum needs to be considered within the range of existing educational approaches and support for this population. VR-HMDs is just one approach amongst a range of others, that may be used by practitioners, teachers, and therapists, and its use should not simply replicate existing practice or be a substitute for human interaction, knowledge, and skills.

ACKNOWLEDGEMENT

This research is supported by Transdisciplinary Research Grant Scheme (TRGS), Ministry of Higher Education (MOHE) and Universiti Kebangsaan Malaysia (UKM), Vot. No: TRGS/1/2020/UKM/02/6/2. We highly appreciate the enormous support received for this research project.

REFERENCES

American Psychiatric Association (2019). Diagnostic and statistical manual of mental disorders (DSM-5®). American Psychiatric Pub.

Bellani, M., Fornasari, L., Chittaro, L., & Brambilla, P. (2011). Virtual reality in autism: state of the art. Epidemiology and psychiatric sciences, 20(3), 235-238.

Bekelis, K.; Calnan, D.; Simmons, N.; MacKenzie, T.A.; Kakoulides, G. (2017). Effect of an Immersive Preoperative Virtual Reality Experience on Patient Reported Outcomes: A Randomized Controlled Trial. Ann. Surg. 265, 1068–1073.

Blyth, C. (2018). Immersive technologies and language learning. Foreign Language Annals, 51(1), 225-232. https://doi.org/10.1111/flan.12327.

Brereton, O.P., Kitchenham, B.A., Turner Budgen, D., Khalil, M., (2007). Lessons from applying the systematic literature review process within the software engineering domain. Journal of Systems and Software 80 (4), 571–583.

Botella, C., Fernndez-lvarez, J., Guilln, V., Garca-Palacios, A., Baos, R (2017): Recent progress in virtual reality exposure therapy for phobias: a systematic review. Curr. Psychiatry Rep.19–42.

Bird, M.L.; Cannell, J.; Jovic, E.; Rathjen, A.; Lane, K.; Tyson, A.; Callisaya, M.; Smith, S. A (2017). Randomized Controlled Trial Investigating the Efficacy of Virtual Reality in Inpatient Stroke Rehabilitation. Arch. Phys. Med. Rehabil.

Brok, W.L.J.E.; Sterkenburg, P.S. (2015). Self-controlled technologies to support skill attainment in persons with an autism spectrum disorder and/or an intellectual disability: A systematic literature review. Disabil. Rehabil. Assist. Technol, 10, 1–10.

Burdea G.C., Coiffet P (2003), Virtual reality technology, John Wiley & Sons, Inc.

Didehbani, N., Allen, T., Kandalaft, M., Krawczyk, D., & Chapman, S. (2016). Virtual reality social cognition training for children with high functioning autism. Computers in Human Behavior, 62, 703-711.

Gal, A., Agam, N., Alchanatis, V., Cohen, Y., Zipori, I., Presnov, E., et al. (2009). Evaluating water stress in irrigated olives: correlation of soil water status, tree water status, and thermal imagery. Irrig. Sci. 27, 367–376. doi: 10.1007/s00271-009-0150-7.

Gernsbacher, M., Morson, M. & Grace, E. (2015). Language Development in Autism. Retrieved from: file:///C:/Users/Owner/OneDrive/PSY320/Autism%20Paper.pdf.

Grossard, C., Palestra, G., Xavier, J., Chetouani, M., Grynszpan, O., David Cohen, D. (2018). ICT and autism care: State of the art. Curr. Opin. Psychiatry, 31, 474–483.

Gorski, P. (2017). Building Virtual Reality Applications for Engineering with Knowledge-Based Approach. Journal of Autism and Developmental Disorders, Vol 8(4); pp. 64 –73. DOI: 10.1515/mper-2017-0037.

Maskey M, Rodgers J, Grahame V, Glod M, Honey E, Kinnear J et al (2019). A randomised controlled feasibility trial of immersive virtual reality treatment with cognitive behaviour therapy for specific phobias in young people with autism spectrum disorder. Journal of Autism Dev Disord 49:1912–1927.

Mallari B, Spaeth EK, Goh H, Boyd BS. (2019). Virtual reality as an analgesic for acute and chronic pain in adults: A systematic review and meta-analysis. J Pain Res 2019 Jul; Volume 12:2053-2085. doi: 10.2147/jpr. s200498.

Mishkind, M.C.; Norr, A.M.; Katz, A.C.; (2017), Reger, G.M. Review of Virtual Reality Treatment in Psychiatry: Evidence Versus Current Diffusion and Use. Curr. Psychiatry Rep. 19, 80.

Mesa-Gresa P, Gil-Gómez H, Lozano-Quilis J-A, Gil-Gómez J-A. (2018). Effectiveness of virtual reality for children and adolescents with autism spectrum disorder: an evidence-based systematic review.

Mohamad UH, Ahmad MN, Benferdia Y, Shapi'i A and Bajuri MY (2021) An Overview of Ontologies in Virtual Reality-Based Training for Healthcare Domain. Front. Med. 8:698855. doi: 10.3389/fmed.2021.698855.

O'Brien, M. G., & Levy, R. M. (2008). Exploration through virtual reality: Encounters with the target culture. Canadian Modern Language Review, 64(4), 663-691.

Parsons, S., & Mitchell, P. (2002). The potential of virtual reality in social skills training for people with autistic spectrum disorders. Journal of intellectual disability research, 46(5), 430-443.

Pulijala, Y.; Ma, M.; Pears, M.; Peebles, D.; Ayoub, (2017). A. Effectiveness of Immersive Virtual Reality in Surgical Training—A Randomized Control Trial. J. Oral Maxillofac. Surg.

Simoes M, Mouga S, Pereira AC, de Carvalho P, Oliveira G, Castelo-Branco M. (2020) Virtual Reality Immersion Rescales Regulation of Interpersonal Distance in Controls but not in Autism Spectrum Disorder. J Autism Dev Disord. Dec;50(12):4317-4328. doi: 10.1007/s10803-020-04484-6. PMID: 32266686.

Tzanavari, S. Matsentidou, C. G. Christou, and C. Poullis, (2015) "User Experience Observations on Factors That Affect Performance in a Road-Crossing Training Application for Children Using the CAVE", Learning and Collaboration Technologies. Technology-Rich Environments for Learning and Collaboration. P. Zaphiris and A. Ioannou (Eds.): Springer Lecture Notes in Computer Science, Vol. 8524, pp. 91-101. ISBN: 978-3-319-07484.

CYBERNETIC PHILOSOPHY OF DIGITAL PUBLIC GOVERNANCE: MODELING RECURSIVE SENSORY SYSTEMS

Konstantin S. Kondratenko
Saint Petersburg State University, Russia

ABSTRACT

This article is devoted to the theoretical aspects of the study of the digital transformation of public governance. The article solves the problem of synthesizing a fairly large number of areas of digital transformation, such as data processing, deliberation, platform communication, project management, etc. The digital transformation model is presented in the form of a recursive sensory system, which is a heuristic for understanding the digitalization of organizations or socio-technical systems. Sensors are the basis for obtaining data in such a system, while each of its components is somehow connected with recursive processes understood in a broad sense - mutual influence, management iterations, isomorphism, etc. Realizing philosophical and cybernetic research methods, the author analyzes in detail the structural components of the Recursive Sensory System and shows their relationship with each other and society in order to describe which heuristic of the Rational Value System is used and to study the features and prospects of the digital transformation of public governance. Solving the lack of theoretical approaches to the above-mentioned processes is an important task of this work.

KEYWORDS

Complex Systems, Recursion, Rational Value System, Recursive Sensory System, digitalization, Cybernetic Philosophy

1. INTRODUCTION

The process of digitalization of public governance truly raises more questions than answers among researchers. The digital transformation of the state is a strange phenomenon where bold experiments in the introduction of high technologies in London, Barcelona and Singapore and the fight against digital inequality - consisting in providing citizens with Internet access coexist. Modern states are ashamed to admit that government bodies have seriously lagged behind businesses, NGOs and other public structures in establishing digital infrastructure. However, salvation came by itself - the era of the Fourth Industrial Revolution has come, and the state, having picked up the flag of technological progress, has hastily begun to patch holes. For example, they have done this by providing citizens with government services in digital forms and by creating user-friendly websites for citizens.

Such a "catch-up" character of digitalization of the state sharply raises the question of the prospects for transformation. Indeed, because public governance cannot simply become more convenient in the digital era (Dunleavy, Margetts et al., 2006) by adding the adjective Algorithmic (Yeung, 2018; Danaher, Hogan, Noone et al., 2017; Aneesh, 2009) to the established and understandable approach of Good Governance. The systemic process of digital transformation is likely to make some elements of the political structure more significant and others less significant. And which part of the state is subject to transformation? What will happen to the entire structure of government? Will the management approach itself change?

A clear drawback, from the point of view of the author of this work, is the lack of a fundamental conceptualization of the research topic. The concept of "complex thinking" by Morin (1992), as well as "systemic thinking" described in the works of Capra (1996), Maturana and Varela (1987), Nicolis and Prigogine (1989) and others, to a greater extent indicates the principle of cognition of the increasingly complex world, including at the expense of technology, and can only serve as a very distant starting point for research. The actor-network theory of Latour, Callon and Law (Latour, 2005; Callon and Latour, 1992; Callon, Law and Rip, 1986) points mainly to the social aspects of technology. Technology is included in the social process as actors or actants, forming socio-technical assemblies together with people. However, the actor-network theory, like any other network theory, takes as the basis of its analysis the actor, a certain material point of the social field, which does not correspond to the goals of this study - to understand what is happening with the actor or with the system of actors. An analysis is needed, not so much an analysis of the processes as an analysis of the structures if we want to set up a thought experiment and present the results of the digital transformation on public governance.

We see a weak starting point within the concepts of "complex thinking", "distributed knowledge", "network society" and many others. While these questions are good for answering the question of how the digital state functions -we are interested in the answer to the question of what the digital state is. Therefore, this work is, first of all, is not so much a political one where the study would be devoted to processes - as it is of a philosophical character since it is interested in entities. From this point of view, researchers usually take techno-optimistic and political-pessimistic positions, dictated, it seems to us, more by feelings such as hope and fear rather than by dry rationality.

The most suitable approach for achieving the objectives of the study seems to be the cybernetic approach, which underlies the processes of management, information exchange and which unites people, machines and society in a single systemic approach. The very concept of "socio-technical system" contains two fields opposed to each other – the social and technical. To get rid of this, we will combine humans and non-humans into one concept "system" and consider what happens to this system. I would also like to note that the systemic approach has proved itself quite successfully in political science - evidenced by the individuals Almond (1988), Easton (1965) and Deutsch (1965). Their concepts - especially the theory of information-cybernetic systems by Deutsch (1965) - will serve as a theoretical basis for this study.

2. RATIONAL VALUE SYSTEMS

The person, society and technology that legitimize the dominance of the modern governance model in this work are variants of Rational Value Systems (RVS; see more details in AUTHOR, 2021). The concept of "value system" is borrowed from psychological literature, where a rare author did not touch on the problem of meaning, incl. Freud (Shape, 1973). The most in-depth studies of meaning as the nucleus of personality have been undertaken within the framework of such areas of activity as humanistic psychology, existential psychology and activity psychology. Meaning, acting in the form of fundamental values, used to use a value relationship; Value Objects (V-objects) appear in the form of meaning (Husserl, 1939). A value relation forms a value representation, or a Value Model (V-model). In order not to consider in detail the topic of value in the structure of a value system, we will simply refer to the works of A. Leontiev (1994), D. Leontiev (2003), Bratus (1988) and Bakhtin (1979).

To a greater extent, the concept of RVS is applicable to a person and society, since the coordinates of their activities are based on semantic, value-based foundations. Machines, apparently, are only on the way to RV-ontology since their architecture is not tied to V-object. The prototype of the RV-machine is the prototypical virtual assistants - with the proviso that their behavior can be described by the assistant model, which will soon become widespread (Internet of things, "big data", "smart city" - all methods of technology are based and actively exploit the helper model). It is unlikely that in the next 20-30 years we will witness successful experiments to create RV-machines that are nothing more than autonomous robots. Using the modeling block described in this article, today's machines execute their commands with the environment according to the "stimulus-response" principle, and can become "smart" machines that analyze the experience of interacting with the environment and executing commands from other machines becoming "smart" technologies. Indeed, the analysis of options for referring to the accumulated experience is the formation of machine intelligence in the literal and figurative sense of the word. However, RV-machines are technologies that have not only a brain, but also a heart.

The main challenge facing software developers today is the need to ensure that the machine can use as much information as possible to improve predictions, decision making and command execution. At the same time, this suggests that the goal of machine intellectualization is not in the formation of V-models, but in the development of governing systems. The question of what kind of systems are these and how their architecture is built will be discussed later on.

It would be logical to present three basic models of behavior of the RVS in relation to the V-object – "subordination", "assistant" and "anticipatory action". These models coincide, in particular, with the structure of abilities described by Shadrikov (2007), consisting of the abilities of an individual (model of subordination), a subject of activity (model of an assistant), developing his abilities "due to the intellectualization of basic mental functions" (Shadrikov, 2007: 57), as well as spiritual abilities of action (model of anticipatory), i.e., ... "methods aimed at knowing other people" (Shadrikov, 2007: 65). The question that follows is: what alternatives to the behavior of V-objects acting as independent systems appear as a result of interaction with RVSs?

It is reasonable to assume that the behavior models of governing systems, acting as V-objects for RVSs, and the RVSs themselves are related according to the principle of reflection, and that the formation of behavior models of governing systems depends on the behavior models of these RVSs, while the activities of governing systems in relation to RVSs is independent. The subordination model is associated with the requirements model, the assistant model - with the task setting model and the anticipatory action model - with the image formation model (Table 1).

Table 1. Relationship between behavior models of RV- and governing systems

Behavior models of RVS	Communication principle	Behavior models of governing systems
Subordination model	**S → R**	Requirements model
Helper model	**Independent activity within the framework of the assigned task**	Target setting model
Leading action model	**Impact on values**	Imaging model

In this case, a separate requirement, task or image formed in the Value Block (V-block) of the RVS must be coordinated with the structures of requirements, tasks or the formation of images of the governing system, as well as the structures of the possibilities for fulfilling requirements, solving problems or perceiving images of the RVSs. First of all, when discussing the sequence for example: the d_n requirement must be consistent with the previous requirements d_{n-1}, d_{n-2},..., with experience in fulfilling the requirements of the RVS md_n, md_{n-1}, md_{n-2},..., and non-standard requirement, i.e. weakly consistent with the previous, but consistent with earlier requirements, should have intermediate requirements to prepare the RVS to meet the requirements. Requirement d must be consistent with the tasks posed, since the formulation and solution of non-trivial problems presupposes the rationalization of requirements by the RVS, and the system can simply refuse to fulfill the requirements that are poorly understood by it, and with the accumulated data in the V-block, since the governing system, which positions itself in a certain way, must correlate the tasks and requirements with the image of oneself in the V-model. The governing system, in other words, having a fairly simple structure proceeding from the expectations of the obeying system, reveals in itself a complex model of governing action, each of which must be correlated with each element of the two systems. However, this complication can only seem to the researcher since the reflecting properties of governing actions in the process of formation and development can change.

The architecture of RVSs is determined by the basic component of their structure - the V-model. It is the V-model that makes the system sense-oriented, associated with values and their implementation, and it is also responsible for the initially undemocratic structure of the system. The V-model is the full-fledged master of the RVS, subordinating the rest of the subsystems to itself - the modeling block, which is responsible for the development of acceptable forms of behavior, the behavioral block and the Rational Block (R-block), which makes decisions and builds communication with the V-object. Moreover, each of these blocks is a completely independent entity that develops through experience (including someone else's experience), skills or abilities which are used for analysis. Suppose, for example, that the R-block is sufficiently developed due to the ability to think critically and the constant need for justification and explanation. How will the system behave with the activation and active expansion of the influence of the V-model? It is not hard to imagine that the R-block will act as a counterweight to the V-model, which will try to keep the entire system from autocracy.

However, there is an important nuance here. Let us estimate, for example, the significance of the V-model in the RVS as 0.7 points in the range (0; 1). Suppose also that the total significance of the R-block components is also 0.7 points, and then we get the desired counterweight. But we did not consider the modeling block and the behavioral block, which can have minimum values, say, 0.1 and 0.2 points, respectively. If we add the weights of the R-block, the behavioral block, and the modeling block, we get 1.0 points versus 0.8 for the V-model. This means that the RVS is more rational than value, and it can act as a governing system in relation to other RVSs. The conclusion that suggests itself from these simple calculations is quite simple: in relation to the V-model, all other subsystems act as a total counterweight.

This explanation also raises a lot of questions. In particular, the question of why we consider the total counterweight. Since the V-model is the most significant component of the RVS, it is capable of subordinating other subsystems to itself, including the R-block. Them, considering this, the next question asked is how to assess the "sufficient development" of the R-block.

These questions are answered with slightly more complex calculations based on simple explanations. First, the counterbalance is fully achieved with more or less similar scales of significance – indeed, it is quite difficult to subdue an equal in strength. For example, this discrepancy in values should not exceed 20%, or there may be gradations - a strong counterweight (a discrepancy of no more than 20%), moderate (no less than 21% and no more than 35%) and a weak counterweight (no less than 36% and no more 50 %). Smaller values will indicate that there is none or too little of a counterweight. Second, it is also necessary to take into account the consistency of the elements with each other, for example, by calculating the average of the score of the two subsystems. If there are four subsystems in an RVS, then there are six connections between them. Each link is "weighted" by the consistency coefficient, and then these values are compared with each other, and the average values of the coefficients associated with the V- and R-blocks, respectively, are compared. An RVS map can provide valuable information about the nature of the system itself (Figure 1).

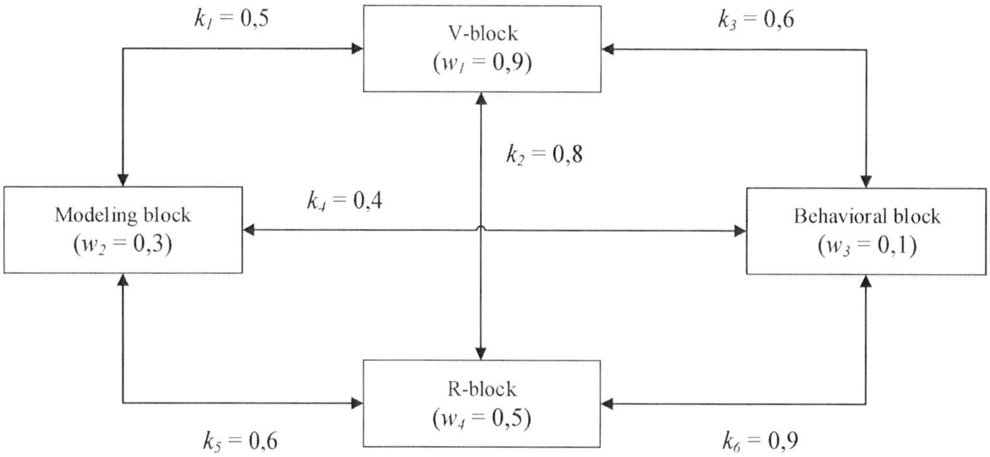

Figure 1. An example of mapping an RVS based on consideration of the balance and consistency of subsystems

In Figure 1, one can see that the RVS is balanced ($w_1 = w_2 + w_3 + + w_4 = 9$) and well matched ($\bar{k}_1 = \bar{k}_2 \approx 6,3$, where $\bar{k}_1$ is the average of the sum of k_1, k_2 and k_3, and $\bar{k}_2$ is the average of the sum of k_4, k_5 and k_6). We also see that the modeling block and the behavioral block are consistent with the R-block rather than with the V-block. This suggests that despite the value of w_4, which is practically half the size of w_1, the R-block is able to build strong connections with the other subsystems. A high value of the weight k_2, in turn, indicates the consistency of the R- and V-blocks. The highest coefficient of consistency is recorded, however, in another connection - between the R- and behavioral blocks, which indicates the developed ability of the RVS to resist intentions that are unreasonable from a rational-behavioral point of view, i.e., to self-control.

These calculations are necessary to establish criteria for evaluating the architecture of the RVS. Based on the criteria of balance (w) and consistency (k), four types of states of RVSs can be distinguished:

- *balanced and consistent* ($W = w_1 - (w_2 + w_3 + w_4) \leq 0,5$ and $K = \bar{k}_1 - \bar{k}_2 \leq 0,5$, i.e. the discrepancy between the weights of the V-block and other subsystems, as well as the average values of the consistency of the links of the RVS and all other links do not exceed the threshold value of 50%). Such systems may well be called *democratic.*

- *balanced and inconsistent* ($W \leq 0.5$ and $K > 0.5$). In such systems, there are several sources of power, therefore, with some reservations, we can call such systems *polyarchic.*

- *unbalanced and consistent* ($W > 0.5$ and $K \leq 0.5$). These can be called *autocratic* systems.
- *unbalanced and inconsistent* ($W < 0.5$ and $K < 0.5$). In such systems, there is a *conflict*.

The state of the RVS, in turn, is the basis for choosing a strategy for the behavior of the RSS.

3. RECURSIVENESS AND GOVERNANCE

The presence of the V-model in the RVS creates conditions for the formation of primary communication in the network of rational systems - the possibility and necessity of subordination to the V-object. Having value presupposes serving this value, i.e., the subordination of the entire system to the one who can manage. Studies by Easton and Dennis (1969), Fromm (1942), Almond (1988) and others show that even in childhood, consciousness goes through the phase of idealization of power. Legitimacy includes an essential personality dimension. In the famous work of Weber (1978), as well as in the works of his researchers and critics (Joosse, 2012, 2017, 2018; Katz, 1975; Reed, 2019; Ritzer, 1975; Schoon and Joseph West, 2017), charismatic leadership is presented as having the most powerful impact on citizens. Value attitudes, therefore, form the basis of the legitimacy underlying the interaction of systems.

Governance systems in this work are presented as Recursive Sensory Systems (RSS) - intelligent machines for governing the values and behavior of RVSs. Their social base is the expectations of citizens, therefore RSS systems are formed according to the principle of reflection, or recursiveness, by which we mean the process of mutual influence of systems on each other, which will be discussed below. Sensors, in turn, are the technical basis for the activity of the RSS. These are the "eyes" and "ears" of the system, with the help of which it receives data on the state of the RVS.

Sensors are receptors in the governing system, which can be nerve endings, people, organizations or technology. The idea of sensors is borrowed from the concept of the internet of things, devices united in an "empire of interconnected things" (Howard, 2015). The sensor network is presented in this paper as the basis for digital governance.

Some authors note two main meanings where researchers of political problems usually put in the term "recursiveness" - general and technical meanings (Townes, 2010). The first meaning is associated with the sociological discussion of the 1980s and is contained in the works of Giddens and Bourdieu: it is a kind of interdependence, mutual influence and mutual generation. Society influences a person, but a person also shapes society. According to Giddens, social actions "are not created by social actors, but are constantly recreated by them through the very means with which they express themselves as actors" (Giddens, 1984: 2) - this is how a sociologist solves the problem of "agent-structure". In turn, Bourdieu's "habitus" acts as a kind of recursive prism between society and man: habitus itself is a product of the environment, but people, through the repertoire of habitus dispositions, can modify social reality. Habitus is a place for interiorization of the external and exteriorization of the internal (Bourdieu, 1972). In the context of digital transformation, the term "recursiveness" gets a new meaning: the recursiveness of, for example, information exchange blurs the roles of "sender" and "receiver", "message" is captured only in the flows of communication between the interacting parties (Crozier, 2007: 7).

The second is technical - the meaning is associated, in particular, with the management of the organization. Each management cycle is nothing more than an iteration of an already existing rule (norm). At the same time, recursiveness implies flexibility in the use of norms for each new case (Tarasenko, Lichutin, 2012). Recursive can also be called a call to a function or project goal in each iteration of project management. This sense of the term fits into the logic of normative and phenomenological constructivism.

Both of these meanings are important for the study of RSSs. But there is another fundamental aspect, which is more connected not with society or government, but with the ontology of recursion. Recursion is a way of *reflecting* reality and modifying reality in models, schemes, projects, etc. In other words, digital governance based on recursion is a machine for cognizing reality, and only secondarily is a conductor of some kind of techno-policy.

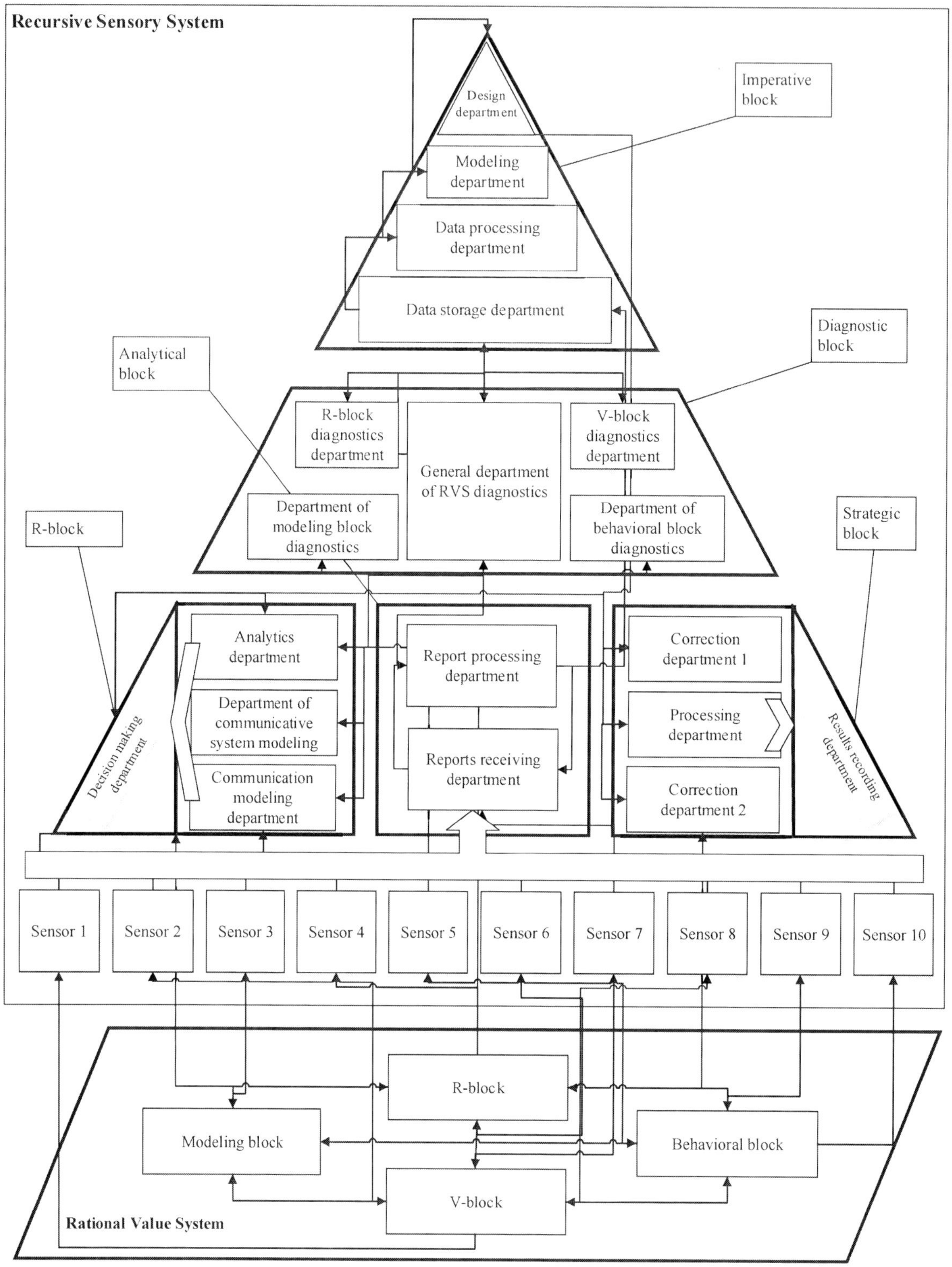

Figure 2. Model of a Recursive Sensory System

4. CONCLUSION

The digitalization of public governance is a process of fundamental transformation of the state. The point is not so much that quantitative changes have a qualitative effect, but rather a change in the foundations of governance. The state is gradually losing the role of "ideologist" and is instead acquiring the role of "researcher". One could argue with this if we recall the examples of the growth of populism in Western Europe and the United States, or the strengthening of the autocracies of Eastern Europe. According to Sedov's law of hierarchical compensation, "the actual growth of diversity at the highest level is ensured by its effective limitation at the previous levels" (Sedov, 1993: 92). However, according to another law - Ashby's law of experience - "information associated with a change in a parameter tends to destroy and replace information about the initial state of the system" (Ashby, 1956: 139).

In the new governance format, scientific expertise, data collection and processing capabilities, communication and joint practices are becoming essential. The presented model of Recursive Sensory Systems describes not so much the "shell" of digital control as layers of shells. The primary layer - the sensor layer - is responsible for receiving signals; the second layer - a data processing layer consisting of components that implement the functions of analysis, processing, communication and discussion - is a more complex structure, since each of the components has feedback with the controlled system; the third layer - the layer of expertise - is associated with the knowledge of the controlled system; finally, the fourth layer - the layer of coordination and decision-making - not only analyzes the operability of the Recursive Sensory System, but also implements the well-known Marx principle, which describes the lack of explanation and the need to influence and change the world.

Any mechanical activity in digital public governance is given to technology. While technology and people jointly participate in the processing of the incoming data, analytical and creative skills will increasingly be highlighted. Each of the bureaucratic departments will have in its structure or will be associated with the departments of monitoring and data analysis, and deep governance seems to be an inevitable process.

Each round of digital transformation will meet with increasingly fierce resistance from citizens. Social conflicts arising from the introduction of new formats of public governance will occur more and more often, without national borders. Technology, as an object of social conflict, and as a result of the conflict itself, loses the status of an actant and becomes a full-fledged actor of social change.

These shocks can be avoided by the balanced operation of the components of the recursive sensory system. Lack of communication and weak involvement of citizens in digital transformation processes appear to be managerial incompetence. Citizens can be more actively involved in the process not only as ordinary people and users, but also as specialists, experts and analysts. The increasing complexity of digital governance implies constant retention of the state from the desire to be isolated from society. Citizens can be included in any governance cycle through technology. Concern for conditions - data openness, joint discussion of projects and initiatives and providing broad opportunities for communication should become the norm, recursively repeated at every stage of governance cycles.

REFERENCES

Almond, G. A., 1988. *Comparative politics today: a world view*. Chicago: Scott, Foresman.

Aneesh, A., 2009. Global labor: Algocratic modes of organization. *Sociological Theory* 27(4): 347-370.

Ashby, W. R., 1956. *An introduction to cybernetics*. New York: J Wiley.

Bourdieu, P., 1972. *Esquisse d'une théorie de la pratique*. Genève: Ed. de Droz.

Bratus, B. S., 1988. *Anomalii lichnosti* [Personality anomalies]. Moscow: Mysl'.

Callon, M., Latour, B., 1992. Don't throw the baby out with the Bath School! A reply to Collins and Yearley. In: Pickering A (ed) *Science as practice and culture*. Chicago: University of Chicago Press: pp. 343-368

Callon, M., Law, J., Rip, A., 1986. *Mapping the dynamics of science and technology: sociology of science in the real world*. Basingstoke: Macmillan.

Capra, F., 1996. *The Web of Life: A New Scientific Understanding of Living Systems*. New York: Anchor Books, Doubleday.

Crozier, M., 2007. Recursive Governance: Contemporary Political Communication and Public Policy. *Political Communication* 24(1): 1-18.

Danaher, J., Hogan, M. J., Noone, C., Kennedy, R., Behan, A., De Paor, A., ... Shankar, K., 2017. Algorithmic governance. *Big Data & Society* 4(2): 1-21.

Deutsch, K. W., 1965. The Nerves of government: models of political communication and control. *The University of Toronto Law Journal.* 16(1): 265-290.

Dunleavy, P., Margetts, H., Bastow, S. and Tinkler, J., 2006. New public management is dead: long live digital-era governance. *Journal of public administration research and theory: J-PART* 16(3): 467-494.

Easton, D. and Dennis, J., 1969. *Children in the Political System: Origins of Political Legitimacy.* New York: McGraw-Hill.

Easton, D., 1965. *A Systems Analysis of Political Life.* New York: Wiley.

Fromm, E., 1942. *The fear of freedom.* London: Routledge & Kegan Paul.

Giddens, A., 1984. *Constitution of Society: An Outline of the Theory of Structuration.* Berkeley: University of California Pres.

Howard, Ph. N., 2015. *Pax Technica: How the Internet of Things May Set Us Free or Lock Us Up.* New Haven, London: Yale University Press.

Husserl, E., 1939. Phenomenology. In: The *Encyclopedia Britannica 14th ed.* Chicago: Encyclopedia Britannica Co: pp. 699-703.

Joosse, P., 2012. The presentation of the charismatic self in everyday life: Reflections on a Canadian new religious movement. *Sociology of Religion* 73(2): 174-199.

Joosse, P., 2017. Max Weber's disciples: Theorizing the charismatic aristocracy. *Sociological Theory* 35(4): 334-358.

Joosse, P., 2018. Countering Trump: Toward a theory of charismatic counter-roles. *Social Forces* 97(2): 921-944.

Katz, J., 1975. Essences as moral identities: Verifiability and responsibility in imputations of deviance and charisma. *American Journal of Sociology* 80(6): 1369-1390.

Kondratenko, K. S., 2020. Elementy teorii racional'no-smyslovyh sistem [Elements of the Rational Value Systems theory]. *Tomsk State University Journal of Philosophy, Sociology and Political Science* 459: 113-118.

Latour, B., 2005. *Reassembling the social: an introduction to actor-network-theory.* Oxford, New York: Oxford University Press.

Leontiev, A. N., 1994. *Filosofiya psihologii: iz nauchnogo naslediya* [Philosophy of Psychology: From the Scientific Heritage]. Moscow: Moscow University Press.

Leontiev, D. A., 2003. *Psihologiya smysla: priroda, stroenie i dinamika smyslovoj real'nosti* [The psychology of meaning: the nature, structure and dynamics of meaningful reality]. Moscow: Smysl.

Maturana, H. R., Varela, F. J., 1987. *The Tree of Knowledge: The Biological Roots of Human Understanding.* Boston: New Science Library.

Morin, E., 1992. *Method: Towards a study of humankind. The nature of nature.* New York: Peter Lang.

Nicolis, G. and Prigogine, I., 1989. *Exploring complexity: An introduction.* New York: W H Freeman.

Ritzer, G., 1975. Professionalization, bureaucratization and rationalization: The views of Max Weber. *Social Forces* 53(4): 627-634.

Schoon, E. W. and Joseph West, A., 2017. From prophecy to practice: Mutual selection cycles in the routinization of charismatic authority. *Journal for the Scientific Study of Religion* 56(4): 781-797.

Sedov, E. A., 1993. *Informacionno-entropijnye svojstva social'nyh sistem* [Information and entropy properties of social systems]. ONS 5: 92-101.

Shadrikov, V. D., 2007. *Mental'noe razvitie cheloveka* [Human mental development]. Moscow: Aspekt Press.

Shape, R. K., 1973. Freud's concepts of meaning. In: Rubinstein B B (ed). *Psychoanalysis and contemporary science 2.* New York, London: Macmillan: pp. 276-303.

Tarasenko, V. V. and Lichutin, A. V., 2012. Rekursivnoe upravlenie [Recursive governance]. *Corporate management.* Available at: https://www.cfin.ru/management/strategy/change/ recursive.shtml (accessed 27 November 2021).

Townes, M., 2010. Usage of Recursive in Political Science. *Political Science & Politics* 43(2): 259-261.

Weber, M., 1978. *Economy and Society. An Outline of Interpretive Sociology.* Berkeley; Los Angeles, CA: University of California Press.

Yeung, K., 2018. Algorithmic regulation: A critical interrogation. *Regulation & Governance* 12: 505-523.

TECHNOLOGICAL USAGE IN DEVELOPMENTAL UNIVERSITIES: A CASE STUDY OF WALTER SISULU UNIVERSITY OF SOUTH AFRICA

Agyei Fosu
Walter Sisulu University Address
Box 19712, Tecoma, East London, 5214, South Africa

ABSTRACT

The contemporary era has changed tools of trade in workplaces and saturated workplaces with more and more technological devices thereby requiring digital competence from prospective workers. This study investigates to establish how Walter Sisulu University (WSU), a developmental university with high population of its students coming from the rural areas of the former disadvantage Black Homelands equipping their future graduates in terms of technological usage for the 4th Industrial Revolution workplaces. The study used mixed-method and purposive sampling approach to gather empirical data from 152 students. Although the findings indicated that the University is providing its students with access and some form of training to use modern digital tools for learning. This was so as respondents (100%) indicated laboratories, library furnished with modern computers and laptops on campuses and 85% of respondents indicating having received some of form technological training from the University. However, there were a cumulative sum of 38.11% representing not really and not at all competence to search in Internet browsing to search information, software and installation on their technological devices calling on the University to do more by inculcating more continuous transferable technological training that will help the students to be abreast and effectively use modern technological devices in the rapid evolving digital globalised workplaces.

KEYWORDS

e-Skills, Eastern Cape, 4th Industrial Revolution Workplaces, Technological Devices, WSU

1. INTRODUCTION

The dawn of the 4th Industrial Revolution engineered by ever increasingly advancement of Internet and technological applications have seen the use and adoption of technology in all spheres of life. As a result of this, workplaces are saturated with technological applications and devices to enhance performances, efficiency, and productivity. Higher education institutions, the producers of specialized labour force to feed the industrial sectors in any nation's economy becomes the breeding grounds where the specialized labour force needs to be equipped with the 4th Industrial Revolution workplaces required skills such as e-skills, research skills, etc., (Damoene, 2003; Kinuthia & Dagaba, 2008). Secondly, the rise of online learning in the mist of COVID-19 pandemic and saturated remote working currently being experienced in higher education institutions globally, the researcher believes it is time to research on aspects of checking how future graduates are been equipped in higher education institutions for the 4th Industrial Revolution workplaces. Hoping that by doing this study, a contribution to the understanding of how future graduates can be gainfully equipped for the technological saturated workplaces associated with the 4th Industrial Revolution.

Jenkins (2009) cautioned and mentioned about the skill set such as mathematical literacy, e-skills and others that will be required in the 4th Industrial Revolution. The author further explained the critical need of e-skills which is described to be individual ability to use technology to improve or enhance their productivity and life. In the sphere of higher education, the infusion of technological educational tools and social media platforms to enhance the core businesses of higher education institutions have brought to light the cautioned and the skill set highlighted by Jenkins (2009). For instance, e-skills has become one of unmentioned admission requirements by higher education institutions in the sense that although e-skill is not a strictly admission requirement normally stated by higher education institutions especially those in developing nations but students in course of their studies may be required to use their institution educational online platform to submit assignments, projects, etc., which require e-skills. Furthermore, the global COVID-19 pandemic and the controlled measures such as online learning, remote working, etc taken by various governments across the globe again has brought to light Jenkins (2009) cautioned and the skill set required for the modern day.

1.1 Description of Research Problem and Background

Eastern Cape Province home to most former disadvantage Black Homelands, makes the province to be one of the poorest provinces in South Africa (Oyelana & Thakhati, 2017). Therefore, access to quality education is critical to socio-economic development of the province. As alluded to by Damoene, (2003); Kinuthia and Dagaba, (2008) that higher education institutions play a critical role in any nation's economy by training and supplying specialized labour forces to support the industrial sector as well offer an opportunity for people to develop their individual capabilities. Developed individual capabilities according to Unterhalter (2013) is a fundamental formation necessary to make individual have confidence and competencies to contribute mean fully to the economic sector which is vital to functioning of a society. Thus, the higher education institutions found within the Eastern Cape province do have a critical role to play towards the Eastern Cape province socio-economic development effort. The National Development Plan (NDP 2030, 2013), highlights the province of Eastern Cape been lowest in terms of number of high schools with computers for teaching and learning thereby resulting in many matriculant from the province entering higher education institutions not having adequate technological literacy levels to cope with the demands of higher education teaching and learning environment.

WSU is a developmental university situated in the Eastern Cape Province. The University produces specialized labour forces such as engineers, medical professionals, etc., who in turn contributes to societal development not only in the Eastern Cape Province but South Africa as a whole. Therefore, it is paramount that WSU graduates are well-equipped for the 4th Industrial Revolution workplaces. It is against this background that this study is been conducted to establish how WSU is preparing its future graduates for the 4th Industrial Revolution workplace.

2. BODY OF PAPER

2.1 Literature Review

In consequence to the global digital revolution, all economic sectors at play in various nation's economy are making rapid changing in terms of integration of technological tools to enhance their business activities to achieve efficiency and higher productivity. This trend is seen globally (Independent Online (IOL) Business Report, February 2013). Acenoglu et al., (2018); Bessen (2018); World Economics Forum (2016) anticipates of a strong possibility of mismatch between the ever-rapid evolving of new technologies and skills availability of workforces. Thereby calling for a systematic, strong, proactive training and re-training of both existing and future workforces to address the issue.

The higher education system, the economic sector responsible for training specialized workforce to feed the industrial sector have in effect seen quite a rapid considerable technological tools of trade transformation and research work on usage of educational technological tools in higher education learning environment. According to these authors (Damoense, 2003; Minnaar, 2011; Seena, 2010; Oxagile 2016) the blending of virtual learning which is enable by integration of educational technologies to supplement the convectional learning in the higher education learning environment are driven by various factors, but the two key factors are

1) the global growing demand of accessing higher education qualifications
2) an opportunity required by potential students to work and at the same time be able to earn their higher education qualification.

Numerous studies assert to the fact that the use of technology by students in their learning presents to them various benefits. According to Kurt (2010) the use of technology by students in early part of their studies assist them in developing a greater thinking order skill. To DePasquale (2003), the use of technology in the early studies of children exposure them to an appropriate use of more advanced applications they will use in their older age. Lin and Yin (2011) concluded in their studies to determine whether the use of WIKI technology by college students would improve their English writing skills that the use of the technology by the students improved their writing skills. Bulut and Delen (2011) concluded in their studies that the use of technology by students improve their studies and therefore must be integrated into classroom curriculum. The authors came to this conclusion when their studies revealed that the participating students who spent a lot of time using technology improve their science and mathematics knowledge as well as in their academic scores. According to Seena (2010) the highly priced skills such as research skills, collaboration, personal information management, etc., in the contemporary era labour market, students developed these skills using technology to self-regulated learning, search information and synthesize to either create a presentation or a project.

2.2 Research Objectives and Questions

The objectives of the study were to establish the following:
1.e-skills and technological devices access of participants before their university education.
2. Participants access to modern technological tools within WSU campuses.
3.Frequencies of using modern technological tools to enforce their studies.
4.Technological training WSU is providing to participants to uplift their e-skills.
The above-mentioned objectives were then turned into research questions for the study.

2.3 Framework that Guided the Study

Van Dijk and Hacker (2003) identified three factors that underline technological inequalities and access to be:

- material access describing it to be the lack of technological tools
- psychological access highlighting it to be computer phobia, unattractiveness, and lack of interest of new technology
- usage access was described by Hacker to be lack of digital opportunities emanating from both material and psychological access.

The study based the assessment of technological usage on Van Dijk and Hacker (2003) usage access from the perspective of material access.

2.4 Research Approach, Design/Method

Mixed-methods (quantitative & qualitative) and non-probability sampling method, namely purposive sampling was used to gather the empirical data from willing students as the study participants from Buffalo City, Queenstown, Ibika and Nelson Mandela Drive Campuses of WSU. The quantitative and qualitative research design took the form of closed and open-ended questionnaire of which both descriptive statistics and thematic analysis were used to analyze the empirical data.

The data gathering for the study was approached as follows:

The researcher is based at Buffalo City Campus, therefore data gathering from other campuses took the form of after telephonic discussion concerning the study and its data gathering with willing colleague lecturers ready to assist, the printed questionnaire was sent to them to gather data from willing students.

A total of 152 fully answered questionnaire were analyzed as the study empirical data.

2.5 Empirical Findings and Discussion

Although an existing set of skill may be the basis from which a new economic revolution may emerge, but the reality is that every economic revolution or shift require an extra new set of skill to add to the existing ones. Based on this, White (2013) highlighted that the 21st century industrial revolution calls for peculiar training of students in developing competence to fit the new shift. Therefore, institutions of learning and training as well as educators needs to teach and train students the new required skills together with the prevailing ones.

The contemporary students require digital competence to make it in the global labour marketplaces.

Literature document the impact of the former apartheid rule system on socio-economic development of previously disadvantage black communities in South Africa and how the impact continues to serve as an impediment in the post-apartheid reconstruction efforts (Rakhometsi 2008; Thobejane 2013; Hale 2010; Keswell 2004). The chosen institution for the current study is a previously disadvantaged black university with majority of its students coming from impoverished rural communities of the former Black Homelands (Ciskei and Transkei) found in the Eastern Cape province. Amidst the misfortune of majority of WSU students, their future careers are embedded in an ever-increasing globalized digital world which require them not only to be specialized in their chosen career paths but as well digital competence. This study therefore investigates the exposure, soft skills training in terms of integration and usage of modern educational technological tools into WSU students learning environment as one of the ways of preparing WSU graduates for the globalized digital workplaces.

The empirical findings of the study are presented below according to the study objectives:

2.5.1 Demographic information of participants

The researcher chose under the demographic information to highlight the percentage of participants from of the selected campus and gender. The researcher believes it is a common knowledge among scholars about age ranges of students in institutions of higher learning and rather shed light on gender as a contribution to try and shed light on the ongoing quest about knowing the girl child education in institutions of higher learning.

Table 1. Demographic information and percentage of participants from the chosen campuses

Gender	Frequency	Percentage
Male	83	54.61
Female	69	45.39
Number of participants from the chosen campuses		
Nelson Mandela Drive	37	24.34
Buffalo City	50	32.89
Ibika	40	26.32
Queenstown	25	16.45

The findings from Table 1 indicated more males (54.61%) participated in the study than females (45.39%). This finding implies there are encouraging girl child education at the institution of higher learning in a region considered to be under-privilege in terms of socio-economic development.

2.5.2 Participants e-skills and technological devices access before university education

Takavarasha et al., (2018) highlights that when it comes to access to digital tools, they not only imply the devices, software, and hardware artefacts but as well how they influence one's ability to learn and use them for engaging in social, business, and educational activities in a responsible and safe ways. Because of My Broadband (2015) findings of Eastern Cape province to be second lowest when it comes to Internet access and digital literacy in South Africa, as a result the study sought to establish the digital literacy of the participants before their university education. The questions and findings are presented in Table 2 below:

Table 2. e-skills of participants before university education

Technological devices own by participants before university education		
Mobile phone	151	99.34
Personal Computer (PC)	46	30.26
Laptop	18	11.84
Tablet	13	8.55
Competence in using standard word processing package, managing files on a computer and smartphone before university education.		
Strongly yes	91	59.87
Not really	38	25
Not at all	23	15.13

Competence in Internet browsing to search information, software downloading and installation on a technological device such as PC, smartphone, etc.		
Strongly yes	94	61.84
Not really	36	23.64
Not at all	22	14.47
Competence in attaching or uploading documents such as files, videos, pictures, etc., to emails and social media platforms before university education.		
Strongly yes	127	83.55
Not really	14	9.21
Not at all	11	7.24

Despite the findings of My Broadband (2015), there is a quite significant level of basic digital literacy among the study participants. For instance, students require basic digital competence such as using standard word processing package to type assignments, projects and being able to upload them unto their institution e-learning platforms as well as being able to download and install their institution e-learning unto their personal technological devices and from Table 2: 59.87% indicated their competence to use standard word processing package, 61.84% also indicated their competence to attach or upload documents on email and social media platforms.

2.5.3 Access to Modern Technological Tools within WSU Campuses

According to Mossberger et al., (2012) speaking from the perspective of digital citizen asserted that students in higher institutions of learning are supposed to have a basic quality of digital literacy because they have access to Internet and technological devices at their institutions of learning. Based on Mossberger et al., (2012) assertion, the study asked respondents in an open-ended question on the questionnaire to mentioned technological devices they have access to on their respective campuses as well as their access to Wi-Fi on campus, hall of residence and stability of connectivity.

All the respondents (100%) indicated laboratories, library furnished with modern computers and laptops on campuses. Also, Wi-Fi access on both campuses and in halls of residence. However, respondents raised an issue of stability of Internet connectivity as only 30% of respondents indicated stability of Internet connectivity on both campus and in halls of residence.

2.5.4 Frequency and how Participants are using Modern Technological Tools to Enforce their Studies

Respondents we asked to indicate the frequency and how they use modern technological tools they have access to in enforcing their studies. Table 3 gives the findings.

Table 3. Frequency and how respondents use modern technological tools to enforce their studies

	Frequency	Percentage	How they use it (In common themes)
Once a week	32	21.05	• use in getting extra lecture notes from the Internet. • Use in watching videos, demonstrations on various subject topics. • Use in getting extra tutorials.
More than once a week	104	68.42	
I don't use	16	10.53	

2.5.5 Technological Training WSU is Providing to Students to Uplift their e-Skills

Respondents were asked as to whether they have received training from WSU to enable them to use modern they have access to on campus.

Majority of respondent (85%) indicated they have received some form of technological training and some of the common mentioned training received were in Microsoft office package, how to use Blackboard and Computer Aided and Digital Design.

3. CONCLUSION

Despite the plight and impediments of students who found themselves in previous disadvantaged schools and regions, they still must compete in the ever-increasing globalised digital workplaces. This study makes its contribution by investigating on how a previous disadvantaged developmental university with majority of students from previously disadvantaged rural areas of South Africa are preparing their future graduates in terms of technological usage for the 21st century industrial digital revolution workplaces. While a quite number of participants have a basic digital skills pre their university education and 85% of participants indicating having received some form of training from WSU, WSU must do more by inculcating more continuous transferable technological training that will help students to be abreast and effectively use modern technological devices in the rapid evolving digital globalised workplaces. The National Development Plan (NDP 2030, 2013) and the work of Oyelana and Thakhati (2017) reveals how the implementation of the former Black Homelands policy by the former apartheid rule continues to hinder socio-economic development in these communities more especially in the Eastern Cape Province. This calls for concerted efforts from higher education institutions found in the province in skilling students from the previously disadvantaged communities for the contemporary era workplaces. Due to financial constraints the study focused on one university.

Future studies can look at other higher education institutions like the vocational training colleges within the provinces.

REFERENCES

Acemoglu D, Restrepo P. 2018. Artificial intelligence and work. *NBER Working Paper No 24196.*

Bessen J. 2018. Automation and jobs: when technology boost employment. *Boston University School of Law*, revised March 2018; http://www.bu.edu/law/faculty -scholarship/working-paper-series/

Bulut, O., and Delen, E. 2011. The relationship between students' exposure to technology and their achievement in science and math. *The Turkish Online Journal of Educational Technology, 10*(3).

Damoense, M. Y. 2003. Online learning: Implications for effective learning for higher education in South Africa. *Australasian Journal of Educational Technology,19*(1).

Hale, F., (2010). The impact of apartheid on the educational endeavours of two missionary agencies, *Studia Historiae Ecclesiasticae* 36(2), 167–185.

Hwang, G. J. 2014. Definition, framework, and research issues of smart learning environments-a context-aware ubiquitous learning perspective. *Smart Learning Environments,1*(1), 4.

IOL Business Report, February 2013. Available: https://www.iol.co.za/business-report/economy/ict-is-a-driving-force-behind-sme-growth-in-africa-1465573[Accessed 23rd June 2021].

Jenkins, H. 2009. *Confronting the challenges of participatory culture: Media education for the 21st century*. Mit Press.

Jochems, W., van Merriënboer, J.J.G. & Koper, R. (Eds.). 2004. *Integrated e-learning: Implications for pedagogy, technology and organisation*, Routledge, London, UK.

Keswell, M. 2004. Education and racial inequality in post-apartheid South Africa, Santa Fe Institute Working Paper, No. 2004-02-008.

Kinuthia, W. & Dagada, R. 2008. E-learning incorporation: an exploratory study of three South African higher education institutions. *International Journal on E-learning,7*(4), 623-639.

Kuo, H. M. 2009. Understanding relationships between academic staff and administrators: An organisational culture perspective. *Journal of Higher Education Policy and Management,31*(1), 43-54.

Kurt, S. 2010. Technology use in elementary education in Turkey: A case study. *New Horizons in Education, 58*(1), 65-76

Lin, W. & Yang, S. 2011. Exploring students' perceptions of integrating Wiki technology and peer feedback into English writing courses. *English Teaching: Practice and Critique,*

10(2), 88-103

Mossberger, K., Tolbert, C. & Mamilton, A. 2012. Measuring digital citizenship: Mobile access and broadband, *International Journal of Communication* 6(2012), 2492–2528.

My Broadband, (2015). Internet access in South Africa: best and worst provinces, viewed n.d., from https://mybroadband.co.za/news/telecoms/127450-internetaccess-in-south-africa-best-and-worst-provinces.html

NDP 2030. 2013. *The future is ours: we must make it work*, viewed n.d., from https://heaids.org.za/site/assets/files/1267/npc_national_development_plan_ vision_2030_-lo-res.pdf

Oyelana,A.A. & Thakhathi, D. R. 2017. Challenges Confronting Amathole District Municipality Managers in Implementing Socioeconomic Strategies in Rural Communities, *Studies of Tribes and Tribals*, 15(1),39-44.

Oxagile. 2016. *History and trends of learning management system.* Available at: contact@oxagile.com [Accessed: 20th November 2021]

Rakhometsi, M.S. 2008. *The transformation of black school education in South Africa,* Unpublished PhD thesis, viewed n.d., from http://www.sahistory.org.za/ sites/default/files/RakometsiMS.pdf

Seena, J. 2010. *An investigation into the impact of e-learning on information technology students at further education and training schools in Mthatha.* Unpublished BTech dissertation. Walter Sisulu University, East London. South Africa.

Takavarasha, S., Cilliers, L. Chinyamurindi, W. 2018. Navigating the unbeaten track from digital literacy to digital citizenship: A case of university students in South Africa's Eastern Cape province, *Reading & Writing* 9(1), a187.

Thobejane, T.D. 2013. History of apartheid education and the problems of reconstruction in South Africa, *Sociology Study* 3(1), 1–12.

Unterhalter, E. 2013. Educating Capabilities. *Journal of Human Development and Capabilities*, *14*(1), 185-188.

Van Dijk, J. & Hacker, K. 2003. The digital divide as a complex, dynamic phenomenon, *The Information Society*, Geneva.

White, G.K. 2013. *Digital fluency: Skills necessary for learning in the digital age*, ACER, Melbourne.

World Economics Forum. 2016.*The future of jobs: employment, skills and workforce strategy for the Fourth Industrial Revolution*, Cologny/Geneva Switzerland January 2016.

COLLEGE/HIGH SCHOOL STUDENTS' CYBERSECURITY CAREER INTEREST

Anthony Joseph[1], Mary Joseph[2] and Tega Ileleji[3]
[1]*Pace University, New York, New York, USA*
[2]*Herbert Lehman College, City University of New York, Bronx, New York, USA*
[3]*Visa Inc, Foster City, CA, USA*

ABSTRACT

This study investigates college and high school students' career interest in cybersecurity. It intends to contribute to the research that explores millennials' and generation Z general lack of interest in the cybersecurity field. The dataset consists of 163 college and high school students who responded to a career interest Science, Technology, Engineering, and Mathematics (STEM) semantic survey adapted to include cybersecurity and computer science. Students rated five sets of paired semantic interest scales: Fascinating to mundane, appealing to unappealing, exciting to unexciting, means nothing to means a lot, and boring to interesting with fascinating, appealing, exciting, means nothing and boring having a rating value of 1. Research showed that there is a shortage of cybersecurity professionals. Yet, millennials are barely interested in a cybersecurity career. The technology workforce is about 90% Whites and Asians and 75% males. Moreover, non-Whites and non-Asians have been leaving the industry workforce because of perceived unfair treatment. This investigation found that the sample of students generally had low cybersecurity and STEM interest, thereby substantiating previous research findings. In addition, there were no statistically significant differences in the mean cybersecurity interest ratings of the various subgroups used in the study: males and females, Whites and Asians and non-Whites and non-Asians, and White and Asian Males and non-White and non-Asian males and females. Therefore, in practical and statistical terms, students' interest in cybersecurity was generally low and essentially the same among the subgroups.

KEYWORDS

Cybersecurity, Career Interest, Generation Z, Millennials, Workforce Shortage

1. INTRODUCTION

Over the years, research has consistently and continually shown that there is a global shortage of cybersecurity professionals (Bittie & Ostrowski, 2020; Furnell et al., 2017; Ileleji & Joseph, 2018; (ISC)^2, 2020[a]; National Academy of Sciences, 2015; Cyberseek, n.d.; Secretary of Commerce & Secretary of Homeland Security, 2018; MacKinnon et al., (2018; Markow & Vilvovsky, 2021; Crumpler & Lewis, 2019). Governments, businesses, and academia demand for cybersecurity professionals outpaced the available supply. At the United States national level, the supply/demand ratio is 0.68 (Cyberseek, n.d.). At the same time, generation Y (or millennials) and generation Z do not appear to have a strong interest in cybersecurity as a career option (Rayome, 2018; (ISC)^2, 2020[b]). Nonetheless, the global cybersecurity workforce increased by an estimated 700, 000 between 2020 and 2021. This narrowed the shortage gap with the highest increases occurring in Germany, Singapore, and the United States. The United States showed a 30% increase suggesting that millennials and generation Zers are warming up to cybersecurity as a career option ((ISC)^2, 2021; Raytheon, 2017; Carnegie Mellon University's Software Engineering Institute, 2019). Typically, millennials and generation Zers entry points into the cybersecurity field are through less technical and more sundry occupational routes than those used by boomers and generation Xers ((ISC)^2, 2021). This may be partially influenced by the cybersecurity labor shortage, and may not be what employers actually want, but rather what they are willing to concede until supply nears demand or the supply shortage is successfully complemented through other means; namely, artificial intelligence and machine learning (IEEE Confluence, 2017; Musser & Garriott, 2021; Pupillo et al. 2021). Despite the evidential improvement in lowering the cybersecurity labor shortage and the increased awareness of the importance of cybersecurity and cybersecurity careers, millennials display risky cyber hygiene and do not show strong interest in cybersecurity careers (Raytheon, 2017; Carnegie

Mellon University's Software Engineering Institute, 2019). In addition, millennials and generation Zers have diverse social interests upon which they place high value and are more interested in working for "companies that share their values" (Raytheon, 2017; Deloitte Global, 2021). Moreover, Carnegie Mellon University's Software Engineering Institute (SEI) (2019) reported that the main causes for the cybersecurity talent shortage were rapid changes in technology, a sizable untapped group of skilled potential employees, exclusionary employer requirements for potential employees, and insufficient awareness of cybersecurity career pathways and opportunities. SEI further stated that the available information on the details of a cybersecurity career is conflicting, confusing, and overwhelming to anyone interested in pursuing a cybersecurity career.

Cybersecurity is by definition a multidisciplinary field ($(ISC)^2$, 2021; McAlaney et al., 2018; Joint Task Force on Cybersecurity Education, 2018; Dawson & Thomson, 2018) that is a computing discipline with elements of engineering, social science, mathematics, and philosophy. However, the social aspects of this socio-technical multidisciplinary field are not sufficiently emphasized (Joint Task Force on Cybersecurity Education, 2018; Dawson & Thomson, 2018) and utilized. More generally, the technology workforce is 90% Whites and Asians and 75% males with women and traditionally underrepresented racial and ethnic minority groups in STEM not sufficiently represented (Scott et al., 2018). These numbers seem to correlate well with those racial/ethnic groups who invest most heavily in and own most of the technology companies. Moreover, non-White and non-Asian males are more likely than other groups of individuals within the technology industry to leave it because of perceived unfair treatment (Scott et al., 2017). Within the cybersecurity workforce, retention is a problem: a consequence of burnout, poaching, inhospitable workplace conditions, insufficient inducements to stay, and voluntary departures (CyberKnights, 2021; Morgan, n.d.). Nevertheless, the convergence of the virtual and physical worlds is quickening, and with it increased vulnerabilities and associated risks, costs to social cohesion, mental health issues, and threats to critical infrastructures (Franco et al., 2022; Optiv, 2019; Markow & Bittle, 2020; Olmstead & Smith, 2017).

This exploratory study addresses the general problem of the shortage of cybersecurity professionals. Carnegie Mellon University's Software Engineering Institute (2019) reported that the main causes for the cybersecurity talent shortage were rapid changes in technology, a sizable untapped group of skilled potential employees, exclusionary employer requirements for potential employees, and insufficient awareness of cybersecurity career pathways and opportunities. Carnegie Mellon University's Software Engineering Institute further stated that the available information on the details of a cybersecurity career is conflicting, confusing, and overwhelming to anyone interested in pursuing a cybersecurity career. This study will explore the aspect of cybersecurity talent shortage that relates to exclusionary employer requirements for potential employees with the focus on females and non-White and non-Asian males. As mentioned earlier, the technology workforce is primarily White and Asian males with underrepresentation of females and racial/ethnic minority groups: Do White and Asian males have greater interest in cybersecurity careers than others? The question that this study is asking is the following: Do females and non-White and non-Asian males have similar interest in cybersecurity as White and Asian males? If the answer is yes, ways to increase the awareness and participation of females and non-White and non-Asian males in the cybersecurity workforce will be suggested to help ameliorate the shortage. To carry out this study, college and high school students' career interest in cybersecurity will be investigated with the aim to contribute to the research literature that explores young people's general interest in cybersecurity with the emphasis on females and non-White and non-Asian males. This investigation was pursued using descriptive statistics in means and standard deviations, correlations, and inferential statistics wherein three pairs of hypotheses were tested on three different subgroups.

2. METHODS AND MATERIALS

The dataset, which consisted of entries from 163 college and high school students from a Northeastern United States metropolitan area, was obtained from a semantic STEM career interest survey (Tyler-Wood, 2010) adapted to include cybersecurity and computer science disciplines. Computer science is combined with the engineering interest to make engineering/computer science career interest, as many computer science departments reside in engineering schools/colleges. For this study, therefore, the semantic career interest survey included Science, Technology, Engineering/Computer Science, and Mathematics, plus Cybersecurity (STEM+Cybersecurity) disciplines. The dataset of 163 samples was divided into subsets: 57 females, 106 males, 76 White and Asian males, 87 females and non-White and non-Asian males, 104 Whites and Asians, and 59 non-Whites and non-Asians. The students rated five sets of paired semantic interest scales: Fascinating to mundane (FM), appealing to unappealing (AUA), exciting to unexciting (EUE), means nothing to means a lot (MNaL), and boring to interesting (BI) in the five STEM+Cybersecurity related disciplinary areas. One was the highest rating for fascinating, appealing, exciting, means nothing, and boring while seven was the highest rating for mundane, unappealing, unexciting, means a lot, and interesting.

To collect the data for the sample of college and high school students, the semantic career interest survey was chosen over interviews and the Likert scale type of survey because interviews were considered to be too time-consuming and costly to obtain the sufficiently large sample size that was desired for this study. While a Likert scale survey would have worked in terms of the sample size and cost, it was deemed to be insufficiently nuanced on the aspects of the descriptiveness of career interest thereby providing less depth of information of how the student survey takers feel than the semantic survey. The sampling strategy used is a mixture of purposive and convenience sampling methods in that a sufficiently large sample size of college and high school students, who had taken at least one course or was taking a course in computing, was desired. Moreover, this sampling method was used to facilitate the division of participants into subgroups large enough for hypotheses testing to be done.

Seventy-three high schools students and 92 college students participated in the in-person survey. The student populations were self-identified as males and females. They were from seven different ethnic/racial groups: Whites, Blacks/African Americans, Spanish/Hispanics/Latinx, Native Americans/Alaskan Natives, Asians, Hawaiians/Pacific Islanders, and Other. The students ranged in ages from about 15 to 60. The high school students were from 12 different high schools with diverse educational foci and a common computer science interest. The college students ranged from baccalaureates to postgraduates; they represented nine different computing related programs Computer Science, Information Technology, Telecommunications and Networks, Professional Studies, and Information Systems, Computer Science and Criminal Justice, Computer Science and Law, Information Technology and Cybersecurity, and Computer Science and Business. Of the 165 surveys collected, one high school and one college survey were removed from the sample because of excessive missing entries. Other tools used in this study included Microsoft Excel used for the calculations of means and standard deviations and as an overall statistical workbook, Mathworks Matlab used for correlation calculations, and SPSS used for the calculations of hypotheses tests.

To gain insight into the robustness of the results obtained through means and correlations of the subgroups, hypotheses testing (Ryan, 2007) were used. It was also used to examine the likelihood of any statistical differences in the interest mean ratings between subgroups. Three pairs of hypothesis tests were conducted under the assumption of no difference in mean ratings: one for fascinating to mundane, appealing to unappealing, and exciting to unexciting (FAE) and the other means nothing to means a lot and boring to interesting (MNB). The hypotheses were tested between White and Asian students and non-White and non-Asian students, male and female students, and White and Asian male students and female and non-White and non-Asian male students.

The semantic survey rating scale values were interpreted as follows: the mean rating values from 1.00 -- 1.99 and from 6.01 -- 7.00 were viewed as high, from 2.00 -- 2.99 and from 5.01 -- 6.00 were viewed as medium, from 3.00 -- 3.99 and from 4.01 -- 5.00 were viewed as low, and 4.00 was considered neutral. In addition, each of the rating scales failed the skewness and kurtosis as well as Lilliefors test for normality. In the Matlab environment, a normal distribution has a kurtosis value of 3.00 and a skewness value of 0.00. Each rating scale FM, AUA, EUE, MNaL, and BI had a kurtosis value of less than 3.00 and a skewness value not equal to 0.00. These results indicate that the five distributions were less prone to outliers and that FM, AUA, and EUE were asymmetrical distributions tending rightward while the MNaL and BI were distributions tending leftward. Moreover, the Lilliefors tests of FM, AUA, EUE, MNaL, and BI rejected the null hypothesis of normality at a significance level of less than 0.001. Because of the lack of normality in the data, Spearman's Rho correlation coefficient was used to measure the strength of the relationship between the pairs of rating scales. The strength of the absolute values of the Spearman's Rho correlations should be interpreted as very weak is 0.00 -- 0.19, weak is 0.20 -- 0.30, moderate is 0.40 -- 0.59, strong is 0.60 -- 0.79, and very strong is 0.80 -- 1.00 (Spearson's Correlation, n.d.).

3. RESULTS

Overall, there were no high mean rating values of 1.00 -- 1.99 and 6.01 -- 7.00 or neutral mean rating values of 4.00 for FM, AUA, EUE, MNaL, and BI rating scales for any of the samples and subsamples. Therefore, the mean rating values observed for each scale were mostly low with few medium ones. As shown in Table 1, science yielded the highest overall low mean rating of 3.14 for FAE (fascinating, appealing, and exciting), relative to the neutral mean rating value of 4.00 and the highest overall medium mean value of 5.42 for MNB (means a lot and interesting). More specifically, there were medium mean rating values for exciting, means a lot, and interesting: 2.89, 5.78, and 5.06 respectively. Cybersecurity yielded the second highest overall mean rating of 3.35 for FAE and the second highest overall mean rating of 4.84 for MNB. These overall mean rating values for cybersecurity indicate low interest. However, MNaL's means a lot mean rating of 5.08 indicates medium interest. Students appeared to be somewhat ambivalent about mathematics and skeptical of technology. For technology, the overall mean rating pattern for FAE and MNB was contrary to that found for science, cybersecurity, and engineering/computer science with FAE and MNB having low mean rating values of 4.62 and 3.64, respectively (see Table 1). These mean rating values are low. Moreover, the cybersecurity standard deviations for the FAE and MNB were the lowest among the disciplines.

Table 1. STEM+Cybersecurity means and standard deviations statistics

Disciplines statistics		Rating scales					Overall ratings	
Category	Statistics	FM	AUA	EUE	MNaL	BI	FAE	MNB
Cybersecurity	Mean	3.42	3.44	3.20	5.08	4.60	3.35	4.84
	Stdev	2.03	1.96	1.78	1.82	1.87	1.92	1.86
Science	Mean	3.27	3.26	2.89	5.78	5.06	3.14	5.42
	Stdev	2.32	2.17	1.85	1.64	2.14	2.13	1.94
Technology	Mean	3.83	5.32	4.71	4.05	3.23	4.62	3.64
	Stdev	2.33	1.80	1.92	2.16	2.04	2.11	2.13
Engineering/Computer Science	Mean	3.32	3.18	5.16	3.60	5.08	3.89	4.34
	Stdev	2.32	2.25	1.92	2.29	2.26	2.35	2.39
Mathematics	Mean	5.12	3.61	3.34	3.68	5.02	4.02	4.35
	Stdev	1.91	2.10	1.89	2.03	2.01	2.11	2.12

Note 1: For the acronym FAE, F represents fascinating of FM (fascinating to mundane) with fascinating =1 and mundane =7, A represents appealing of AUA (appealing to unappealing) with appealing =1 and unappealing = 7, and E represents exciting of EUE (exciting to unexciting) with exciting =1 and unexciting = 7; and for MNB, MN represents means nothing of MNaL (means nothing to means a lot) with means nothing =1 and means a lot = 7 and B represents boring of BI (boring to interesting) with boring =1 and interesting = 7.

Note 2: 1.00--1.99 & 6.01-7.00 is high; 2.00-2.99 & 5.01-6.00 is medium; 3.00-3.99 & 4.01-5.00 is low; and 4:00 is neutral

The correlations between the rating scales for cybersecurity relative to the other STEM disciplines showed that the distribution of correlation coefficient values between rating scales of cybersecurity and science were quite similar (see Table 2 a & d). The correlations between AUA and FM, EUE and FM, and between EUE and AUA were positive and at least strong, and notably very strong between AUA and FM. The correlation between BI and MNaL was positive and moderate. On the contrary, the correlation coefficient values between MNaL and FM, MNaL and AUA, MNaL and EUE, BI and FM, BI and AUA, and between BI and EUE were negative and generally moderate; they were weak between MNal and FM (0.2546) and between MNaL and AUA (0.2279) for science. Moreover, these correlation values associated with cybersecurity and science were statistically significant at the 0.01 level. In each case, the probability values were less than 0.0000. In addition, mathematics and technology had a somewhat similar pattern in correlation values between pairs of rating scales (see Table 2 b & c). For example, the correlation values of 0.0613 and (0.0616) between AUA and FM for mathematics and technology respectively were weak and not statistically significant at the 0.05 level indicating that statistically the scales were not correlated. In addition, the correlations between EUE and AUA for these two disciplines were strong and statistically significant with values of 0.7619 and 0.6005 respectively. Other similarities in the rating scales for mathematics and technology were apparent in the moderate and statistically significant negative correlation values of (0.4271) and (0.4384), respectively, between BI and AUA as well as the weak and statistically significant positive correlation values between BI and FM and the negative correlation values between BI and EUE (see Table 2 b & c).

Table 2 (a-e). STEM+Cybersecurity Spearman's rho correlations between rating scales

(a) Science correlations

	FM	AUA	EUE	MNaL
AUA	0.8466**			
EUE	0.7423**	0.7243**		
MNaL	(0.2546)**	(0.2279)**	(0.4237)**	
BI	(0.5803)**	(0.5336)**	(0.5205)**	0.4557**

(b) Technology correlations

	FM	AUA	EUE	MNaL
AUA	(0.0616)			
EUE	(0.3766)**	0.6005**		
MNaL	0.6685**	(0.2132)**	(0.4337)**	
BI	0.3908**	(0.4384)**	(0.3210)**	0.5686**

(c) Mathematics correlations

	FM	AUA	EUE	MNaL
AUA	0.0613			
EUE	(0.1526)	0.7619**		
MNaL	(0.2025)**	0.6839**	0.6626**	
BI	0.2317**	(0.4271)**	(0.3255)**	(0.3152)**

(d) Cybersecurity correlations

	FM	AUA	EUE	MNaL
AUA	0.8606**			
EUE	0.7662**	0.8002**		
MNaL	(0.4583)**	(0.4385)**	(0.4345)**	
BI	(0.5732)**	(0.5430)**	(0.4430)**	0.5609**

(e) Engineering/computer science correlations

	FM	AUA	EUE	MNaL
AUA	0.9047**			
EUE	(0.5438)**	(0.4912)**		
MNaL	0.6832**	0.6461**	(0.5284)**	
BI	(0.6190)**	(0.6471)**	0.5085**	(0.4207)**

** indicates a p-value is less than the significance level of 1%

* indicates a p-value is less than the significance level of 5%

Note: FM (fascinating to mundane) with fascinating =1 and mundane =7, AUA (appealing to unappealing) with appealing =1 and unappealing = 7, EUE (exciting to unexciting) with exciting =1 and unexciting = 7, MNaL (means nothing to means a lot) with means nothing =1 and means a lot = 7, and BI (boring to interesting) with boring =1 and interesting = 7.

When correlations were taken across the disciplines of STEM+Cybersecurity, it was found that the correlation coefficient values between the rating scales of science and cybersecurity had the most distinctly consistent pattern while those of mathematics and technology had the largest set of very weak to weak coefficients. The other eight sets of correlation values between the rating scales of science and mathematics, science and engineering/computer science, science and technology, mathematics and engineering/computer science, mathematics and cybersecurity, engineering/computer science and technology, engineering/computer science and cybersecurity, and technology and cybersecurity had less distinctly consistent patterns of correlation coefficient values. For example, except for the correlation coefficient value of 0.1444 between MNal and FM, the remaining 24 correlation values of science and cybersecurity rating scales were statistically significant at least at the 0.05 level. In fact, only three of the 24 correlation values were significant at the 0.05 level: those were between MNaL and AUA, MNaL and EUE, and between MNaL and BI. However, none of the correlation coefficient values was strong or very strong; they ranged in values from very weak to moderate. Moreover, the correlation coefficients between science and cybersecurity rating scale pairs FM and FM, FM and AUA, FM and EUE, AUA and FM, AUA and AUA, AUA and EUE, EUE and FM, EUE and AUA, and EUE and EUE were moderate, statistically significant, and positive while those between the pairs FM and MNaL, FM and BI, AUA and MNaL, AUA and BI, EUE and MNaL, and EUE and BI were weak, statistically significant, and negative. The arithmetical sign of the correlations between the pairs MNaL and FM, MNaL and AUA, MNaL and EUE, BI and FM, BI and AUA, and BI and EUE were very weak to moderate statistically significant negative values whereas the correlations between the pairs MNaL and MNaL, MNaL and BI, BI and MNaL, and BI and BI were very weak to moderate and statistically significant positive values. The patterns of the correlation coefficient values obtained between the rating scales for mathematics and technology were unlike those obtained for science and cybersecurity in arithmetical signs, strengths, and statistical significance. For example, 13 of the correlation coefficient values were very weak and 14 were statistically significant at least at the 0.05 significance level.

Focusing on the details of students' interest in cybersecurity, it was found that the subgroups (males, females, Whites and Asians (WA), non-Whites and non-Asians (non-WA), White and Asian (WA) males, and non-White and non-Asian (non-WA) males) almost completely have low interest in cybersecurity with only slight variations in their mean ratings (see Table 3). Although non-WA somewhat appeared to indicate having the lowest interest in cybersecurity as evidenced by the mean ratings and the correlation coefficient values, the Independent-Samples Mann-Whitney U test (see Table 4) did not reject the null hypotheses of no difference in the distributions of FAE mean ratings and the distributions of MNB mean ratings for WA and non-WA, females and males, and WA males and females and non-WA males. Moreover, while FAE and the associated paired rating scales FM, AUA, and EUE for each subgroup was low, between 3.00 -- 3.99, which indicates that students' perception of cybersecurity being fascinating, appealing, and exciting was low, MNB and the associated paired rating scales MNaL and BI for the subgroups had instances of lows and mediums with BI being low for each subgroup and MNB and MNaL being low for some subgroups and medium for others (see Table 3). In fact, the males and WA males' subgroups perceptions of cybersecurity as reflected in MNB and MNaL were medium; the WA's subgroup perception of cybersecurity was also medium for MNaL.

Table 3. Cybersecurity means and standard deviations statistics

Subgroup statistics		Rating scales					Overall ratings	
Category	Statistics	FM	AUA	EUE	MNaL	BI	FAE	MNB
Overall	Mean	3.42	3.44	3.20	5.08	4.60	3.35	4.84
	Stdev	2.03	1.96	1.78	1.82	1.87	1.92	1.86
Females	Mean	3.56	3.63	3.30	4.65	4.39	3.50	4.52
	Stdev	1.82	1.81	1.60	1.89	1.73	1.74	1.81
Males	Mean	3.34	3.33	3.14	5.31	4.72	3.27	5.01
	Stdev	2.14	2.03	1.87	1.75	1.95	2.01	1.87
WA	Mean	3.38	3.25	3.04	5.17	4.66	3.22	4.92
	Stdev	2.09	1.95	1.74	1.80	1.89	1.93	1.86
Non-WA	Mean	3.49	3.76	3.47	4.92	4.49	3.58	4.70
	Stdev	1.94	1.94	1.82	1.87	1.86	1.90	1.87
WA Males	Mean	3.39	3.21	3.04	5.37	4.76	3.21	5.07
	Stdev	2.19	2.02	1.87	1.79	1.98	2.03	1.90
Females & Non-WA Males	Mean	3.44	3.63	3.33	4.83	4.46	3.47	4.64
	Stdev	1.89	1.89	1.69	1.82	1.78	1.82	1.81

Note 1: WA stands for Whites and Asians; stdev = standard deviation

Note 2: For the acronym FAE, F represents fascinating of FM (fascinating to mundane) with fascinating =1 and mundane =7, A represents appealing of AUA (appealing to unappealing) with appealing =1 and unappealing = 7, and E represents exciting of EUE (exciting to unexciting) with exciting =1 and unexciting = 7; and for MNB, MN represents means nothing of MNaL (means nothing to means a lot) with means nothing =1 and means a lot = 7 and B represents boring of BI (boring to interesting) with boring =1 and interesting = 7.

Furthermore, Table 3 shows that females' mean rating values were slightly lower than that of males with standard deviations that were generally smaller, denoting less spread about the mean. The highest mean rating values recorded were for WA males, which were generally slightly higher than the ratings of WA, and the mean ratings of WA were essentially the same as that of the males. Non-WA had arguably the lowest mean ratings; these ratings were slightly lower than those of females over the five rating scales with higher standard deviation values.

Table 4. Hypothesis tests summaries

No.	Null Hypothesis	Test	Sig.[a,b]	Decision
1	The distribution of FAE mean ratings is the same across the White and Asian student group and the non-White and non-Asian student group.	Independent-Samples Mann-Whitney U Test	0.176	Retain the null hypothesis.
2	The distribution of MNB mean ratings is the same across the White and Asian student group and the non-White and non-Asian student group.	Independent-Samples Mann-Whitney U Test	0.416	Retain the null hypothesis.
3	The distribution of FAE mean ratings is the same across the students' gender groups.	Independent-Samples Mann-Whitney U Test	0.343	Retain the null hypothesis.
4	The distribution of MNB mean ratings is the same across the students' gender groups.	Independent-Samples Mann-Whitney U Test	0.055	Retain the null hypothesis.
5	The distribution of FAE mean ratings is the same across the White and Asian Males student group and the Females, Non-White, and Non-Asian Males student group.	Independent-Samples Mann-Whitney U Test	0.300	Retain the null hypothesis.
6	The distribution of MNB mean ratings is the same across the White and Asian Males student group and the Females, Non-White, and Non-Asian Males student group.	Independent-Samples Mann-Whitney U Test	0.083	Retain the null hypothesis.

a. The significance level is .050; b. Asymptotic significance (Sig.) is displayed

Note: For FAE, F represents fascinating of FM (fascinating to mundane), A represents appealing of AUA (appealing to unappealing), and E represents exciting of EUE (exciting to unexciting); and for MNB, MN represents means nothing of MNaL (means nothing to means a lot) and B represents boring of BI (boring to interesting).

The pattern of the correlation coefficient values was consistent across the six subgroups (females, males, females and non-WA, WA males, WA, and non-WA) in terms of sign, general strength, and statistical significance. However, the correlation values' strength and statistical significance vary among the subgroups with the greatest difference in strength and statistical significance observed between WA and non-WA. Nonetheless, across all the subgroups the correlation value with the strongest strength was observed between AUA and FM. Seven of the 10 correlation coefficient values for WA were strong to very strong whereas the remaining three correlation coefficient values were moderate, and related to MNaL. This correlation finding for WA was essentially the same as that for WA males in strength with slight variations in values. Both WA and WA males' correlation coefficient values were statistically significant at the 0.01 level. In contrast, there was no very strong correlation value for non-WA, but there were two strong values with the stronger being 0.7651 between AUA and FM and two moderate values with the stronger being 0.5900 between EUE and FM. The remaining six correlation coefficient values for non-WA were very weak (one) to weak (five). These six values were associated with MNaL and BI and each of their relationships with FM, AUA, and EUE. Eight of the 10 correlation values for non-WA were statistically significant at least at the 0.05 level. In terms of correlation strength and the statistical significance of the correlations, the non-WAs were more consistent with the females' subgroup than any other subgroup.

4. DISCUSSION

This investigation shows that the sampled college and high school students do not have a strong interest in a cybersecurity career. Their interest is essentially low. In fact, their interest in STEM careers is low to medium. Moreover, inferential statistics showed that the difference in interest in cybersecurity careers between males and females, between Whites and Asians and non-Whites and non-Asians, and between White and Asian males and others were not statistically significant. This finding of low interest in cybersecurity answers the posed question, *Do females and non-White and non-Asian males have similar interest in cybersecurity as White and Asian males?*, in the affirmative.

Although career interest in STEM+Cybersecurity disciplines was low and the differences in the mean ratings were generally marginal, only science had higher mean ratings than cybersecurity. Moreover, the ratings of cybersecurity and science were more consistent with each other than among any other pairs of STEM+Cybersecurity disciplines as reflected in their disciplinary individual and cross correlation analyses. Furthermore, while White and Asian males appeared to have the highest interest in cybersecurity, this was essentially very marginal, and when the relatively smaller numbers of females (57) and non-Whites and non-Asians ((59) are taken into consideration there might be no obvious difference in interest in cybersecurity among the subgroups. Besides, the inferential statistics showed that there were no statistically significant differences in the mean ratings of males and females, Whites and Asians and non-Whites and non-Asians, and White and Asian males and females and non-White and non-Asian males. This suggests that the students regardless of gender and racial/ethnic grouping largely felt the same way about cybersecurity.

The finding of low interest in this study supports some earlier studies findings of low interest among millennials and generation Zers. For example, references Rayome (2018), $(ISC)^2$ (2020^b) and Raytheon (2017) reported on the low interest that millennials and generation Zers have in cybersecurity as a career and attributed it to a general lack of awareness and media stereotypes. In fact, millennials and generation Zers perceived cybersecurity workers more negatively than baby boomers and generation Xers ($(ISC)^2$, 2020^b). While the survey respondents of ($(ISC)^2$, 2020^b) generally viewed a cybersecurity career positively, they did not see themselves in it; one of the reasons given for this lack of interest is the relative "high cost of entry." The high cost of entry (Cyberseek, n.d.; $(ISC)^2$, 2020^b) could remain a problem for some aspirants unless the extra cost for the multiple certifications and the continual professional development that might be needed beyond a college degree for career entry, currency, productivity, and advancement are clearly justified with returns on the investments. The survey results of this study clearly showed that high school and college students who consisted mainly of generation Zers and millennials generally have a low interest in STEM with more interest in cybersecurity than in mathematics and technology disciplines and related career areas that they would likely have more exposure to and awareness of. Moreover, most of the students were attending high schools with computer science programs or taking either a college computing course or program. Furthermore, many emerging fields such as blockchain, data science, artificial intelligence, cloud computing, and Internet of things as well as the established engineering fields including computer science and software engineering pay salaries competitive with that received by cybersecurity professionals, and yet have a lower cost of entry as there is not as much emphasis on the myriad certifications supplementing a college degree. Given the societal role of cybersecurity in the safeguarding of cyberspace and that a 100% secure cyberspace is not practical even if achievable, it becomes incumbent on cybersecurity professionals to ensure that cyberspace is protected from all cyber-attacks to avoid a potentially serious breach with costly economic and reputational consequences. If a cybersecurity work environment is not well-staffed with workers scheduled with their well-being in mind, these workers could become over-worked and highly stressed given their responsibilities for protecting more and more of peoples' lives that are being put online. This will likely further exacerbate general workplace experiences of high levels of stress, anxiety, and burnout reported by millennials and generation Zers (Deloitte Global, 2022). In addition, the low interest of millennials and generation Zers in STEM may have more to do with their vocational interest being somewhere else and they might need more encouragement to become STEM-centric and cybersecurity focus. According to National Center for Education Statistics' 2019-2020 Postsecondary Education report (National Center for Education Statistics, 2022) of the approximately three million associate and baccalaureate level degrees conferred in 2019-2020, the majority was for business and health related professions. At the associate level, only 8% were conferred for STEM and at the baccalaureate level 21% were conferred for STEM with 6% being for engineering and 5% being for computer and information sciences and support services. Therefore, millennials and generation Zers interest might be mainly outside of STEM related disciplines including cybersecurity.

An implication of the finding of this study is that a vibrant, diverse, inclusive, equitable, productive, and creative cybersecurity workforce that is accepting of others is feasible. A barrier to this cybersecurity workforce could be the exclusionary employer requirements for potential employees (Carnegie Mellon University's Software Engineering Institute, 2019), which would need to be removed to accommodate an increased number of females and non-White and non-Asian males. Scott et al (2018) reported that the more racial/ethnic and gender diverse a company is the more profitable it is with a greater customer base and market value. In the current cybersecurity landscape, this general lack of racial/ethnic and gender diversity likely translate into the inability to identify some attacks because of insufficient knowledge of cultural and social skills (Dawson & Thomson, 2018; McAlaney et al, 2018) that are built into the attack. Two of the reasons why the cybersecurity workforce are mainly Whites and Asians may have more to do with management and familiarity rather than productivity, product quality, customer service, and profit (($(ISC)^2$, 2020[b]; Scott et al., 2018). Because of the general lack of representative role models and mentors, females and non-White and non-Asian males need ample awareness of the cybersecurity field and the different job opportunities and career pathways, opportunities for advancement, flexible work schedules, and opportunities to switch from one cybersecurity pathway to another. In addition, suitable incentives, motivators, and advocacies (Carnegie Mellon University's Software Engineering Institute, 2019; McAlaney et al., 2018; Optiv, 2019; Haney & Lutters, 2019) might be needed at different junctures in the educational pipeline from pre-K to post-graduate school and beyond. These interventions are likely necessary to encourage and recruit the most suitable candidates to the cybersecurity field and to promote good cyber-hygiene in cyberspace (McAlaney et al., 2018; Olmstead & Smith, 2017) because over 90% of the cybersecurity troubles that organizations experience are user based while another 43% are from employees (Franco et al., 2022). Additionally, cybersecurity could borrow additional interventions from nursing to ameliorate the friction caused by the high cost of entry and competition from other fields in its attempt to attract females and non-White and Asian males, and other underrepresented groups. Like cybersecurity, nursing has a high cost of entry: a college degree and a license to practice supplemented with professional developments to remain current. Nursing also appears to have a diversity problem as it pertains to providing good health care for all (Phillips & Malone, 2014). There are programs in nursing that successfully recruit and retain individuals who are from underrepresented minorities (Phillips & Malone, 2014). The strategies employed in these programs that could be beneficial throughout the cybersecurity ecosystem include the following: support and encouragement before enrollment in a college program as well as ongoing support throughout college; supportive environments; cultural competency training; working with community partners; counseling; academic and financial support; mentoring; teacher, peer, and social support; and appropriate role models. Furthermore, upper management needs to make diversity integral to its organization's strategic plan with clear measurable objective outcomes that link diversity to product quality, customer service, productivity, and profit.

5. CONCLUSION

This study shows that millennials and generation Zers have a low interest in cybersecurity and more generally STEM related disciplines. It adds to the findings of earlier studies that reported on millennials' and generation Zers' low interest in cybersecurity. Although the samples of females as well as non-Whites and non-Asians were relatively smaller than males and Whites and Asians, there were no essentially practical or statistically significant differences in the mean interest ratings of the subgroups. With the high cost of entry into cybersecurity, insufficient readily available information about it, and competition from other high paying STEM fields such artificial intelligence and data science, it is not sufficiently compelling to pursue careers in cybersecurity, especially among those who are females or members from racial/ethnic minorities. This paper further uses data on the number of degrees conferred in the United States in 2019-2020 to show that students' career interests were mainly in business and health related fields, not STEM related fields. However, the small sample sizes of some of the subgroups, notably females and non-whites and non-Asians, should be viewed as a limitation of this study.

REFERENCES

Bittie, S and Ostrowski, S. (2020, November 9). CyberSeek™ Helps Organizations Address Growing Cybersecurity Staffing Challenges. *Burning Glass Technologies*. https://www.burning-glass.com/cyberseek-helps-organizations-address-growing-cybersecurity-staffing-challenges/

Carnegie Mellon University's Software Engineering Institute. (2019, February). Cybersecurity Career Paths and Progression. *Department of Homeland Security's Cybersecurity and Infrastructure Security Agency (CISA*. https://niccs.cisa.gov/sites/default/files/documents/pdf/cybersecurity%20career%20paths%20and%20progressionv2.pdf?trackDocs=cybersecurity%20career%20paths%20and%20progressionv2.pdf

Crumpler, W. and Lewis, J. (2019, January). The Cybersecurity Workforce Gap. *Center for Strategic and International Studies*. https://www.csis.org/analysis/cybersecurity-workforce-gap

CyberKnights. (2021). What Fuels Cybersecurity Retention, and Reduces Employee Attrition, is Achievable? https://www.cyberknights.us/what-fuels-cybersecurity-retention-and-reduces-employee-attrition-is-achievable/

Cyberseek. (n.d.). *Cybersecurity Supply/Demand Heat Map*. Retrieved February 10, 2022, from https://cyberseekus.herokuapp.com/heatmap.html

Dawson, J. and Thomson, R. (2018, June). The Future Cybersecurity Workforce: Going Beyond Technical Skills for Successful Cyber Performance. *Frontiers in Psychology*, 9. https://doi.org/10.3389/fpsyg.2018.00744_Reference [23]

Deloitte Global. (2021). A Call for Accountability and Action: The Deloitte Global 2021 Millennial and Gen Z Survey. *Deloitte*. https://www2.deloitte.com/content/dam/Deloitte/global/Documents/2021-deloitte-global-millennial-survey-report.pdf

Deloitte Global. (2022). Striving for Balance, Advocating for Change: The Deloitte Global 2022 Gen Z and Millennial Survey. *Deloitte*. https://www2.deloitte.com/content/dam/Deloitte/cn/Documents/about-deloitte/deloitte-2022-genz-millennial-survey-en-220601.pdf

Franco, E. et al. (2022, January). The Global Risks Report 2022, 17th edition. *World Economic Forum*. https://www.weforum.org/reports/global-risks-report-2022

Furnell, S. et al. (2017, February). Can't Get the Staff? The Growing Need for Cybersecurity Skills. *Computer Fraud & Security*, 2017(2), pp. 5-10.

Haney, J. and Lutters, W. (2019, June 20–22). Motivating Cybersecurity Advocates: Implications for Recruitment and Retention. *Proceedings of the 2019 Association for Computing Machinery (ACM) Special Interest Group on Management Information Systems (SIGMIS) Conference on Computers and People Research*. Nashville, TN, US, pp. 109-117. https://tsapps.nist.gov/publication/get_pdf.cfm?pub_id=927445

IEEE Confluence. (2017). Artificial Intelligence and Machine Learning Applied to Cybersecurity. *IEEE Industry Engagement Committee and Syntegrity Three-Day IEEE Confluence, 6-8 October 2017*. https://www.ieee.org/about/industry/confluence/feedback.html

Ileleji, T, & Joseph, A. (2018, October 22-23). Cybersecurity Talent Shortage and High School Students' Career Interests. *Proceeding of ninth Annual International Conference on Computer Science Education: Innovation and Technology (CSEIT 2018)*. Singapore, Singapore, pp. 105-112.

Joint Task Force on Cybersecurity Education. (2018, February). Cybersecurity Curricula 2017: Curriculum Guidelines for Post-Secondary Degree Programs in Cybersecurity. *Association for Computing Machinery*, Ver. 1. https://dl.acm.org/doi/book/10.1145/3184594

International Information System Security Certification Consortium (ISC)^2. (2021). A Resilient Cybersecurity Profession Charts the Path Forward: (ISC)^2 Cybersecurity Workforce Study, 2021. [https://www.isc2.org/Research/Workforce-Study

International Information System Security Certification Consortium (ISC)^2. (2020^a). Cybersecurity Professionals Stand Up to a Pandemic: (ISC)^2 Cybersecurity Workforce Study. https://www.isc2.org/-/media/ISC2/Research/2020/Workforce-Study/ISC2ResearchDrivenWhitepaperFINAL.ashx?la=en&hash=2879EE167ACBA7100C330429C7EBC623BAF4E07B

International Information System Security Certification Consortium (ISC)^2. (2020^b). How Views on Cybersecurity Professionals are Changing and What Hiring Organizations Need to Know: The 2020 (ISC)^2 Cybersecurity Perception Study. https://www.isc2.org/Research/Perception-Study

MacKinnon, M. et al. (2018, July). The Changing Faces of Cybersecurity - Closing the Cyber Risk Gap. *Deloitte and Toronto Financial Services Alliance*. https://www2.deloitte.com/ca/en/pages/risk/articles/the-changing-faces-of-cybersecurity.html

Markow, W. and Vilvovsky, N. (2021, March). Securing a Nation: Improving Federal Cybersecurity Hiring in the United States. *Burning Glass Technologies*. https://www.burning-glass.com/research-project/cybersecurity-securing-nation/

Markow, W. and Bittle, S. (2020, October). Protecting the Future: The Fastest-Growing Cybersecurity Skills. *Burning Glass Technologies.* https://www.burning-glass.com/wp-content/uploads/2020/10/Fastest_Growing_Cybersecurity_Skills_Report.pdf

McAlaney, J. et al. (2018, November). Behaviour Change: Cybersecurity. *The British Psychological Society.* https://www.bps.org.uk/news-and-policy/changing-behaviour-cybersecurity

Morgan, S. (n.d.). The 2019/2020 Official Annual Cybersecurity Jobs Report. *Herjavec Group.* https://www.herjavecgroup.com/2019-cybersecurity-jobs-report-cybersecurity-ventures/

Musser, M. and Garriott, A. (2021, June). Machine Learning and Cybersecurity: Hype and Reality. *Center for Security and Emerging Technology (CSET) at Georgetown University.* https://cset.georgetown.edu/wp-content/uploads/Machine-Learning-and-Cybersecurity.pdf

National Academy of Sciences (2015). *Cybersecurity Dilemmas: Technology, Policy, and Incentives: Summary of Discussions at the 2014 Raymond and Beverly Sackler U.S.-U.K. Scientific Forum.* Washington, DC: The National Academies Press. https://doi.org/10.17226/21833

National Center for Education Statistics. (2022). Undergraduate Degree Fields: Condition of Education. *U.S. Department of Education-Institute of Education Sciences.* https://nces.ed.gov/programs/coe/indicator/cta

Olmstead, K. and Smith, A. (2017, March). What the Public Knows About Cybersecurity. *Pew Research Center.* https://www.pewresearch.org/internet/2017/03/22/what-the-public-knows-about-cybersecurity/

Optiv. (2019, October). How to Get Into Cybersecurity, Even Without a Technical Background.: https://www.optiv.com/insights/discover/blog/how-get-cybersecurity-even-without-technical-background

Phillips, J. and Malone, B. (2014, January-February). Increasing Racial/Ethnic Diversity in Nursing to Reduce Health Disparities and Achieve Health Equity. *Association of Schools and Programs of Public Health*, Public Health Report, 129(2), pp. 45-50. https://doi: 10.1177/00333549141291s209

Pupillo, L. et al. (2021, May). Artificial Intelligence and Cybersecurity Technology, Governance and Policy Challenges. *Centre for European Policy Studies (CEPS).* https://www.ceps.eu/ceps-publications/artificial-intelligence-and-cybersecurity-2/

Rayome, A. (2018, May 8). Only 9% of Millennials are Interested in a Cybersecurity Career. *TechRepublic.* https://www.techrepublic.com/article/only-9-of-millennials-are-interested-in-a-cybersecurity-career/

Raytheon. (2017). Securing Our Future: Cybersecurity and the Millennial Workforce. https://www.raytheon.com/sites/default/files/2017-12/2017_cyber_report_rev1.pdf

Ryan, T. (2007). *Modern Engineering Statistics*. John Wiley & Sons, Hoboken, New Jersey, USA.

Scott, A. et al. (2018, February). The Leaky Tech Pipeline: A Comprehensive Framework for Understanding and Addressing the Lack of Diversity across the Tech Ecosystem. *Kopar Center for Social Impact.* https://www.kaporcenter.org/the-leaky-tech-pipeline-a-comprehensive-framework-for-understanding-and-addressing-the-lack-of-diversity-across-the-tech-ecosystem/

Scott, A. et al. (2017, April). Tech Leavers Study. *Kapor Center for Social Impact.* https://www.kaporcenter.org/tech-leavers/

Spearson's Correlation (n.d.). http://www.statstutor.ac.uk/resources/uploaded/ spearmans.pdf

Secretary of Commerce and Secretary of Homeland Security. (2018, May 30). Report to the President on Supporting the Growth and Sustainment of the Nation's Cybersecurity Workforce: Building the Foundation for a More Secure American Future. *United States Department of Commerce and the United States Department of Homeland Security..* https://www.nist.gov/itl/applied-cybersecurity/nice/resources/executive-order-13800/report

Tyler-Wood, T. et al. (2010). Instruments for Assessing Interest in STEM Content and Careers. *Journal of Technology and Teacher Education*, 18(2), pp. 341-363.

LEVERAGING SOCMINT: EXTRAPOLATING CYBER THREAT INTELLIGENCE FROM RUSSIA-UKRAINE CONFLICT

Bipun Thapa
Marymount University, 2807 N Glebe Rd, Arlington, VA 22207, USA

ABSTRACT

The paper aims to derive Cyber Threat Intelligence (CTI) from the Russia-Ukraine conflict with the help of Social Media Intelligence (SOCMINT), a framework that emanates reasoning from voluntarily available public information—using open-source tools and APIs, datasets created are assessed through topic modeling, thematic analysis (word cloud), Logit function, and neural network classification. The topic modeling and word cloud failed to provide consequential intelligence due to weak datasets - censorship and integrity remain big concerns. Logit function supplied statistically significant features that were influential in the outcome of the tweets, and MLP, a neural network classifier, yielded 91% accuracy when identifying tweet alliance (Russia or Ukraine).

KEYWORDS

Social Media Intelligence, Cyber Threat Intelligence, Topic Modeling, Machine Learning, Opinion Mining, Logit

1. INTRODUCTION

There is a vital necessity for robust Cyber Threat Intelligence (CTI) due to evolving attacks(Husari et al., 2019), which can derail normalcy for organizations and governments. Accumulating threat intelligence is generally a prelude to solution discovery; thus, organizations and vendors are more acquiescent in sharing their knowledge and collaborating responses, albeit with varying efficacy(Sauerwein et al., 2021). Social media intelligence, an extension of Open-source intelligence (OSINT), provides a new conduit to intelligence for CTI (Omand et al., 2012)by sifting through social media data to measure the pulse of the cyber security posture.

Various vendors postulate(*Using Social Media (SOCMINT) in Threat Hunting*, n.d.) the integration of SOCMINT in the existing OSINT ecosystem where the latter provides context to the cyber threats in question like time, location, and trends, though for SOCMINT to be consequential, it has to be complete, accurate, relevant, and timely. From a defender's perspective, employing SOCMINT capabilities is another tool in the arsenal to protect the assets(Kropotov & Yarochkin, 2019).

Easily deployable Application Programming Interfaces (APIs)(*Getting Started — Tweepy 4.2.0 Documentation*, n.d.)for social media(*APIs for Scholarly Resources | Scholarly Publishing - MIT Libraries*, n.d.), and usability of programming languages like Python(Lakshmi, 2018) and its burgeoning libraries enable the construction of customized CTI frameworks. More notably, the open-source and inexpensive resource rollout acts as a catalyst for the adoption.

The storyline of the Russia-Ukraine conflict has been Ukraine's admirable resistance(Khan, 2022) to all vanguards, including its effective social media management to educate the global diaspora and combat Russian cyber attacks (*Ukraine's Digital Ministry Is a Formidable War Machine | WIRED*, 2022). Willing digital volunteers have assisted in fighting for Ukraine to win the 'information war'(McLaughlin, 2022), and Western media often tout that Ukraine is 'winning' the social media battle(Trouillard, 2022).

With the Russia-Ukraine conflict as the context, this exploratory research attempts to extricate cyber-related intelligence to find a meaningful understanding of the subject. The research aims to satisfy the SOCMINT concept by curating the dataset from publicly available cybersecurity information (Twitter), analyzing it, and synthesizing its essence.

President Volodymyr Zelenskyy and President Vladimir Putin are the lead thespians of the conflict from each side, and by leveraging their following on Twitter, the cyber-related dataset with ample entries will be curated. The datasets will be the basis of analysis; however, it is not established to prove the active hypothesis that Ukraine is winning the social media war. An intermittent outage of social media in Russia would undersample tweets, and hence it would not deliver a realistic foundation for such commentary(Milmo, 2022).

Therefore, this research only attempts to create a repeatable, modular framework that is a proof-of-concept for the SOCMINT-infused CTI framework and potentially derive valuable experiences. This research endeavors to address two research questions:

RQ1 - Can Cybersecurity intelligence be derived from Twitter data?

RQ2 -What independent variables are important alliance indicators for Ukraine and Russia, and can they correctly predict the alliance?

In Section 2, the paper investigates contemporary work on the subject, followed by Section 3, which describes the methodology of the research. Section 4 presents the results and analyzes to discover limitations and future scope (Section 5), and finally, Section 6 summarizes the essence of the findings.

2. RELATED WORK

As part of the intelligence-gathering family, SOCMINT, coined by Sir David Omand and fellow researchers(Ivan et al., 2015), is a combination of tools that leverages social media tools to uncover meaningful intelligence to aid the investigation. OSINT or open-source intelligence is often put in the same category as SOCMINT, but one key attribute separates the two, the latter can analyze both public and private data, whereas the former is strictly focused on publicly available data(*Social Media Intelligence*, n.d.).

Assessing the varying emotions of participants is plausible due to the plentiful crowd-sourced data, real-time perspicuity, and supposing intents from diverse groups. Eclectic techniques are inscribed in intelligence lexicons; Signals Intelligence (SIGINT), Imagery Intelligence (IMINT), and Geomatics (GEOINT) are pertinent examples. The SOCMINT, by nature, could capture a combination of the intelligence above collectively; a single entry could have an image, geographical information, and textual intelligence(Mahood, 2015).

To successfully identify perished Russian soldiers in Ukraine, Clearview AI, a US-based company, scrapped social media images to match the pictures of dead soldiers as a courtesy to the ailing family(Dave, 2022). Thailand maintains a dedicated task force to continuously monitor the public dissent towards monarchy and political groups, with a system structured to reward the whistleblowers(*Three Surveillance Technologies That Protesters Need to Know about - IFEX*, 2019). Egypt, around mid-2014, with suspicious premises to determine 'security hazards', facilitated technologies to monitor social media with insidious purpose. Venezuela imprisoned numerous people by observing social media discord that was harmless in the other jurisdictions, like posting dollar rates or personal opinions differing from the ruling body.

'Kansas City No Violence Alliance', an initiative to comprise a predictive instrument for future offenders, uses social media intelligence in its model(*Social Media Intelligence*, n.d.). Squeaky Dolphin, a presumed GCHQ product, allegedly compromised data cables to monitor comments about prominent British personalities through YouTube and Facebook content(Kelion, 2014). China's social credit system incorporates payment delinquency, public habits, non-compliance with local laws, and social media behavior to control individuals' access to society(Kobie, 2019).

To improve security posture and confidence, or conversely, to create chaos and uncertainty, the use cases of SOCMINT are equally applicable because, for the most part, the analysis is conducted on public data. SOCMINT sources are disorderly and informal from tweets, blogs, forum posts, chats, or any avenues available(Forrester & Hollander, 2016). In addition, excellent open-source network visualization tools, intelligence gathering APIs, and forensic instruments make it easier to collect information.

TWINT is an OSINT tool that utilizes scraped data from Twitter for specific criteria like username or hashtags to comprehend ongoing trends(Kropotov & Yarochkin, 2019). The major tech companies allow APIs to connect to their environment, for example, Tweepy for Twitter and PRAW for Reddit, which makes data analysis easy, albeit they impose rate limits so the insights cannot be visualized in their entirety(*Code Snippets — Tweepy 3.5.0 Documentation*, n.d.).

At the onset of the Russia-Ukraine conflict, considerable hacktivists and the cyber army have taken sides though there is no way to validate the claims. Such parties induce cyber warfare, attacks on supply-chain, DDoS to major banks and government sites, and data breaches to expose tactics. Recorded Future, an intelligence-gathering entity(Vail, 2022), explored the available information to understand the alliance as illustrated in Table 1 below with their corresponding Twitter handles, except that most Russian groups have had their account suspended due to increasing violations of community guidelines. This member list, although it may or may not be comprehensive, was synthesized from credible news sources and intelligence-gathering sites like Recorded Future, Anomali, and ThreatQuotient.

Table 1. Cyber Group Alliance

Group	Alliance	Twitter Handle
Anonymous	Ukraine	@YourAnonOne
IT Army of Ukraine	Ukraine	@ITarmyUA
Belarusian Cyber Partisans	Ukraine	@cpartisans
Secjuice	Ukraine	@Secjuice
Conti leaks	Ukraine	@ContiLeaks
RedBanditsRU	Russia	@RedBanditsRU
Sandworm	Russia	unknown or suspended
Freecivilian	Russia	unknown or suspended
Digital Cobra Gang (DCG)	Russia	unknown or suspended
Zatoich	Russia	unknown or suspended

Ordinary Linear Squares(OLS) Regression model can be incorporated to understand the factors that impact Twitter behavior(Costa et al., 2021); could variables like length of the tweet, mentions or hashtags be significant when predicting the alliance by calculating its statistical significance? In a study of one hundred Twitter users analyzing online behavior, with an accuracy of 75.13 percent, the authors were not only able to predict their preferences but their deeper suites of personalities like openness and agreeableness(Mahajan et al., 2022). Value systems or personal beliefs could be an important predictor of why someone engages in retweets and are sometimes as effective as the traditional machine learning models like Random Forest, XGB, and Logistic Regression(Kakar et al., 2021).

The presence of pro and anti-Kremlin bots are plentiful, and both parties were invariant in promoting their desired accounts on Twitter with one exception - pro-Kremlin's source of truth was their state media, and anti-Kremlin derived their content from areas that could not be controlled by the state(Stukal et al., 2019). Infodemic on social media is another challenge in social media as the dissemination of information is quite rapid, and expert sources and non-expert sources have no different impacts(Wang et al., 2021).

There is some argument in the literature that while SOCMINT provides some situational awareness in the aftermath of the event(Dover, 2020), it is incapable of predicting the immediate threats and creating a significant foundation to yield intelligence(McLoughlin et al., 2020). In addition, there are privacy issues even with publicly available data that could be ripe for misuse, and adversaries like Al-Shabaab are highly active on social media providing an intelligence-gathering framework for their use as well, which makes SOCMINT a double-edged sword(Momi, 2021).

The use cases of SOCMINT are abreast in the literature - however, organizing data and deducing intelligence is a difficult task due to the dynamism of data, lack of validity in the public domain, and possibilities of duplication making it somewhat risky to base the foundation. Nevertheless, we are addicted to public perceptions, and having some synthesis of the opinions could be valuable. The Russia-Ukraine conflict is the most impactful event to start in 2022, and surveying the cyber public opinion is beneficial with available tools to measure the pulse of space provides value.

3. METHODOLOGY

In this exploratory research design, the first step is to curate the dataset from Twitter to address the research questions. President Zelensky and President Putin lead the conflict from each side; therefore, numerous alliances have backed their cause. In this case(*Tweet Object | Docs | Twitter Developer Platform*, n.d.), two separate country-specific datasets are generated with the same columns. The Twitter data is highly malleable, and #hashtags in its ecosystem provide sorting and aggregation(Laucuka, 2018) of information - a topic modeling mechanism that encapsulates similar themes.

A synthesis of popular hashtags originating from the presidents and alliances is selected by analyzing Twitter intelligence. The factor analysis reduced the hashtags to the ones that provided explicit support. The research aims to uncover cyber-related intelligence from the dispute; therefore, war-related, popular hashtags paired with cyber-related hashtags are blended to create the most relevant dataset for each country. Table II describes the mapping of hashtags to the countries.

Table 2. Cyber Hashtag Harvesting

Country	Support		Cyber
Ukraine	#ukraine,#ukraine #standwithukraine #ukrainewar #ukraina #ukrainerussiawar, #strongertogether, #helpukraine		#cyber #cyberattack #cybersecurity #hacking #cyberwar #cyberwarfare
Russia	#kremlin #moscow #istandwithrussia	#IStandWithPutin	#cyber #cyberattack #cybersecurity #hacking #cyberwar #cyberwarfare

Python is a powerful programming language due to its rich libraries. Tweepy, twint, and snscrape are effective APIs that pull tweets based on the prescribed criteria - in this instance, Tweepy was used due to its official alignment with the company, although rate-limiting is a nuance. The feature selection or the columns extracted from a tweet are listed below in Table 3 below.

Table 3. Feature Dictionary

Columns Extracted	Definition	Data Type
AU1R0	Alliance, R =0, U = 1	Boolean
Time Hour	Time of the tweet	Int
Followers Count	users following user	Int
Tweets	Content of the tweet	String
Length	Total characters of the tweet	Int
Location	Location for this account's profile	String
Statuses Count	Tweets issued by the user	Int
Friends Count	Users this account is following	Int
Favorites Count	Number of Tweets this user has liked in the account's lifetime.	Int
Account creation	Data account was created	Int
Retweet Count	Retweeted by other users	Int
Favorite of the tweet	Favorite of the tweet	String
Account Verified	Account confirmed by Twitter	Boolean
Listed Count (public list)	Users adding people to their list	Int

The dataset curated will be pre-processed to eliminate repetitive tags, empty columns, and integrate one-hot encoding to convert the categorical variables to numerical for model integration. Latent Dirichlet Allocation and integration of Logit helped to understand which explanatory variable is statistically significant when predicting the alliance of the tweets. The alpha or p-value for statistical significance is 0.05; any independent variable yielding this value or lower is presumed to be statistically significant (negative or positive) to the dependent variable or, in other words, influences alliance(Sperandei, 2014). Figure 1 illustrates a graphical representation of the methodology.

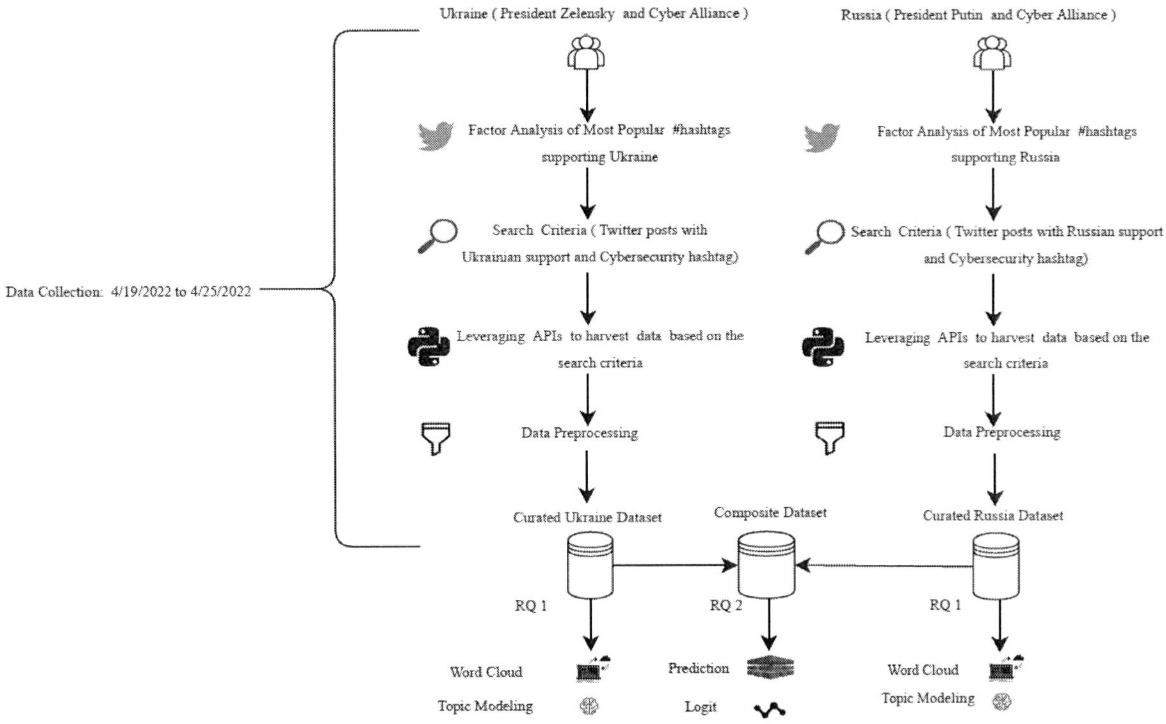

Figure 1. Methodology Overview

4. RESULTS AND DISCUSSIONS

The framework with specific criteria as defined in Figure 1 harvested 2017 tweets that aligned with Ukraine and 1007 that supported Russia. If any of the missing values from the columns were missing, the entire row of data was eliminated.

4.1 RQ1 - Can Cybersecurity intelligence be derived from the Tweets?

Contrary to the initial supposition, confining data harvesting to strict criteria where the tweets had to mention programmed cybersecurity hashtags with Russia-Ukraine in the background, the intelligence yielded was not denotative. Most tweets mentioned very little about cybersecurity-related topics. An assumption that the tweets would discover pertinent information about vulnerabilities, risks, and threats due to escalating conflict was inaccurate. The intelligence gathered mostly revolved around the actors of the wars.

No meaningful intelligence was deduced. Figure 2 and Figure 3 assemble thematic visualizations of the conflict for each alliance. Both Ukrainian and Russian datasets were devoid of cyber intelligence and looked fairly similar in essence, although their hashtags advertised cyber-inclinations - the research gathered random data from the Twittersphere, but the unruly and uncorroborated tweets from randomly users are of major concern.

Figure 2. Word Map of Ukrainian Tweets

Figure 3. Word Map of Russian Tweets

Another way to explore textual content is by LdaMOdel(Blei, 2003), which uses a generative probabilistic model to classify discrete data. For the Ukrainian tweets, the texts were classified into three topics and three subtopics, as shown in Figure 4. Likewise, Figure 5 shows the classification of Russian tweets

```
[([(0.06697432, 'invasion'), (0.043968353, 'amp'), (0.042402808, 'forces')],
  -1.663477528864579),
 ([(0.044847447, 'new'), (0.032542273, 'civilians'), (0.027998375, 'amp')],
  -16.6106552919199),
 ([(0.06433899, 'russian'), (0.0438365, 'west'), (0.043834955, 'ukraine')],
  -17.382324496238436)]
```

Figure 4. Topic Modeling (LDA) Ukrainian Tweets

```
[([(0.020311186, 'russia'), (0.019370638, 'russian'), (0.008891063, 'war')],
  -2.188174247546721),
 ([(0.022913601, 'war'), (0.008763276, 'amp'), (0.008728733, 'russian')],
  -2.295054839347346),
 ([(0.0145660825, 'russian'), (0.013962858, 'amp'), (0.010483739, 'ukraine')],
  -2.489858479269049)]
```

Figure 5. Topic Modeling (LDA) Russian Tweets

Both models did not exhibit cyber content relevance, most topics were rudimentary and devoid of cyber-specific topics, further confirming that cybersecurity intelligence was negligible.

4.2 RQ2 - What Independent Variables are Important Alliance Indicators for Ukraine and Russia, and can they Correctly Predict the Alliance?

A composite dataset was created with Russia and Ukraine entires; the Ukrainian affiliation was denoted '1' or 'True' in the 'AU1R0' column and '0' and 'False' for Russian affiliation. Logit Regression, a binary classification model with conditional probability(Taboga, n.d.) was used to exhibit the relationship between 'AU1R0', a dependent variable, and the numerous explanatory variables. 'AU1R0' denotes the alliance - if a tweet has a '1' value in this variable, it means the tweet was explicitly supporting Ukrainian initiatives in cyberspace.

Multicollinearity using Variance Inflation Factors (VIF) was used to eliminate competing features. In Table IV below, 'Time Hour', 'Length', and 'Account Creation', whose VIF score is greater than 4, are assumed to be noise in the modeling - hence eliminated to explain the dependent variable, 'AU1R0'. After iterating the model, it produced a high p-value for 'Followers Count' and 'Account Verified', which were deemed insignificant to the model, so it was eliminated as well.

Table 4. Multicollinearity Assessment (VIF)

Features	VIF Score
Time Hour	5.72
Followers Count	1.15
Length	4.52
Statuses Count	1.60
Friends Count	1.24
Favorites Count	1.65
Account creation	10.15
Retweet Count	1.00
Favorite of the tweet	1.43
Account Verified	1.63
Listed Count (public list)	1.25

If the p-value is ≤ 0.05, then the independent variable is significant, and thus, will impact the direction of the dependent variable, negatively or positively. As shown in Figure 6, the p-value of 'Favorite of the tweet', 'retweet_count', 'Friends Count', 'Listed Count', 'Favorites Count', and 'statuses_count' are important independent variables in predicting the Ukrainian alliance.

A multilayer perceptron (MLP) is a readily deployable feedforward neural network often used in structured data that doesn't require intensive computing(Popescu et al., 2009). They are part of the neural network, with one or more hidden layers where classification or prediction is conducted on the output layer. An extremely flexible algorithm(Brownlee, 2016), in this case, is used to solve a binary class problem regarding the Ukrainian alliance.

```
Optimization terminated successfully.
         Current function value: 0.337289
         Iterations 14
                           Logit Regression Results
==============================================================================
Dep. Variable:                  AU1R0   No. Observations:                 3024
Model:                          Logit   Df Residuals:                     3018
Method:                           MLE   Df Model:                            5
Date:                Wed, 04 May 2022   Pseudo R-squ.:                  0.4699
Time:                        20:23:00   Log-Likelihood:                -1020.0
converged:                       True   LL-Null:                       -1924.1
Covariance Type:            nonrobust   LLR p-value:                     0.000
=========================================================================================
                            coef    std err          z      P>|z|      [0.025      0.975]
-----------------------------------------------------------------------------------------
Favorite of the tweet    -0.4528      0.043    -10.505      0.000      -0.537      -0.368
retweet_count             0.1235      0.011     11.007      0.000       0.102       0.146
Friends Count             0.0002   4.37e-05      5.174      0.000       0.000       0.000
Listed Count             -0.0042      0.001     -4.336      0.000      -0.006      -0.002
Favorites Count        1.024e-05   3.02e-06      3.390      0.001    4.32e-06    1.62e-05
statuses_count         1.707e-05   2.88e-06      5.933      0.000    1.14e-05    2.27e-05
=========================================================================================

Possibly complete quasi-separation: A fraction 0.32 of observations can be
perfectly predicted. This might indicate that there is complete
quasi-separation. In this case some parameters will not be identified.
```

Figure 6. Logit for Tweets

A standard evaluation metric is deployed for the model after 80% of the data was dedicated to training and 20% to testing, where precision, recall, f1-score, and accuracy are calculated. The model infused with MLP was highly accurate in predicting the '1' or Ukrainian alliance with an average of over 90% in all metrics, while '0' was slightly lower but still respectable. An undersampling of '0' could have hurt the model's learning capabilities.

Table 5. Evaluation Metrics

	Precision	recall	f1-score	support
0	0.89	0.84	0.86	198
1	0.92	0.95	0.94	407
accuracy			0.91	605
macro avg.	0.91	0.89	0.90	605
weighted avg.	0.91	0.91	0.91	605

Figure 7 presents a visualization of the evaluation while True Negative is 166, False Positive is 32, False Negative is 20, and True Positive is 387 for the given data.

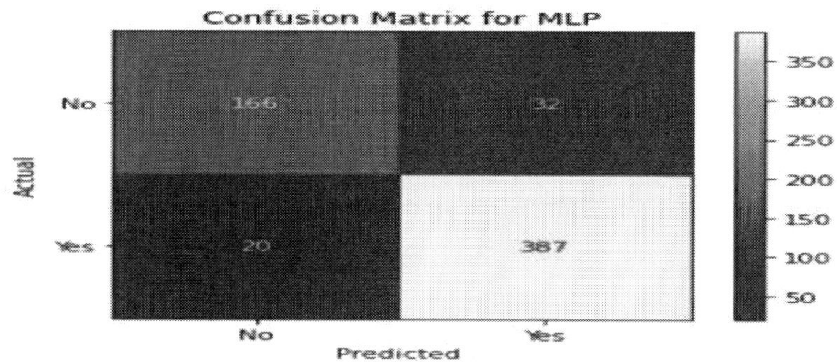

Figure 7. Confusion Matrix

5. LIMITATIONS AND FUTURE SCOPE

The research was unable to find a consequential CTI that was beneficial. Nevertheless, cybersecurity remains an interdisciplinary field, encapsulating all aspects of lives. One of the glaring limitations of the research was the probity of data collection - what makes social media rich in value is its abundance and availability but at the same time anyone can post anything and the dubiety of diverse jurisdictions makes the data collection process unfair. The data was collected randomly in a relatively short period, and while Ukraine-centric data was plentiful, ongoing Twitter restrictions meant Russia-originated/supporting data was rare and often censored. Some trending hashtags at the onset of the conflict were removed, further missing out on the intelligence. The research tracked hashtags to find cybersecurity-centric tweets, but often it was futile in discovering useful content.

SOCMINT is in its infancy and here to stay; refined methodologies could yield more benefits for the future. Rather than chasing hashtags, a careful pre-selection of influential users that exhibit sincere cybersecurity content should be traced using graph and network theory. An understanding of the relationships from that sub-group will be more prudent in discovering intelligence as opposed to random samplings. Censorship is a huge issue for academics that want to leverage the power of public data - the current Twitter ownership change, in which the new owner will allow leniency towards data freedom to provide unbounded access, is a supposition that could provide research aspirations. Alternatively, creating synthetic data based on artificial intelligence could provide alternate data fuel.

6. CONCLUSION

Whilst the availability of data is tempting, various issues remain to uncover intelligence - questionable integrity and sporadic availability remain key issues. The public data is owned by private companies and often their policies dictate the intelligence-gathering. A quick stoppage of data flow can derail a framework. When data is available, a refined approach is required to parse the information, as free data is not always the most valuable data. Nevertheless, the functional APIs, low open-source programming, and seamlessness of data discovery make the CTI integration appealing. This research provided sequential steps on building an intelligence-gathering framework whilst incorporating machine learning techniques to make it further discernible. Although it didn't provide substantial value to the Russia-Ukraine conflict largely due to weak data, a better part of the framework is dependable and can be replicated for future experiments.

REFERENCES

APIs for Scholarly Resources | Scholarly Publishing—MIT Libraries. (n.d.). Retrieved April 18, 2022, from https://libraries.mit.edu/scholarly/publishing/apis-for-scholarly-resources/

Blei, D. M. (2003). *Latent Dirichlet Allocation*. 30.

Brownlee, J. (2016, May 16). Crash Course On Multi-Layer Perceptron Neural Networks. *Machine Learning Mastery*. https://machinelearningmastery.com/neural-networks-crash-course/

Code Snippets—Tweepy 3.5.0 documentation. (n.d.). Retrieved April 13, 2022, from https://docs.tweepy.org/en/v3.5.0/code_snippet.html

Costa, C., Aparicio, M., & Aparicio, J. (2021). Sentiment Analysis of Portuguese Political Parties Communication. *The 39th ACM International Conference on Design of Communication*, 63–69. https://doi.org/10.1145/3472714.3473624

Dave, P. (2022). *Ukraine uses facial recognition to identify dead Russian soldiers, minister says | Reuters*. https://www.reuters.com/technology/ukraine-uses-facial-recognition-identify-dead-russian-soldiers-minister-says-2022-03-23/

Dover, R. (2020). SOCMINT: A shifting balance of opportunity. *Intelligence and National Security*, *35*(2), 216–232. https://doi.org/10.1080/02684527.2019.1694132

Forrester, B., & Hollander, K. den. (2016). *The role of Social Media in the Intelligence Cycle*.

Getting started—Tweepy 4.2.0 documentation. (n.d.). Retrieved November 1, 2021, from https://docs.tweepy.org/en/stable/getting_started.html#models

Husari, G., Al-Shaer, E., Chu, B., & Rahman, R. F. (2019). Learning APT chains from cyber threat intelligence. *Proceedings of the 6th Annual Symposium on Hot Topics in the Science of Security - HotSoS '19*, 1–2. https://doi.org/10.1145/3314058.3317728

Ivan, A. L., Iov, C. A., Lutai, R. C., & Grad, M. N. (2015). SOCIAL MEDIA INTELLIGENCE: OPPORTUNITIES AND LIMITATIONS. *SOCIAL MEDIA INTELLIGENCE*, 7.

Kakar, S., Dhaka, D., & Mehrotra, M. (2021). Value-Based Retweet Prediction on Twitter. *Informatica*, *45*(2). https://doi.org/10.31449/inf.v45i2.3465

Kelion, L. (2014). *Snowden leaks: GCHQ 'spied on Facebook and YouTube'—BBC News*. https://www.bbc.com/news/technology-25927844

Khan, I. (2022). *Zelenskyy Humanizes Ukraine's Plight in His Social Media Messaging—CNET*. https://www.cnet.com/news/politics/zelenskyy-humanizes-ukraines-plight-in-his-social-media-messaging/

Kobie, N. (2019). *The complicated truth about China's social credit system*. The Complicated Truth about China's Social Credit System. https://www.wired.co.uk/article/china-social-credit-system-explained

Kropotov, V., & Yarochkin, F. (2019). *Basic Social Media Intelligence (SOCMINT) Tools To Help Fight Disinformation*. http://www.mikekujawski.ca/2019/02/25/basic-social-media-intelligence-socmint-tools-to-help-fight-disinformation/

Lakshmi, J. V. N. (2018). Machine learning techniques using python for data analysis in performance evaluation. *International Journal of Intelligent Systems Technologies and Applications*, *17*(1/2), 3. https://doi.org/10.1504/IJISTA.2018.10012853

Laucuka, A. (2018). Communicative Functions of Hashtags. *Economics and Culture*, *15*(1), 56–62. https://doi.org/10.2478/jec-2018-0006

Mahajan, R., Mahajan, R., Sharma, E., & Mansotra, V. (2022). "Are we tweeting our real selves?" personality prediction of Indian Twitter users using deep learning ensemble model. *Computers in Human Behavior*, *128*, 107101. https://doi.org/10.1016/j.chb.2021.107101

Mahood, Lc. M. (2015). *SOCMINT: Following and Liking Social Media Intelligence*. 25.

McLaughlin, J. (2022). *Social media volunteers aim to help Ukraine win the information war: NPR*. https://www.npr.org/2022/03/17/1087137578/social-media-volunteers-aim-to-help-ukraine-win-the-information-war

McLoughlin, L., Ward, S., & Lomas, D. W. B. (2020). 'Hello, world': GCHQ, Twitter and social media engagement. *Intelligence and National Security*, *35*(2), 233–251. https://doi.org/10.1080/02684527.2020.1713434

Milmo, D. (2022). *Russia blocks access to Facebook and Twitter | Russia | The Guardian*. https://www.theguardian.com/world/2022/mar/04/russia-completely-blocks-access-to-facebook-and-twitter

Momi, R. (2021). *SOCMINT: Social Media Intelligence a New Discipline? - Grey Dynamics*. https://www.greydynamics.com/socmint-social-media-intelligence-a-new-discipline/

Omand, D., Bartlett, J., & Miller, C. (2012). Introducing Social Media Intelligence (SOCMINT). *Intelligence and National Security*, *27*(6), 801–823. https://doi.org/10.1080/02684527.2012.716965

Popescu, M.-C., Balas, V. E., Perescu-Popescu, L., & Mastorakis, N. (2009). *Multilayer Perceptron and Neural Networks*. *8*(7), 11.

Sauerwein, C., Fischer, D., Rubsamen, M., Rosenberger, G., Stelzer, D., & Breu, R. (2021). From Threat Data to Actionable Intelligence: An Exploratory Analysis of the Intelligence Cycle Implementation in Cyber Threat Intelligence Sharing Platforms. *The 16th International Conference on Availability, Reliability and Security*, 1–9. https://doi.org/10.1145/3465481.3470048

Social Media Intelligence. (n.d.). Privacy International. Retrieved April 12, 2022, from http://privacyinternational.org/explainer/55/social-media-intelligence

Sperandei, S. (2014). Understanding logistic regression analysis. *Biochemia Medica*, *24*(1), 12–18. https://doi.org/10.11613/BM.2014.003

Stukal, D., Sanovich, S., Tucker, J. A., & Bonneau, R. (2019). For Whom the Bot Tolls: A Neural Networks Approach to Measuring Political Orientation of Twitter Bots in Russia. *SAGE Open*, *9*(2), 215824401982771. https://doi.org/10.1177/2158244019827715

Taboga, M. (n.d.). *Logistic classification model (logit or logistic regression)*. Retrieved May 4, 2022, from https://www.statlect.com/fundamentals-of-statistics/logistic-classification-model

Three surveillance technologies that protesters need to know about—IFEX. (2019). https://ifex.org/three-surveillance-technologies-that-protesters-need-to-know-about/

Trouillard, S. (2022). *Following the Ukraine war – and fighting it – on social media*. https://www.france24.com/en/europe/20220308-following-the-war-in-ukraine-%E2%80%93-and-fighting-it-%E2%80%93-on-social-media

Tweet object | Docs | Twitter Developer Platform. (n.d.). Retrieved April 19, 2022, from https://developer.twitter.com/en/docs/twitter-api/v1/data-dictionary/object-model/tweet

Ukraine's Digital Ministry Is a Formidable War Machine | WIRED. (2022). https://www.wired.com/story/ukraine-digital-ministry-war/

Using Social Media (SOCMINT) in Threat Hunting. (n.d.). Retrieved April 18, 2022, from https://www.anomali.com/blog/using-social-media-socmint-in-threat-hunting

Vail, E. (2022). *Russia or Ukraine: Hacking groups take sides—The Record by Recorded Future*. https://therecord.media/russia-or-ukraine-hacking-groups-take-sides/

Wang, H., Li, Y., Hutch, M., Naidech, A., & Luo, Y. (2021). Using Tweets to Understand How COVID-19–Related Health Beliefs Are Affected in the Age of Social Media: Twitter Data Analysis Study. *Journal of Medical Internet Research*, *23*(2), e26302. https://doi.org/10.2196/26302

EXPECTATIONS OF SOFTWARE DEVELOPMENT EDUCATION: STUDENTS VS PROFESSIONALS

Janet Liebenberg
North-West University, South Africa

ABSTRACT

ICT departments at universities are challenged to provide software development education that meets the expectations of both the students and the industry to deliver students with applicable professional competencies and behaviours. The question is whether a gap exists between the expectations of students and professionals. This study in South Africa reports on the views of software development students and software development professionals and the results were compared to determine if the views of students regarding professional competencies and behaviours are compatible with the views of professionals. The results revealed a gap in views regarding industry involvement and matching views regarding the importance of soft skills. Recommendations are made to universities and the industry to narrow the expectation gap between students and professionals.

KEYWORDS

Education, Competencies, Industry, Students, Software Professionals, Expectation Gap

1. INTRODUCTION

The overall levels of technical and soft skills needed from software developers are not sufficient to meet demand, not only in South Africa, but also worldwide (BusinessTech, 2022, Connolly, 2013). Therefore, more software development students need to be attracted and retained, but will students take courses that do not meet their expectations? Students desire courses and projects that provide them with applicable professional competencies and skills. At the same time, the software development (SD) industry expects students to be educated in courses and projects that are professionally relevant and that prepare them well for the workplace (Moreno et al., 2012), but are these two perspectives compatible?

Numerous studies have investigated SD education at university level, but have been mostly concerned with ways to meet the needs of industry. Reisman (2004) argues that studies in higher education often neglect to collect data from two of higher education's most important players, namely faculty and students. It is therefore essential that the view of higher education's largest stakeholder, namely the students, should also be considered. There is little doubt that discovering what motivates individuals and rewarding them for what they find important are key to successfully recruiting and retaining talent (Bunton and Brewer, 2012). In light of the shortage of skilled software developers, the present study investigated the compatibility of the expectations of students and the expectations of professionals regarding professional competencies and behaviours in SD education. The research questions were:

- What are SD students' views on professional competencies and behaviours?
- What are professional software developers' views on professional competencies and behaviours?
- What is the compatibility between the views of students and the views of professionals regarding professional competencies and behaviours in SD education?

The results could help universities (SD educators, curriculum developers) and the SD industry (employers, corporate trainers) to form a picture of, on the one hand matching views and on the other hand differing views regarding SD professional competencies and behaviours. University and industry can relate the results of the study to fill the expectation gap in the academic preparation of future software developers and improve the employability of software developer graduates. It can therefore result in more relevant SD education with regard to students, as well as the software industry and might contribute to meeting the demand for skilled software developers.

2. LITERATURE BACKGROUND

2.1 The Student-Centric View

Students in current university classes are described by a great number of writers as Generation Z or the Net Generation (aged about 10 to 27) - mainly born 1995 to the early 2010's in the US and Canada. These young people have grown up with computers and the Internet (one device can do everything) and therefore they have a natural aptitude and high skill levels when using new technologies (Oblinger et al., 2005). The Net generation has unique modes of communication, learning preferences, social choices, and entertainment preferences defined by their early exposure to technology (Saiedian, 2009). These students are driven by different motivations, learning styles, characteristics, skills sets and social concerns than the previous generation. For universities and colleges it is pertinent to learn to adapt their courses, programs, processes, environment and initiatives to meet the needs of these students (Seemiller and Grace, 2016).

Tapscott (2008) found that the Net Geners share eight norms: Freedom, customization, scrutiny, integrity, collaboration, entertainment, speed and innovation, and contends that education must evolve to meet students' changing demands and to teach students what they need to know to thrive in an information-based economy. In addition, Liebenberg and Pieterse (2016) found when they investigated the career goals of Generation Z programming students that job security and work/life balance ranked top for these students.

Seemiller and Grace (2016) explain regarding Generation Z as learners: "*Gen.Z demands an education that will be useful and relevant in getting a job after graduation, a learning that will help fill the toolbox with applicable knowledge and skills for the workforce, an education that will be a utility toward their eventual career (in the « real world »). Therefore they have a preference for experiential learning and will seek out courses that teach the critical skills employers want*".

2.2 The Industry-Centric View

Industry is, similar to the university, challenged to attract and retain skilled employees (Hein, 2016). The research of Bullen et al. (2009) examined workforce trends in IT-provider companies and identified problems in the following areas: graduates who are not trained in areas that the industry is seeking; thin pipeline for specific technical skills; increasing pressure to source IT capability; and lag in university responsiveness to the needs of the industry.

Several studies suggest a gap between the knowledge and skills demanded by the industry and the knowledge and skills gained by graduates of university computing courses. Gallagher et al. (2010) found that a mix of skills is essential for IT professionals but the skills most critical are non-technical skills, such as project management, business-domain knowledge and relationship skills. There is no globally accepted definition of non-technical skills and different terms being used are soft skills, key competencies, enabling skills, generic skills, core skills, essential skills, necessary skills, and Critical Cross-field Outcomes (Kechagias, 2011, SAQA, 2000). Although a large number of definitions may be found in the literature on the concept of skill, most emphasize that all skills are capable of being learned and developed and involve the appropriate and observable performance of particular types of activity and tasks (Kechagias, 2011).

Several studies found that professionals in the computing field lack non-technical or soft skills, such as communications skills with co-workers and customers (negotiation or giving presentations), client-facing capabilities, leadership skills and relationship skills (Calitz et al., 2015, Moreno et al., 2012, Radermacher et al., 2014). Hazzan and Har-Shai (2013) found that almost all problems associated with software development processes are connected to people and are rooted not in technological aspects, but in the expression of soft skills. Consequently, industry recognizes the importance of soft skills in software development processes, not only at the managerial level, but also among software team members.

2.3 The Role of the University

Courses with a primary emphasis on current technology in which most of the knowledge will become obsolete as the technology does are a major challenge in the education of software developers (Topi et al., 2010). Employees who perform well in scientific and technical concepts but not in supplementary skills such as managing and working with others will be of much less value to organizations in the long run (Taylor, 2016).

Pressures arising from the changing character of software and from external pressures on educational institutions will require changes in what software developers are taught and how they are taught (McKinnon and McCrae, 2012). To effectively fill this gap between the knowledge and skills demanded by the industry and the knowledge and skills gained by graduates of university computing courses, it would be necessary, on the one hand, to guarantee that the educational programmes provide the knowledge required for the job profiles suggested by industry and on the other hand, to ensure that this knowledge is taught in a manner enabling future professionals to correctly tackle the problems that they will face during their professional career (Bothe et al., 2009, Loftus et al., 2011).

Technical skills are a part of most educational curricula, but soft skills need further emphasis in the university curricula (Robles, 2012). Hazzan et al. (2020) explain that soft concepts and skills are especially important today as they are dominant in the teaching of computer science in schools and the university, as well as influence software development processes in the industry. These facts require computer science students and teachers, computer scientists, and software engineers to acquire, in addition to disciplinary and technical knowledge, additional skills: both social (such as teamwork and getting and giving feedback), cognitive (e.g., thinking in terms of different levels of abstraction), and organizational (e.g., familiarity with different kinds of organizations)

The Joint Task Force on Computing Curricula: Association for Computing Machinery (ACM) and IEEE Computer Society (2013) emphasizes that the education that students receive must adequately prepare them for the workforce in a more holistic way than simply conveying technical facts. Students will, through the general university experience, acquire some soft skills and personal attributes (e.g., patience, time management, work ethic), but for the rest of the skills, provision must be made through specific curricula. Soft skills should be learned and grasped gradually, based on students' engagement, active learning, and reflection (Hazzan and Har-Shai, 2013).

It is the expectation of both the students and the industry to deliver graduates with applicable professional competencies and behaviours. It is the role of the university to provide an education that meets the expectations of students and industry. The question is whether a gap exists between the expectations of students and industry professionals.

3. METHODOLOGY

This quantitative study was conducted in South Africa and consisted of two groups of participants, namely SD students and SD professionals. In this section, the demographics of the participants, as well as the data collection and analysis are discussed.

3.1 Participants

3.1.1 Students

The target population was four academic year groups of the SD classes at a university in South Africa and 386 questionnaires were posted as an assignment on the e-learning system to these students. The number of usable responses received totalled 297, making for an overall response rate of 76.9%.

Table 1 provides a summary of the biographic data. The gender profile is typical of most CS classes with only 25% of the respondents being women. The participants included 276 BSc undergraduate students and 21 students enrolled for the subsequent BSc (Hons) in CS and IS. Most of the undergraduate students follow the three-year BSc in IT and CS program in the School of CS and IS. A subsequent one-year BSc (Hons) in CS and IS is offered and this degree gives access to a Master's degree in CS.

3.1.2 Professionals

A convenience sample of 995 professional software developers in South Africa was taken and the respondents were members of the following groups of the professional networks LinkedIn and MyBroadband: Software and Web Developers in South Africa, SA Developer.NET and C# Developers/Architects. They were personally contacted via e-mail and requested to complete the anonymous online survey. The link of the survey was also forwarded by some of the respondents to their colleagues for completion. In addition, five managers at software houses were contacted and they sent the link of the survey to the software developers in their company. The number of usable responses received totalled 214, which indicates a response rate of around 21%, although an exact figure cannot be determined.

Table 2 provides a summary of the biographic data. The gender profile is a concern but not surprising with only 8% of the respondents being female. The age profile indicates that 12% of the respondents are young, Generation Z computing professionals, about 74% are Millenials (ages 25-39) and a mere 14% are from Generation X (ages 40-59).

In terms of the respondents' education 49% of them are in possession of a CS/IS degree or degrees, with another 22.5% having related degrees. It is not uncommon in software development to find people with few qualifications with 4.5% of the respondents having only a high school certificate. The work experience of respondents indicates that 70.5% of them have more than five years' work experience.

Table 1. Profile of students (n=297)

Criteria	Categories	Number (%) of students
Gender	Male	222 (75%)
	Female	75 (25%)
Academic Year	1	145 (49%)
	2	76 (26%)
	3	55 (19%)
	4 (Hons)	21 (7%)

Table 2. Profile of professionals (n=214)

Criteria	Categories	Number (%) of respondents
Gender	Male	196 (92%)
	Female	18 (8%)
Age category	18-24	25 (12%)
	25-29	64 (30%)
	30-39	94 (44%)
	40-49	28 (13%)
	50-59	3 (1%)
	>=60	0 (0%)
Education	High School Certificate	10 (4.5%)
	Certification	22 (10%)
	National diploma	30 (14%)
	CS/IS degree(s)	104 (49%)
	BSc/BCom	38 (18%)
	Engineering degree	10 (4.5%)
Work experience (in years)	0-4	63 (29.5%)
	5-9	62 (29%)
	10-14	51 (24%)
	15-19	22 (10%)
	20-29	13 (6%)
	30-39	3 (1.5%)
	>=40	0 (0%)

3.2 Data Collection, Instrument and Analysis

3.2.1 Students

An initial list of questions was developed by both writing new items and adapting items from available surveys, such as ROSE (Schreiner and Sjøberg, 2004) and the South African Qualifications Authority's (SAQA, 2000) list of "Critical Cross-field Outcomes". Once the initial questions were generated, they were sent to three industrial and two academic experts to refine the instrument. Feedback from this pilot study served as the basis for correcting, refining, and enhancing the questions and it resulted in a questionnaire with a pool of 26 items. The first section of the questionnaires gathered information on the biographic data of the respondents as shown in Table 2.

The questionnaire was further divided into two domains (See Appendix A). The first domain "In class" with 14 items enquired on their perceptions of required competencies and behaviours in their SD classes, such as industry experience of lecturers. The second domain "My career" had 12 items and gathered data on their future career, such as what is expected from a good software developer. The first and second domains were accompanied by a five-point Likert response scale from 1 (Strongly disagree) to 5 (Strongly agree). The domain "In class" had a comment box where students could list topics that interest them and the domain, "My career", included a question: "In your first job after graduation, which job would you ideally want?" with a list of 36 choices from which a maximum of three alternatives could be selected.

3.2.2 Professionals

The questionnaire that was used for the SD students was used as a point of departure for the questionnaire for the SD professionals. The first section of the questionnaire gathered information on the biographic data of the respondents as shown in Table 2. The questionnaire was further also divided into the same two domains (see Appendix A).

3.2.3 Students and Professionals

Factor analysis was used to investigate the 26 items in more detail to reduce the variables into a smaller number of factors. The 511 responses were examined using principal components factor analysis and the 26 attitude items yielded 4 interpretable factors and 6 items were being handled as single research variables. Factors were named according to their main context. A Cronbach's α coefficient was calculated for each of the factors and was found as Table 3 shows, to be reliable ($\alpha \geq 0.60$).

Table 3. Factors* (with reliability coefficients) and single variables

Factors	**Cronbach's alpha (α)**
Critical outcomes required in courses	0.849
Knowledge of course requirements	0.606
Positive attitude towards work and colleagues	0.800
Emotional/social skills required	0.760
Single variables	
Software developers needed	
Industry experience of lecturers	
Project work	
People from industry teaching	
Good examination results	
Neat and tidy appearance	

** See Appendix A for the items in each factor and descriptive statistics*

Basic analysis of quantitative data was done by calculating the mean values and standard deviation of each of the 10 variables. The statistical tests used in the analysis varied as necessary to match the metric being analysed. When the results of the interaction analysis are reported, only the significant interactions or primary effects will typically be discussed. A convenience sample instead of a random sample was used therefore the p-values will be reported for the sake of completeness but will not be interpreted.

4. RESULTS AND DISCUSSION

In this section, important data for each of the research questions are considered.

4.1 Students' and Professionals' Views

In Table 4 the results of the two domains "In class" and "My career" are shown. For the students, the mean values of five of the 10 variables are relatively high. Students feel that a positive attitude towards work and colleagues is very important. Students also feel strongly about the need for emotional and social skills in the workplace and they agree that more software developers are needed in South Africa. The only variable showing a lower mean is that students feel that it is not too important for people from industry to come in and teach some of their classes. Further analysis showed that the more advanced the students were in their studies, the more importance they attached to industrial experience and knowledge brought to them in their classes.

The comment box where students listed topics that interest them produced varied results but the comments included: HCI/user interfaces, Game development, Data analysis via Excel; Statistical analysis systems. These comments might be an indication that some students have noticed the increasing demand for Data Scientists (Dippnall, 2021)

The top five ideal jobs from 36 choices for the software development students are software developer, computer game designer, data scientist, systems analyst and software engineer. It is worth noting that game design and development are among the top jobs for these students although South Africa has a very small video game development industry. However, Covid-19 rapidly changed the world of work with remote work becoming the norm rather than the exception to the rule.

Table 4. Students' and Professionals' views

	Students		Professionals	
	Mean*	Std. Deviation	Mean*	Std. Deviation
Critical outcomes required in courses	4.051	0.717	4.346	0.44
Emotional/social skills	4.247	0.628	3.778	0.679
Good examination results	3.616	1.153	3.033	1.041
Industry experience of lecturers	3.776	0.94	4.425	0.752
Knowledge of course requirements	3.859	0.694	3.201	0.596
Neat and tidy appearance	3.753	1.225	2.79	1.112
People from industry teaching	2.29	1.231	4.369	0.769
Positive attitude towards work and colleagues	4.501	0.53	4.41	0.477
Project work	4.02	0.954	4.322	0.734
Software developers needed	4.207	0.875	4.322	0.89

* *Likert-style responses were ranked from 1 to 5 respectively*

For the professionals, like for the students the results of the two domains "Educational background of new recruits" and "Career" are shown. For the professionals, Table 4 shows that the mean values of six of the 10 variables are relatively high. The professionals feel that it is very important for lecturers to have industry experience in their armour. The professionals also feel strongly about a positive attitude towards work and colleagues, and they agree that people from industry should come in and teach some of the university classes. The only variable showing a lower mean is that professionals feel that it is not too important for software developers to have a neat and tidy appearance.

4.2 Students vs Professionals

The results of the students and professionals were compared to determine if the views of students are compatible with the views of professionals. Differences were analysed with a T-Test and Table 5 shows significant differences in means between 6 variables. Four variables reveal that the mean values of the students are significantly higher than the mean values of the professionals and for only two variables the professionals had significantly higher means. The top section of Table 5 shows the results of the variables where significant differences were found and the rest of the variables with insignificant differences follow in the bottom section.

Two variables in the top section of Table 5 showed large practically significant differences between the students' views and the professionals' views. The professionals felt very strongly that it is important for people from industry (themselves) to assist in teaching the students whereas the students felt that it is not necessarily that important. However, the closer the students came to entering the workplace, the more they valued industrial experience brought to them in their classes. McKinnon and McCrae (2012) state that undergraduate students commonly do not think about their employability skills until they are about to graduate and the students in this study confirmed that notion. The students on the other hand felt significantly more strongly than the professionals that the critical outcomes (teamwork; self-organized; information collection and evaluation; communication skills; science and technology use; worldview of related systems) are important for students to learn.

There were also four variables showing a medium practically significant difference between the students' and professionals' views. Three of the four variables (Neat and tidy appearance; Emotional/social skills; Good examination results) indicated greater importance for the students than the professionals. It is noteworthy that numerous studies, including this study, indicated the importance of soft skills for the industry and now this study found that soft skills (emotional/social skills and critical outcomes) have even greater importance for students. The fourth variable Industry experience of lecturers is much more important to the professionals than to the students and it is not surprising since this finding concurs with the finding that the professionals also view the variable that people from industry should come in and teach some classes as essential.

Table 5. Students vs Professionals

	Mean of Students (n=297)	Mean of Professionals (n=214)	Effect size (d)	p
People from industry teaching	2.290	4.369	1.690**	<0.001
Critical outcomes required in courses	3.859	3.201	0.948**	<0.001
Neat and tidy appearance	3.753	2.790	0.786*	<0.001
Emotional/social skills	4.247	3.778	0.691*	<0.001
Industry experience of lecturers	3.776	4.425	0.691*	<0.001
Good examination results	3.616	3.033	0.506*	<0.001
Knowledge of course requirements	4.051	4.346	0.411	<0.001
Project work	4.020	4.322	0.316	<0.001
Positive attitude towards work and colleagues	4.501	4.410	0.171	<0.050
Software developers needed	4.207	4.322	0.129	>=0.050

* *medium practically significant difference (d>=0.5)*
** *large practically significant difference (d>=0.8)*

The variable where the students' views are most compatible with the professionals' views (d<0.2) is Software developers needed. The three remaining variables (Knowledge of course requirements; Project work; Positive attitude towards work and colleagues) had such small practically significant differences that it can also be said that the views of the students are compatible with the views of the professionals regarding these variables.

5. CONCLUSIONS AND RECOMMENDATIONS

The above results paint a picture for all the stakeholders of SD education of, on the one hand matching expectations and on the other hand differing expectations of software students and software professionals. The university and industry can therefore utilize this rare insight to improve SD education. The results of this study can on one hand contribute towards universities improving education in the relevant professional competencies and behaviours and on the other hand employers' ability to recruit candidates that fit their short- or long-term needs. In addition, an improved link between education institutions and industry is built.

The current professional competencies and behaviours important to students and the software industry are presented, which can assist educators in identifying areas where students may not measure up to the expectations of industry companies and in improving the curriculum at their universities to better prepare them for their future careers. The matching views inform the universities of the important professional competencies and behaviours that should receive particular attention because in the process both the students and the industry's requirements are satisfied.

There is a mismatch between the students' and the professionals' views regarding industry involvement in teaching. The university and industry must work together to provide 'real-world' group projects from students' first year of study. Students' expectations and conceptions of SD careers seem somewhat misguided when the results are contemplated. Students should gain experience in the industry much earlier in their education to eliminate possible misconceptions regarding their future career.

Students regard soft skills as very important and the importance for the industry of soft skills is proved in this study and numerous other studies. It should be a clear indication to universities to pay attention to the development of the soft skills of their students. It should furthermore be an indication to the software industry that their new recruits agree with them on the importance of soft skills, and it might also reflect students' awareness of their own weaknesses and skills gaps.

Industry-university collaboration should be launched and maintained to establish two-way knowledge and skill exchanges which will result in joint research and education projects and will keep educational offerings grounded in professional practice. It is essential to work towards compatible views and expectations regarding software development education between industry and higher education's largest stakeholder, namely the students.

This study investigated expectations of students and professionals regarding SD education and future work can therefore be to investigate the reality - how universities teach and what students can really learn compared to their already known expectations.

REFERENCES

Bothe, K., Budimac, Z., Cortazar, R., Ivanovic, M. & Zedan, H. 2009. Development of a modern curriculum in software engineering at master level across Countries. *Comput. Sci. Inf. Syst.,* 6, 1-21.

Bullen, C. V., Abraham, T., Gallagher, K., Simon, J. C. & Zwieg, P. 2009. IT workforce trends: Implications for curriculum and hiring. *Communications of the Association for Information Systems,* 24, 9.

Bunton, T. & Brewer, J. Discovering workplace motivators for the millennial generation of IT employees. Proceedings of the 1st Annual conference on Research in information technology (RIIT '12), 2012 New York, NY. ACM, 13-18.

Businesstech. 2022. The list of critical skills needed in South Africa right now. Available: https://businesstech.co.za/news/business/555526/the-list-of-critical-skills-needed-in-south-africa-right-now/.

Calitz, A., Cullen, M. & Greyling, J. South African alumni perceptions of the industry ICT skills requirements. Proceedings of the 44th annual conference of the Southern African Computer Lecturers' Association (SACLA), 2015 Johannesburg. University of Witwatersrand, 36-47.

Connolly, B. 2013. IT worker shortage continues as jobs remain unfilled. *CIO* [Online]. Available: http://www.cio.com.au/article/454650/worker_shortage_continues_jobs_remain_unfilled [Accessed 31 Oct 2020].

Dippnall, S. 2021. Why businesses need to start upskilling with data science. *ITWEB* [Online]. Available: https://www.itweb.co.za/content/dgp45qa6yRKvX9l8.

Gallagher, K. P., Kaiser, K. M., Simon, J. C., Beath, C. M. & Goles, T. 2010. The requisite variety of skills for IT professionals. *Communications of the ACM,* 53, 144-148.

Hazzan, O. & Har-Shai, G. Teaching computer science soft skills as soft concepts. Proceeding of the 44th ACM technical symposium on Computer science education, 2013. 59-64.

Hazzan, O., Lapidot, T. & Ragonis, N. 2020. *Guide to teaching computer science: An activity-based approach*, Springer.

hein, R. 2016. How to conquer recruiting, retention and IT skills challenges. *CIO* [Online]. Available: https://www.cio.com/article/240225/how-to-conquer-recruiting-retention-and-it-skills-challenges.html [Accessed 20 May 2022].

Joint Task Force on Computing Curricula: Association For Computing Machinery (ACM) And IEEE Computer Society 2013. *Computer science curricula 2013: curriculum guidelines for undergraduate degree programs in computer science,* New York, NY, USA, ACM.

Kechagias, K. 2011. Teaching and assessing soft skills. SCST.

Liebenberg, J. & Pieterse, V. 2016. Career goals of software development professionals and software development students. . *Computer Science Education Research Conference (CSERC 2016).* Pretoria, South Africa: ACM.

Loftus, C., Thomas, L. & Zander, C. Can graduating students design: revisited. Proceedings of the 42nd ACM technical symposium on Computer science education, 2011. ACM, 105-110.

Mckinnon, S. & Mccrae, J. 2012. Closing the gap: preparing computing students for employment through embedding work-related learning in the taught curriculum. *Industry and Higher Education,* 26, 315-320.

Moreno, A. M., Sanchez-Segura, M.-I., Medina-Dominguez, F. & Carvajal, L. 2012. Balancing software engineering education and industrial needs. *Journal of systems and software,* 85, 1607-1620.

Oblinger, D., Oblinger, J. L. & Lippincott, J. K. 2005. *Educating the net generation*, Boulder, Colo.: EDUCAUSE, c2005. 1 v.(various pagings): illustrations.

Radermacher, A., Walia, G. & Knudson, D. Investigating the skill gap between graduating students and industry expectations. Companion Proceedings of the 36th international conference on software engineering, 2014. 291-300.

Reisman, S. 2004. Higher education's role in job training. *IT Professional,* 6, 6-7.

Robles, M. M. 2012. Executive perceptions of the top 10 soft skills needed in today's workplace. *Business communication quarterly,* 75, 453-465.

Saiedian, H. 2009. Software engineering challenges of the "Net" generation. *Journal of systems and software,* 82, 551-552.

Saqa 2000. *The national qualifications framework and curriculum development*, South African Qualifications Authority.

Schreiner, C. & Sjøberg, S. 2004. Sowing the seeds of ROSE. Background, rationale, questionnaire development and data collection for ROSE–a comparative study of studens' view of science and science education. *Acta Didactica,* 4.

Seemiller, C. & Grace, M. 2016. *Generation Z goes to college*, John Wiley & Sons.

Tapscott, D. 2008. *Grown up digital: How the net generation is changing your world HC*, McGraw-Hill.

Taylor, E. 2016. Investigating the perception of stakeholders on soft skills development of students: Evidence from South Africa. *Interdisciplinary journal of e-skills and lifelong learning,* 12, 1-18.

Topi, H., Valacich, J. S., Wright, R. T., Kaiser, K., Nunamaker Jr, J. F., Sipior, J. C. & De Vreede, G.-J. 2010. IS 2010: Curriculum guidelines for undergraduate degree programs in information systems. *Communications of the Association for Information Systems,* 26, 18.

APPENDIX A

Factors	Items	Mean* of Students (n=297)	Mean* of Professionals (n=214)
Critical outcomes required in courses	***In the software development classes:*** ... students should learn to work with others as a member of a team or group.	4.174	4.383
	...should require from students to organise and manage themselves effectively.	4.119	4.374
	... students should learn to collect and critically evaluate information.	4.020	4.491
	... students should learn to communicate effectively, both verbally and in writing.	3.997	4.360
	... students should learn to use science and technology effectively.	4.010	4.201
	... students must be able to demonstrate an understanding of the world as a set of related systems by recognising that problem-solving contexts do not exist in isolation.	3.976	4.266
Knowledge of course requirements	Students know what software developers do in the workplace.	3.839	2.696
	I know what the outcomes are for a software development degree.	3.710	3.495
	The university courses have high expectations of students.	3.945	3.178
	The instruction in the software development classes is relevant.	3.952	3.435
Industry experience of lecturers	Lecturers should have industry experience.	3.776	4.425
Project work	Projects play an important role in the education of students.	4.020	4.322
People from industry teaching	People from industry should be brought into software development classes	2.290	4.369
Good examination results	To be a good software developer you have to have a good set of exam results	3.616	3.033
Positive attitude towards work and colleagues	***To be a good software developer you have to:*** ... have a good attitude including a willingness to listen and to take instructions	4.559	4.477
	... be prepared to work hard and to learn (a thirst for knowledge)	4.527	4.509
	... have good time-management skills	4.332	4.397
	... have respect for others	4.447	4.425
	... have a desire to succeed (realistically ambitious)	4.532	4.229
	... have a preparedness to take responsibility	4.600	4.423
Emotional/social skills	... have modern leadership skills like self-confidence and a preparedness to lead by example	4.135	3.645
	... have the ability to relate well to and to build relationships with others (emotional intelligence)	4.214	3.831
	... have at least some idea of what career direction one wish to take	4.269	3.836
	... have a reasonable level of general knowledge	4.369	3.793
Neat and tidy appearance	... have a neat and tidy appearance	3.753	2.790
Software developers needed	More software developers are needed in the country	4.207	4.322

* *Likert-style responses were ranked from 1 to 5 respectively*

THE EMERGENCE OF LIMINAL CYBERSPACE – CHALLENGES FOR THE ONTOLOGICAL WORK IN CYBERSECURITY

Jukka Vuorinen and Ville Uusitupa
University of Jyväskylä, Finland

ABSTRACT

This philosophy-oriented paper examines cybersecurity and its ontological work in relation to spaces which are created by conventional perimeter security model and Zero Trust model. We argue that security works by a code of inclusion and exclusion, e.g., an individual user seeking access is either included or excluded in relation to the system. Therefore, cybersecurity divides the space through employing the code of inclusion/exclusion which directly affects the agency of users. We examine how the growing complexity of network environment makes information and cybersecurity to struggle with the simplicity of the inclusion/exclusion code. The simplified bifurcation is held by maintaining a strict order of the space for included users (i.e., how users and devices can behave once they are let in). Furthermore, we analyse the emergence of liminal spaces that contain both included and excluded actors. Liminal spaces, which have increased during the pandemic era, provide an intriguing spot through which security can be examined in terms of what it does, how it works out the ontological status (included/excluded) of its subjects.

KEYWORDS

Spatiality, Liminality, Ontological Work, Zero Trust, User-centric Cybersecurity

1. INTRODUCTION

Cybersecurity enables or halts users depending on whether the user is identified, authenticated, and authorised. In other words, the agency of the user depends on the decision of information security in terms of inclusion and exclusion. The bifurcation of inclusion/exclusion has spatial consequences. Cybersecurity divides the space between the inside and the outside. The former is the region in which the use of system takes place and is controlled and managed by information security policies, whereas the outside is mainly the unknown environment that has to be blocked (Vuorinen and Tetri 2012). This type of bifurcation can be carried out in different manners, but they still perform the same inclusion/exclusion code. The conventional perimeter security model, which develops security relying on the physical metaphors (e.g., castle walls, doors, see Weaver and Weaver 2008), has been criticised not being fit for the mobile or remote use of systems (e.g., Campbell 2020, Pieters 2011, Rose et al. 2020). Despite the "de-perimeterisation" efforts, security still works with the same code of inclusion and exclusion (insides and outsides emerge). The ontological work of cybersecurity – attempt to find out the "being" or "becoming" of user, what or who is that – differs in these different spaces. The challenge is thrown in by emergence the liminal spaces that are not exclusively inside or outside. For example, a user (authorized insider) can use their own device (unmanaged outside element) in the organisational network environment. In addition, working remotely from home (outside the local network of the office) refers to such a liminal space as well. The liminal space contains elements of insides out outside, which challenges the essential bifurcation of inclusion and exclusion – the crucial discourse and practice in the field.

In this paper, we analyse the significance of spatiality for cybersecurity. In the course of history, spaces and security have formed a significant pair. For example, the analysis of spatiality from plague towns to prisons and mobile controls can be found (Deleuze 2017, Foucault 2007a, 2007b). However, in the case of cybersecurity, space does not provide a place of internment, but it is a fluid and divisible space of transformation filled with different actors. The Covid-19 pandemic has changed the arrangement of spaces in which we work. Notably, the spaces – different sites of use – and data gathered from the sites, such as geolocation, IP address and timestamps, provide important information, which is used to separate the compliant users from the suspicious ones. We analyse space in liminal space in terms of ontology and two different cybersecurity models, the conventional perimeter security model and Zero Trust model. The latter has gained popularity in recent years and claims to tackle the information security problems relating to remote work although it increases complexity (Bertino 2021). We analyse the models in terms of "ontological work", which pertains to being and becoming. "What is an actor?" forms an ontological question that resides at the heart of cybersecurity. Ontological work relates to how an actor is identified and authenticated. What is the significance of space in this ontological work that defines the position of individual user? Importantly, we do not seek to determine which of the models is better, but we focus on the analysis of space and ontological work. By doing this, we can have a better understanding of the environment, in which the individual user seeking their autonomy acts.

We begin by analysing the essential information security code of inclusion/exclusion and its spatial ramifications. We examine the spaces that the code organises and analyse how these regions work. As we have explored the bifurcation of space, we analyse how the two security models treat their spaces. In addition, we describe the emergence of liminal – mixed – space that has become the dominating space for cybersecurity to operate in the pandemic environment.

2. BIFURCATION – THE IDEAL PURIFICATION OF SPACE

2.1 The Essential Dualistic Code

Information security is based on the idea of inclusion. A user who signs on to service goes through a process of inclusion. Inclusion pertains to the processes of identification, authentication, and authorization: who the user is, and what are the privileges given. With a chosen method, information security algorithms analyse whether a user is the one they claim to be. A username with a shared secret (a password), or a token that the user has (a mobile phone with a particular number, a key in case of a door), or what a user is like (a biometric fingerprint scanner) can be used for identification and authentication. This is the essential ontological work of cybersecurity (Vuorinen 2014; Vuorinen and Tetri 2012). Granting access to a system means that the user can go over a barrier – a door is opened. As the (virtual or physical) perimeters are crossed, the status of the user changes from an unknown outsider into a known insider. If a user is not identified and authenticated, then, of course, access is denied. This reveals the counterpart of inclusion: exclusion. Evidently, solely the particular users are let in while the other are excluded. This demonstrates the dualistic code by which information security works: allowed/denied.

The code follows strictly simple binary logic leaving no room between the digits zero and one, on and off, allowed and prohibited. There is no partial access. From the administrative point of view, a user is either allowed to see information or it remains disclosed. Surely, all modern information systems that are used by multiple users have different layers of security such as granular user accounts. Simply, for example, students at a university cannot access each other's accounts as the user accounts are isolated from each other. The accounts are parallel but simultaneously inclusive/exclusive. In addition, there are vertical user rights from a user to an administrator and a root, which can be organized different ways to create scalable layers of security (Hong and Kim 2016). Nonetheless, the code follows the same dualistic logic.

With the attempts to define or describe the dimensions of information security, the triad of confidentiality, integrity, and availability can be mentioned (e.g., Agarwal and Agarwal 2011, Dhillon and Backhouse 2001, Samonas And Coss 2014). The binary logic of the inclusion/exclusion method can be understood in relation to these terms. The confidentiality and availability procedures function in terms of inclusion and exclusion. Integrity refers to the persistent form. For example, a file should hold its order (e.g., a hash) while being in storage – i.e., it should remain the same. If the file loses its order, becomes different, it is not secure or useful. In other words, it can be included (trusted) only if it holds its initial form – integrity. Otherwise, it is useless and becomes excluded.

2.2 The Code Divides and Cleans the Space

The fundamental dualistic code creates bifurcating spaces. Let us examine more closely what the spatial ramifications of the code are. At the ideal level, the dual category system (the code) bifurcates the space in which it is applied. In terms of information security, the ideal spatial consequence is a split of space into an orderly and controlled safe region of inside and outside that is a volatile, vibrant, uncontrollable, and possibly hostile exterior (Vuorinen and Tetri 2012). The outside is the (virtual/cyber) world of chaos that goes beyond the organised inside. Such divisions are not merely ideal but practical in some cases. For example, in perimeter security model uses information about location as a way of further inclusion (Weaver and Weaver 2008, Rose et al. 2020). If a user is within network perimeter, access can be given to all user resources within network.

Bifurcations go beyond security. "Inside" is defined by its order that springs from the desire of the holder – administrator, root, managers (Vuorinen and Tetri 2012). Insides and order are mundane. To clean a table is to exclude dirt. Mary Douglas (2003), a British anthropologist, makes a classic note on dirt; matter becomes dirt by its relation to other objects. For example, food on the plate is not dirt but as it falls on clothes, bedsheets, or on the floor, it instantly becomes undesired dirt. We want to emphasise that what is considered clean and dirty is defined by the desired order of the inside. In terms of information security, this means that inside is constituted on the inclusion of desired actors whether these were users, software or hardware. With regard to the desire and organisation, information security policies denote the desire of the organisation. All the actors and activities that are compliant with the policy are clean, proper and orderly. Cleanliness is based on the absence of noise. Noise can be understood in terms of systems here (Serres 2007), to a disturbing actor that distorts the logic of the system. This way, we can argue that information security threats are actors (e.g., hackers, malicious code, misuse of devices), that are incompatible with the order of inside. The danger is constituted by the position and effects that the threat actor would cause within. The actor can be harmless in another place – just as food on the plate instead of a floor.

The order of the inside is twofold. It concerns the relations of inside actors and, in addition their inner order (e.g., software and even thoughts of users). Firstly, the order pertains to the interconnective (and often spatial) arrangement of actors, including hardware, software, and users. Here, the question is of relations: which actors are allowed to connect, which actors can communicate and on which terms, which actors with specific parameters can read or/and write (see Rose et al. 2020). However, the order is also about engineering and managing the space in which the connections emerge. For example, using a desktop computer has spatial and virtual significance: where the system is used, which physical facilities are used, how they are cooled, how the power supply is protected. Furthermore, the specificity of locations allows hardware and software to be manipulated physically on the spot if such activity is needed. In other words, the order is about organising the relations of users, devices, software and data through a set of controls. Secondly, the order extends to the inner relations of these actors; each device is updated and made compliant, information security policies are imposed on users, the data is backed up. Staff can be rushed into security education programs. Here, we have arrived at the heart of security awareness campaigns that seek to grasp and influence the subjectively lived and experienced – phenomenological – world within the users. In terms of research, information security is compelled by the idea to make people behave in a particular manner (e.g. Alias 2019, Safa et.al. 2016; Vroom and von Solms 2004). It and its controls are employed to serve the desire of the organisation. However, simultaneously the organisation is bound to feed the security machine that consumes the energy of the system (Vuorinen and Tetri 2012).

3. SPACE AS A PROVIDER OF CERTAINTY

The constitution of an orderly inside seeks to gather spatial information. Spatial information refers to the firm knowledge of and about the space used. For example, spatial information can be developed as a part of situational awareness. For example, in a controlled environment – such as a well-managed facility for using IT resources – it is possible to gather data about the ordinary network traffic – where the packets travel, with what frequency, from which points – and then to define the baseline of that activity. This way, the order of the inside is made visible. The baseline describes the tempo-spatial rhythm that the users and devices with their routines create. It anticipates the future in terms of what to expect what it should be like and provides a canvas against which to compare all the future traffic. Baseline describes the cosy rhythmic hum of the inside that can give a warm and fuzzy feeling of security. If there are deviations, then flags rise. This provides an opportunity for further analysis of the case. The outside, on the other hand, is beyond control. Indeed, the external network can be probed, and information of situational awareness can be shared between trusted parties, but it cannot be managed. In terms of the inside, the perimeter is the surface of contact towards the outside; by analysing it, something of the outside can be known, but it cannot be captured entirely. In other words, the outside environment is too large to be known. Security establishes the bifurcation by organising the inside in the most impeccable way possible with the current resources. Inside stands out because of its order.

The rich knowledge of the site creates spatial certainty that brings about stability. When the baseline is known, it can be used as a tool of identification. At least in the conventional perimeter network security architecture, the company network space works as an identification. For example, inside resources are accessible for users in the company network IP-address range. An IP address is not a bullet-proof method of identification. However, in addition to IP, all the signals that the user emits directly or through the side channels provide partial evidence. Likewise, hiding such information makes the identity disappear. The anonymity of Tor-network is partly based on the fact that every user looks similar; no uniqueness can be extracted. In a conventional company network, user behaviour and familiar signals do not matter in terms of trusting in the user's identification as it is the filtering system at the perimeter that is trusted. In other words, when ordinary user behaviour is known, it can be used as supporting evidence of identification. It should be noted that information security is interruptive by its nature (Vuorinen & Tetri 2012). Security procedures tend to interrupt the user (or the system) and give order words: place the finger on the scanner, give your credentials, restart the machine to update it. The baseline makes information security quieter and more unintrusive. It can hold the identity of a user without asking it constantly if the environment is controlled. The routine activity can be used as a part of the continuous authentication of users.

Spatial stability – working at the same site with the same devices – makes it possible to employ more controls. A stable environment provides certainty about the space, as the resources can be accessed, analysed and managed with group policies, which can be in a social (discursive) or software format (group policy). The more controls there are or more the stability of the environment is trusted, the more the order is strengthened. We have described an ideal case of the inside. Ideally, it is known and dominated by the desired order that establishes the difference between the system and its external environment. However, in practice, the insides are filled with movement, distortion and noise. There are dark corners within. Moreover, devices fail, and users do not comply. Notably, the inside lives on the resources of outside. There is no energy within – everything has to be imported inside. This is aligned with the general system theories: The systems (also social systems) differentiate themselves in relation to their environment (Luhmann 1989). Thus, the inside is merely reorganised and controlled outside (see also Vuorinen & Tetri 2012). The users are outsiders that become insiders through the filtering process. The devices can include dirty firmware. Updates can be buggy. In addition, a device can be entirely managed, but certainty about user's identity can be questioned through user and entity behaviour analytics. Constantly, the outside is within the inside region, but the exterior is present in the form of resources that are put in order and arranged according to policies. Information security policies are probably imported as well. Nonetheless, organisations are used to dealing with such import procedures.

Hitherto, as we have described "inside" as an orderly and controlled space, we have referred to it as an ideal type in the field of information security. How the goals of information security are achieved is a different question. Security is never about control of everything, but its target is limited: the confined inside region. All the security measures seek to establish and reinforce the division between inside and outside. The security methods seek to prevent the external actors from accessing the system in the first place – "keep the dirt out". However, the practitioners that fight the security threats probably agree with Michel Serres's (2007, 86) argument of work "is a struggle against noise", but noise always finds its way within. Thus, it is important to know whether there the system is safe or compromised – "yet, look for the dirt within".

Zero Trust holds that "there is no implicit trust granted to assets or user accounts based solely on their physical or network location (i.e., local area networks versus the internet) or based on asset ownership (enterprise or personally owned)" (Rose et al. 2020, ii.) Currently, it seems that practitioners are keen to be careful and assume that systems are infiltrated. In terms of bifurcation, dirt is already within. Importantly, in all the above cases (whether the inside is trusted or not), in case an external actor is found, it is sought to be removed, excluded. It always comes down to "authorized and approved subjects (combination of user, application (or service), and device) can access the data to the exclusion of all other subjects (i.e., attackers)" (Rose et al. 2020, 4). In simple terms, purity is sought. Thus, bifurcation has become a dominating idea of security, in a sense it is a trope that is recognisable in security seeking environments. The trope is present in the field as metaphors of castles (e.g., Weaver and Weaver 2008), but also in the terms such as green zones (a safe region surrounded by a more uncontrolled space), the cyber kill chain (Lockheed Martin's terminology) which underlines outside region as a source of advanced persistent threats. The bifurcation is present in every firewall and login procedure; they establish the inside and outside.

Although the binary code of security and the bifurcation make it possible to think in terms of pure/impure, safe/contaminated, the Zero trust paradigm implies that the practical securing work does not use such binary thinking in such a way that the inside would be trusted by default. This is because the paradigm assumes that the attacker is already present in the inside region. The inside is seemingly in order but on basis of that order no trust is build. The inside means nothing in terms of trust as "an enterprise-owned environment is no different—or no more trustworthy—than any nonenterprise-owned environment" (Rose et al, 2020, 1). Instead, it is about risks which refer to the possibility of an incident and the severity of lost or damaged assets (Pieters 2011). Risk varies in terms of a threat, a system and an asset. For example, to lose a mundane shopping list is quite different from losing a social media account that brings food to the table. The risk involves unknown, even unthought and differs from the dualistic code. Risk is a continuum. It can increase or decrease.

Overarchingly, to form our argument at a more abstract level, we conclude that this struggle against actors that are incompatible with the order of inside – struggle against impurities – is, in fact, ontological work which requires resources. This is to say that the questions of being and becoming have to be answered. For example, an actor that behaves inconsistently becomes suspicious. An ontological question arises: what is that? Can it be a threat? What can it become? In other words, is that dangerous, i.e. incompatible with the inside order? As this work of finding out the "real" being of the actor requires work, there are two different strategies to spare resources. The conventional network architecture deals with the ontological work at the border of the system. To filter – for example, asking credentials – is to work with the question of ontology. As the filtering process is completed, users can be given further security privileges within the organisational network – i.e., user becomes inherently trusted after the filtering process. However, Zero Trust paradigm assumes users to be compatible with the order temporarily. An insider is an insider as long as new privileges are needed or something violating the baseline occurs. For example, by authenticating successfully to a service the user is temporarily verified as an insider. Yet, if the same user proceeds to download an excessive number of files, which is considered as a deviation compared to the baseline, then the user might be forced to authenticate themselves again and prove they are still the same insider they claimed to be before. The ontological question, what is that, needs to be answered. The verdict comes in the dualistic form but the ontological work behind the verdict is about probability and risk. In other words, ontological work seeks to find out what an actor is and then translate it on the level of binary logic (threat or not).

4. THE PANDEMIC AND SPATIAL UNCERTAINTY

The covid-19 pandemic has changed the working environments, as organisations have shifted to the mode of remote work. Indeed, this is a significant change from the information security point of view. The users are no longer in the gentle embrace of an office environment that would enable the use of managed devices in a sensorrich environment. Instead, the insiders are out there in the volatile exterior. In terms of inside/outside binary, the users are in a liminal space; out there in a suspicious environment but not yet totally out of control. Control is partly lost as the signals of behaviour fade into the depths of the outside world, which baselines cannot capture. The canvas of comparison has been torn into pieces. More importantly, diminishing spatial control means losing spatial certainty that no longer translates into ontological certainty. Consequently, a question arises: how the problem of uncertainty is solved? The answer seems clear, as more ontological work is obviously done. However, this time the inner space is not available for analysis and the outside space is too vast and general; thus, the user and the device in use becomes the subject of analysis. The analysis itself is ontological work; it seeks to banish the suspicion. In Zero Trust environment that suspicion is aroused constantly.

The loss of control and change in space adds a new mixed code parallel to the conventional binary code of information security. The space of security becomes a liminal space that mixes inside order and outside elements. In Table 1 we have described the contradictory logic of liminal space and how the models react to it.

Table 1. Spatiality, models, and ontological work

Spatial dimension	Actors	Ontological work at the level of inclusion and exclusion	Ontological work carried out in Perimeter security model	Ontological work carried out in Zero Trust Model
Outside	Chaotic actors, hackers, malicious code, but also useful resources, and insiders that have logged out.	Actors need to be defined in terms of combability with the inner order. The insider users and compatible software needs to be recognised and allowed access (availability). Other actors must be excluded (confidentiality).*	Outside is not trusted – sought to be excluded.	Outside is not trusted – sought to be excluded.
Inside	Authorized users, devices and other resources	Users and devices that must be kept in order in relation to the information security policies. Insiders must be enabled to work, to achieve the goal of confidentiality, integrity, and availability (CIA). **	The insiders are enabled at the perimeter. Once actor is within the ontological work can be delegated partly to the spatial location. All the actors that have passed the perimeter check are trusted.	The insiders are trusted only in momentary manner. Microperimeters and continuous analysis of user behaviour and context becomes the basis of categorizing users to insiders and outsiders. The privileges cannot be inherited from borderline control i.e., treats inside as an outside.
Liminal; mixed inside and outside	Authorised users (insiders) reside in uncontrolled spaces such as homes and using unmanaged devices. Cloud services, legacy IT and Shadow IT.	The same as * and ** above.	The model has difficulties in dealing with unmanaged devices. The only possibility is to extend the office base into the outer world through trusted gateways. It cannot handle "use your own device" requirement. Ontological work is increased at the border – e.g., multiphase authentication.	Works in the similar manner as above: as if users and devices were constantly coming inside from outside. Constant Re-checking.

For sure, the shift in the pandemic era has not been such dramatic as it may seem. Firstly, cloud services have been widely used in modern organisations before the covid. Perhaps, remote work opportunities were not utilised with such volume as in lockdown periods, but the option for such a way of acting was available. This means that the pandemic did not create new problematisations but rather it emphasised and boosted some problematisations. For example, the Zero Trust security model promotes constant ontological work as it promotes doubt and paranoia in the form of continuous checking on the identity of the user. It means that there is no permanent inside, but an insider status is only given temporarily. The inside is certainly organised orderly: security policies are applied, and insiders are granted access. However, spatial or temporal trust is not inheritable in Zero Trust paradigm (Rose et al. 2020). Trust fades away with time and spatial changes. For example, a certain period of passivity or a change of IP address is considered suspicious. For a user, working in a Zero trust environment is a constant becoming of a trusted user. Indeed, information security is considered a process (making of security) in a conventional network security context. Still, Zero Trust rejects "being" – the stability of status – and embraces becoming, which refers to vanishing trust. Trust is momentary; multifactor authentication is temporary. Zero Trust is about continuous renewing and rechecking. The continuous process reminds of Gilles Deleuze's (2017) description of control society. There are no places of rest, there is no final destination; everything is mobile; there are no specific spaces for controls, but they are everywhere. In a disciplinary society, on the other hand, there are particular spaces of surveillance and control, which are limited spatially and temporally.

In the pandemic era, some organizations have had the urge to stick with high control without compromises. For example, if an organisation is dealing with sensitive data, maintaining a high level of control can be a top priority. In these cases, the spatial inside space becomes stretched, which leads to the question of information security topology. The inside space is merely extended into the homes. This means, for example, solely using strictly controlled and managed devices with secured connections. In the pandemic environment such a requirement is recognised in the field (Bernard &Nicholson& Golden, 2020). The office moves home and works over VPN. However, there cannot be total control over, who is using, are the home office doors locked and computers locked. This of course has a decreasing effect on spatial certainty which emphasises the position of the user again. In a sense, users are simultaneously inside and outside. Zero trust framework treats the inside space in the same manner: as if outside actors were within the organisational space all the time.

The cloud services are used on daily basis in various organizations. There are different cloud-based services from infrastructure to software, but all of these outsource – partly at least – information security management. While the services can be administered and managed in a restricted way, physical access to the infra is out of the question. Cloud services are not within an organization, but neither they are outside. Being partly controlled by a trusted third party these services lie in a third space beyond inside and outside, i.e., in the liminal. An example of cloud services in the liminal space approaching the chaotic outside is the so-called Shadow IT, typically the cloud storages and other SaaS used by organizations' business departments or individual users without the consent and management of the centralized IT and security functions.

An intriguing example is yet to be provided by legacy IT that is found problematic in the field. Legacy IT is outdated machinery that cannot be updated to meet the current standards. Now as seen above the covid pandemic has turned the organizations and users to move towards the outside, taking a position in the liminal space; slightly out of control yet able to hold some sort of recognisable order. The legacy IT cannot transform itself into the liminal space but stays an outside element even in a controlled environment. Initially an insider slips away as the space changes and slowly it becomes an incompatible element with the inner and desired order. It has become a threat.

5. CONCLUSION

An individual as a user is subjected to the ontological work of information security, which seeks to answer what the user is and eventually decides whether the individual should be allowed to use the system in the requested way. In other words, information security affects the agency of individual users. In practice, information security is bound to work by the dualistic code of inclusion and exclusion. Consequently, the code bifurcates space into inside and outside regions. The former is sought to be organised orderly by the desire of its owner. The latter, on the other hand, is out of control in the sense that the chaotic outside world can be observed but cannot be managed. If a system were totally isolated – without a connection with the outside – maintaining the order of inside could perhaps be uncomplicated. However, the inside region requires outside as a resource pool from which to draw energy. This is to say that the two sides, almost always, are in connection with each other. Security requires energy. However, it turns that energy into the form of interruptions as it works ontologically (e.g., a requirement of credentials is an interruption). To save energy and to avoid unnecessary interruptions, the bifurcation of space can be harnessed for ontological work. Thus, to answer the problem of inclusion and exclusion, the perimetric network security model inherently trusts (includes) all the users within the organisational network. E.g., spatial information such as an IP address can be used as a tool of authentication. In other words, the perimetric network security model trusts the filtering system that resides at the borderline and gives privileges for everyone within the perimeter. However, if the connections between the inside and the outside increase and gain strength, liminal spaces, which mix outside and inside elements, emerge. The liminal spaces become significant in terms of information security as they defy the dualism of the code by which security works. In simple terms, the liminal spaces distort the spatial order of information security. Therefore, it also makes harnessing spatial information for ontological work difficult. For example, if a user is not inside the perimeter, then using that information for inclusion/exclusion decisions is impossible. With the rise of remote work – and the covid-19 pandemic made the remote work common practice for a number of organisations – the liminal spaces have become dominating form of information security space. Zero Trust information security model transforms the topology of information security. The space is still divided between inside and outside, but now there are spots of inside scattered around the outside. Secured gateways connect the inside to the spots, but these liminal space actors are trusted only momentarily. This means that the ontological work has increased; thus, the subjects are more likely to be interrupted. Or they do not have the option to move inside the space that would be trusted. In terms of the autonomy of an individual, in Zero Trust model, the conditions that make autonomy possible are questioned more frequently.

REFERENCES

Agarwal, A. and Agarwal, A., 2011. The security risks associated with cloud computing. *International Journal of Computer Applications in Engineering Sciences*, *1*, pp.257-259.

Alias, R.A., 2019. Information security policy compliance: Systematic literature review. *Procedia Computer Science*, *161*, pp.1216-1224.

Bertino, E., 2021. Zero Trust Architecture: Does It Help?. *IEEE Security & Privacy*, *19*(05), pp.95-96.

Campbell, M., 2020. Beyond zero trust: trust is a vulnerability. *Computer*, *53*(10), pp.110-113.

Deleuze, G., 2017. *Postscript on the Societies of Control.* Routledge, Abingdon.

Dhillon, G. and Backhouse, J., 2001. Current directions in IS security research: towards socio-organizational perspectives. *Information systems journal*, *11*(2), pp.127-153.

Douglas, M., 2003. *Purity and danger: An analysis of concepts of pollution and taboo*. Routledge.

Foucault, M., 2007a. *Discipline and punish: The birth of the prison*. Duke University Press, Durham.

Foucault, M., 2007b *Security, territory, population: lectures at the Collège de France, 1977–78.* Springer, Heidelberg.

Hong, J., Kim, D. 2016. Towards scalable security analysis using multi-layered security models. Journal of Network and *Computer Applications*, 75, pp.56–168.

Luhmann, N. (1989) *Ecological communication.* University of Chicago Press, Chicago.

Pieters, W., 2011. Representing humans in system security models: An actor-network approach. J. *Wirel. Mob. Networks Ubiquitous Comput. Dependable Appl.*, 2(1), pp.75-92.

Samonas, S. and Coss, D., 2014. The CIA strikes back: Redefining confidentiality, integrity and availability in security. *Journal of Information System Security*, *10*(3).

Safa, N.S., Von Solms, R. and Furnell, S., 2016. Information security policy compliance model in organizations. *computers & security*, *56*, pp.70-82.

Serres, M., 2007. *The Parasite*. Minnesota University Press, Minnesota.

Vroom, C. and Von Solms, R., 2004. Towards information security behavioural compliance. *Computers & security*, *23*(3), pp.191-198.

Vuorinen, Jukka. Parasitic *Order Machine – A Sociology and Ontology of Information Securing*. Annales Universitatis Turkuensis, Turku (2014).

Vuorinen, J. and Tetri, P., 2012. The order machine–The ontology of information security. *Journal of the Association for Information Systems*, *13*(9), pp. 695–713.

Weaver R, Weaver D., 2008. *Guide to tactical perimeter defense: becoming a security network specialist*. Thomson/Course Technology, Boston.

BRIDGING THE DIGITAL COMPETENCE GAP: TELL US WHAT YOU NEED

Sandra Santos[1], Margarida Lucas[2] and Pedro Bem-Haja[3]
[1]*Centre for Research in Higher Education Policies (CIPES), Matosinhos, Portugal*
[2]*Department of Education and Psychology, CIDTFF – The Research Centre on Didactics and Technology in the Education of Trainers, University of Aveiro, Portugal*
[3]*Department of Education and Psychology, CINTESIS – The Center for Research in Health Technologies and Services, University of Aveiro, Portugal*

ABSTRACT

Digital competence is one of the eight key competences for lifelong learning. In a generic way, it can be defined as the set of knowledge, skills, attitudes, abilities and strategies that make citizens capable of using digital technologies in a creative, critical, meaningful and responsible way for different personal and professional activities.
In the specific case of Higher Education (HE) students, the idea prevails that, because they were born in the digital age, they are digitally competent. However, several studies show that this is not the case, indicating that, in general, students find it difficult to capitalize on the potential of digital technologies for their personal, academic and professional development, which has an impact on their employability and work prospects.
The purpose of this study is threefold: i) to examine the digital competence of a group of HE students (N=411); ii) to understand where their digital competence needs to be improved and updated, and iii) to determine which modality students choose to respond to their needs. The study is conceptualized through the European Digital Competence Framework for Citizens (DigComp) and results show students score lower in relation to the competence areas Digital content creation, Safety and Problem solving. The competence areas where students reached higher scores correspond to Information and data literacy and Communication and collaboration. Students identified higher needs of improvement in relation to the competence areas Safety, followed by Problem solving and Digital content creation. The preferred modality for digital competence development was online tutorials, followed by compulsory curricular units and optional curricular units. It is hoped that this study will contribute to inform the design of HE strategies that can better support the development of students' digital competence.

KEYWORDS

Assessment, Digcomp, Digital Proficiency, Higher Education, Lifelong Learning, Training Needs

1. INTRODUCTION

Digital competences are becoming more and more relevant to meet the demands of contemporary and future learning, working and socialization environments (e.g. European Political Strategy Centre, 2016; Gonzalez Vazquez *et al.*, 2019; Ehlers, 2020). Higher Education (HE) students, therefore, should gather a set of digital competences that lead them to succeed throughout their academic training course and facilitate their integration and progression in the job market. In the context of a knowledge society, the role of HE institutions has been widely debated: on the one hand, they are the foundation for the acquisition of technical competences, specific to the different academic fields, necessary to enroll in a profession; on the other hand, there are other core competences, such as digital competences, that facilitate lifelong learning and help prepare for changing labor markets (Jørgensen, 2019). An acknowledgment of the key role of HE institutions in the digital ecosystem comes also from the Digital Education Action Plan (2021-2027), an European Union policy initiative aimed at supporting a sustainable and effective adaptation of the education and training systems of EU Member States to the digital age (European Commission, 2020). It is assumed that HE students from this new generation are digitally competent. However, it seems that, although they are capable of using digital devices for their daily needs, they seem to struggle to use them to foster their personal, academic and professional development (Margaryan, Littlejohn and Vojt, 2011; Corrin *et al.*, 2018).

This paper seeks to identify students' competences and training needs in five digital competence areas, namely, 1- Information and data literacy, 2- Communication and collaboration, 3- Digital content creation, 4- Safety and 5- Problem solving, and to determine which modality students would choose to respond to their needs. It is expected that this study will contribute to inform the design of HE strategies that can better support the development of students' digital competence and bridge the gap. This survey is particularly relevant for the distance teaching and learning process dynamization, which has accelerated the demand for higher levels of digital competences. It is, therefore, important for HE institutions to understand whether their students are able to learn through a digital modality and, if not, how they can support the promotion of digital competences without compromising the courses' curricular components (Crawford *et al.*, 2020).

2. BACKGROUND

In a generic way, digital competence can be defined as the set of knowledge, skills, attitudes, abilities and strategies that make citizens capable of using digital technologies in a creative, critical, meaningful and responsible way to achieve goals related with work, employability, learning, leisure and participation in society, independently or with others (Ferrari, 2013; Ilomäki *et al.*, 2016). It is both a mind-set and a survival kit offering benefits at social, economic, political, health and cultural levels (Eshet, 2012) and a key driver for greater social inclusion, employability, competitiveness in the job market and economic growth (Broadband Commission Working Group on Education, 2017; European Commission, 2019).

The rapid digitalization of society and the labor market make it imperative for graduates to acquire basic and advanced digital skills to actively and effectively participate in the digital sphere (European Commission, 2020). The ability to find, manage and store digital information securely, communicate, collaborate and share online responsibly, create, manage and protect digital identities efficiently, find solutions to problems using digital tools and services effectively or engage with innovative technology, such as artificial intelligence or robots are perceived as essential to respond to the challenges of a highly-digitalized economy (European Commission, 2019). In addition, such skills are reported to be essential for better academic engagement and study enthusiasm, which is a concern for worldwide HE institutions (Bergdahl, Nouri and Fors, 2020; Heidari *et al.*, 2021). Digital competences are also transversal to the development of other skills, such as lifelong learning, decision making, problem solving, or efficacy to deal with change, innovation and the unknown (Eshet, 2012; Broadband Commission Working Group on Education, 2017). These skills may determine who stays ahead and who falls behind in such a demanding society and world of work (OECD, 2019).

Digital competences are considered fundamental in the recent frameworks developed by the European Commission, such as "The European Entrepreneurship Competence Framework" (EntreComp) or "The Digital Competence Framework for Citizens" (DigComp) (Carretero, Vuorikari and Punie, 2017). The later framework is a reference for the assessment and the development of initiatives to foster digital competence both at European and Member State level. It entails a total of 21 specific digital competences distributed across five competence areas, namely, Information and data literacy, Communication and collaboration, Digital content creation, Safety and Problem solving. For each of the 21 competences, eight proficiency levels are defined and identified through a descriptor that may form the basis for the construction of digital competence assessment instruments and inform intervention strategies according to the identified needs.

The idea that HE students are naturally highly proficient in what digital competences may concern has been debated and deconstructed. Even though this new generation has the abilities to deal with technological devices and tools in their daily lives, they seem to have difficulties in using them to enhance their personal and professional wellbeing. Also, although HE students are frequently designated as "digital natives", there seems to be an heterogeneity in their digital competences' profile (European Commission, 2020). The most recent European skills and jobs survey (Cedefop, 2018) shows that about 85% of all jobs in Europe need at least a basic digital skills level. However, even though some studies have shown that students perform positively in some areas of digital competence, it does not mean that they are fully competent (López-Meneses *et al.*, 2020). Results from studies focusing on the development of digital skills in HE demonstrate students' lack of skills for searching, selecting and treating information (Santos, Azevedo and Pedro, 2013; Strømsø and Bråten, 2014) or lack of confidence in sharing contents and managing personal identity and privacy (Lupton, Oddone and Dreamson, 2019; Martzoukou *et al.*, 2020). A recent study with Chinese HE students aimed at assessing their digital competence with an instrument inspired by the

DigComp model also evidences that even though most students had a positive perception regarding their level of digital competence, especially in information and data literacy, communication and collaboration, some competences needed improvement to deal with the increasing tasks' complexity, such as creating digital content and programming, and problem-solving when facing technical problems and understanding of technological trends (Zhao *et al.*, 2021). Similar results were previously found in two other surveys on digital competences based on the DigComp model, with HE students from Spanish and Italian universities (Llorent-Vaquero, Tallón-Rosales and Monastero, 2020), and from three HE institutions in Scotland, Ireland and Greece (Martzoukou *et al.*, 2020). The safety competence area has also been identified as lacking improvement, even though students' perceptions are not unanimous. Some studies report positive perceptions on this competence area (Zhao *et al.*, 2021), while others identify this area as one reaching lower levels of confidence among HE students (Gallego Arrufat, Torres-Hernández and Pessoa, 2019).

Overall, research seems to suggest that there is the potential and the necessity for HE students to improve their digital competence to deal with more complex information and increase their performance on digital tasks. These data also make it evident that there are some performance nuances on digital competences, depending on the competence area. Therefore, a comprehensive approach on HE institutions regarding digital competence would start from an evaluation of students' training needs with reliable instruments based on the descriptors of digital competence. Such an assessment would form the basis for an outline of the intervention strategy to help students overcome their handicaps and diminish disparities in the digital competence.

As such, the present study aims at i) examine the digital competence of a group of HE students; ii) understand where their digital competence needs to be improved and updated, and; iii) determine which modality students would choose to respond to their needs. For these purposes, the study employs an instrument grounded in the DigComp model.

3. METHODOLOGY

3.1 Sample, Instrument and Procedure

In the Spring semester 2020, an online survey was launched at the University of Aveiro, to collect data on students' readiness for online learning, focusing on their digital competences. Respondents, representing a wide range of academic fields, agreed to participate and agreed with the use of their data for research.

The survey items were inspired by the DigComp Framework. At the end of the survey, students' responses were mapped against the first six proficiency levels proposed by the framework. To each level, a rule for cut-off scores was calculated. These (as well as the survey items) were confirmed as valid in a previous study (Lucas *et al.*, 2022). The cut-off scores attributed to each proficiency level correspond to: below 16 (A1), between 16 and 29 (A2), between 30 and 45 (B1), between 46 and 61 (B2), between 62 and 76 (C1) and above 76 (C2).

Data analysis was performed using R. To assess the differences in means among the different digital competences, a repeated measures ANOVA was performed, followed by multiple comparisons with Holm correction. To verify differences in the proportion of students who admitted the need to improve competence areas, binomial analyses were performed. These analyses compare, for each area, the proportional differences of students who reported "No" need for improvement with the proportion of students who reported a need by choosing the "Yes" option. Regarding the preferred modality for digital competence development, the difference in proportions between each answer was evaluated through a chi-squared analysis followed by multiple comparisons with Bonferroni adjustments on the p value.

4. RESULTS

4.1 Proficiency Scores by Competence Area

Figure 1 shows the proficiency scores of the sample in each of the five competence areas.

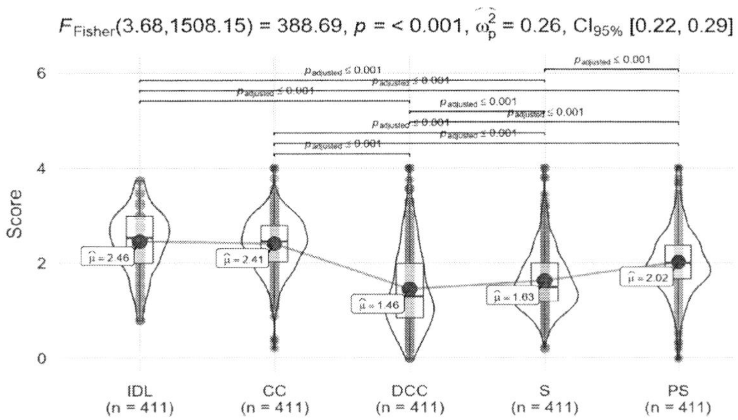

Note: IDL = Information and data literacy; CC = Communication and collaboration; DCC = Digital content creation; S = Safety; PS = Problem solving.

Figure 1. Proficiency scores in each competence area

The lowest scores of digital competence are found in "Digital content creation" (M=1.46), "Safety" (M=1.63) and "Problem solving" (M=2.02). Higher scores were reached in "Information and data literacy" (M=2.46) and "Communication and collaboration" (M=2.41) competence areas. Inferentially, the repeated measures ANOVA show significant differences among competence areas (see inference on top of Figure 1). Multiple comparisons with Holm correction also indicate significant differences among competence areas with the exception of the comparison between "Information and data literacy" and "Communication and collaboration" (p values adjusted with Holm correction are displayed in Figure 1).

4.2 Competence Areas in Need of Improvement

Figure 2 shows the flows of the competence areas students identified as in need of improvement.

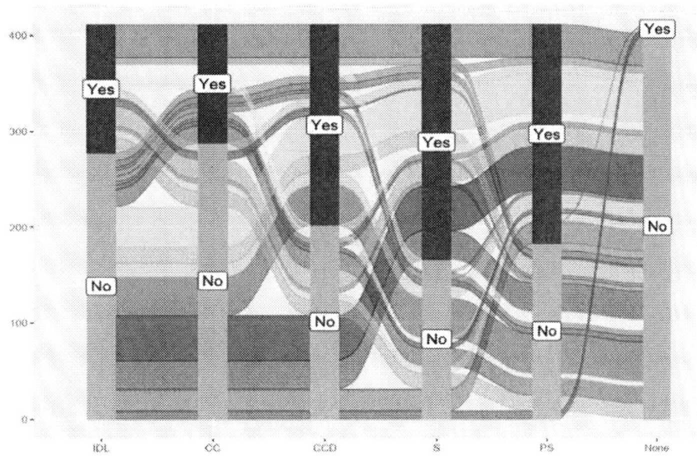

Note: IDL = Information and data literacy; CC = Communication and collaboration; DCC = Digital content creation; S = Safety; PS = Problem solving.

Figure 2. Alluvial flows of competences to be improved

The binomial analysis of the competence areas in need of improvement is displayed in Table 1. Through the descriptive analysis of the results, it is possible to verify that the proportion of students wanting to improve competences is higher in relation to the competence areas "Safety", "Problem solving" and "Digital content creation". However, inferentially, statistically differences in terms of proportions, were only obtained for the competence areas "Safety" and "Problem solving". At the same time, for the competence areas "Information and data literacy" and "Communication and collaboration", there is a larger proportion of students stating not needing to improve competences in these areas. The difference between those wanting and not wanting to improve competences in these areas achieved statistically differences. Only nine students admitted not needing to improve any competence/competence area. A closer look at the alluvial diagram allows us to verify that the thickest flow only trespasses the competence areas "Safety" and "Problem solving".

Table 1. Binomial analysis of competences to be improved

Competence areas	Y/N	N	Proportion	*p*	95% CI Lower	95% CI Upper	BF_{10}
Information and data literacy	No	277	0.6740	<.001	0.6263	0.7191	6.247e00+9
	Yes	134	0.3260	<.001	0.2808	0.3737	6.247e00+9
Communication and collaboration	No	288	0.7007	<.001	0.654	0.745	3.548e+13
	Yes	123	0.2993	<.001	0.255	0.346	3.548e+13
Digital content creation	No	202	0.4915	0.76730	0.4421	0.5409	0.06549
	Yes	209	0.5085	0.76730	0.4590	0.5578	0.06549
Safety	No	166	0.4039	<.001	0.3560	0.4531	125.93872
	Yes	245	0.5961	<.001	0.5469	0.6439	125.93872
Problem solving	No	183	0.4453	0.02986	0.3965	0.4947	0.72405
	Yes	228	0.5547	0.02986	0.5052	0.6034	0.72405
None	No	402	0.9781	<.001	0.9588	0.9899	1.520e+103
	Yes	9	0.0219	<.001	0.0100	0.0411	1.520e+103

Note: H_a proportion $\neq 0.5$

4.3 Preferred Modality for Digital Competence Development

Figure 3 illustrates the preferred modality selected by students to develop their digital competence.

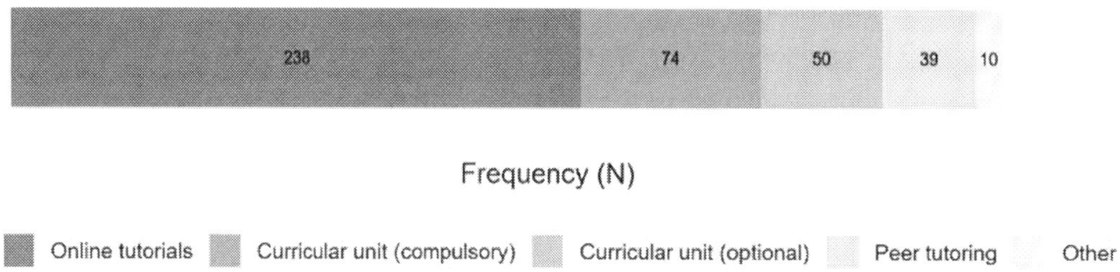

Figure 3. Preferred modality for digital competence development

It is possible to see that the preferred modality chosen by students to develop their digital competence correspond to "Online tutorials", followed by "Curricular unit (compulsory)", "Curricular unit (optional)" and "Peer tutoring". The last option selected by students correspond to "Other" and suggestions include the modalities: workshops, MOOCs, webinars and short training courses.

The Chi-squared test (Goodness of fit) shows that there was no homogeneous distribution among response options, χ^2 (4)=394.85, p<.001. Pairwise binomial tests with Bonferroni adjustment show significant statistical differences between all the response options with the exception for the comparisons between the "Curricular unit (compulsory)" and "Curricular unit (optional)" (adjusted p=.384), and between the "Curricular unit (optional)" and "Peer tutoring" responses (adjusted p=.999).

5. DISCUSSION AND CONCLUSION

The goals of this study were threefold: i) to analyze the digital competence level of a sample of HE students; ii) to identify which digital competences need to be improved, and; iii) to verify which modality of training students prefer to develop their lacking or low proficiency digital competences.

In response to the first goal, the results suggest that higher proficiency levels were reported for "Information and data literacy" and "Communication and collaboration" competence areas and the lowest scores were found in relation to competence areas "Digital content creation", followed by "Safety" and "Problem solving". These results are similar to those found in literature (Gallego Arrufat, Torres-Hernández and Pessoa, 2019; Llorent-Vaquero, Tallón-Rosales and Monastero, 2020; Martzoukou *et al.*, 2020; Zhao *et al.*, 2021). The results found in "Safety" competence area were as expected, considering the majority of the revised studies, contradicting only the results obtained by Zhao and colleagues (2021).

Regarding the second goal, the results show a significant large proportion of students that consider they do not need to improve "Information and data literacy" and "Communication and collaboration" competences and a significant large proportion of students wanting to improve "Safety" and "Problem solving" related competences. Even though "Digital content creation" competence area was the one where lower proficiency levels were found, there is not a significant proportion of students wanting to improve their competences in this area. As competences included in this area relate to programming, creating and modifying digital content and applying copyright licenses to it, students might not find these competences as valuable or necessary for their learning and future professional experience as the other competence areas.

The preferred modality for students' training and development of digital competence (third goal) corresponds to "Online tutorials", followed by the modalities that encompass the integration of training in curricular units, either compulsory or optional.

Overall, this study results are in line with literature and reports reviewed that highlight not only the existence of competence areas where HE students showed higher proficiency levels than others, but also the potential and the need of HE students to improve their digital competences, thus joining the body of research that challenges the myth that they are digital natives.

These results also raise the attention for the relevance of performing an analysis of needs to better guide the construction of responses aimed at bridging the gaps in students' specific lacking or needing improvement competences. An evaluation of students' training needs, with reliable assessment instruments based on the descriptors of digital competence makes it possible for HE institutions to decide on which efforts should be made to include digital competences' training in their agenda and meet students' and employers' needs concerning the basic and advanced skills fundamental to succeed in the different spheres of life. Some challenges may, however, be encountered. Delivering the same training solution to all the students seems simpler than providing context-specific tools and resources. However, a 'one-size- fits-all' approach is self-defeating, as students benefit the most when enrolled in individualized, flexible, improvement strategies. Planning and providing training programmes designed to meet the needs of target beneficiaries is a factor that could determine the success of the implemented strategies (Broadband Commission Working Group on Education, 2017).

This study may also inform HE institutions on the modality of intervention that better suits students' interests and preferences and that, simultaneously, may face less obstacles for the implementation of training programs. Online tutorials may be a cost-effective solution both for students, who may learn new competences at their pace, and institutions, considering that it does not require a massive change in the curricular plan, nor additional resources to teach digital competences face-to-face. It also makes it possible for this training opportunity to reach a wider proportion of students as there are no geographic or social-limitations (Ferrari, 2012). In any event, HE institutions may use the collected data to help them weight the advantages and disadvantages of the strategies they find effective to improve their students' digital competences, in line with the Digital Education Action Plan for the digital transition.

Some limitations in this study should be also regarded. This study sample is composed of Portuguese speaking students from a single university, therefore the results cannot be generalizable to other HE institutions, either in Portugal or in the European context. It is possible that students' competences needing improvement may be different in other contexts or the preferred modality for competences' development may diverge. Considering that students' competences may diverge depending on the HE institution typology (e.g., vocational education and training institutions *versus* cooperative HE institutions) (Wild and Schulze Heuling, 2020), future efforts should be made to assess HE students' digital competences in other institutions with different challenges and opportunities, including universities and polytechnic institutes, and to conduct comparative studies with students from different European HE institutions. Another limitation relates to the data collection with a self-report instrument, which may not reflect students' true digital competences, as there is the tendency of them to overestimate their competence level. Future studies should be based on the observation of students' performance to control for the subjectivity of measures based on personal perceptions.

Despite the limitations, this study is an original contribution for the study of the Portuguese HE students' digital competences that suggests some paths to promote those competences and meet the needs of students, employers, and education stakeholders, thus laying the foundations for an inclusive and equitable education on digital competences.

ACKNOWLEDGEMENT

This work is financially supported by National Funds through FCT – Fundação para a Ciência e Tecnologia, I.P. under the projects UIDB/00194/2020 and PTDC/CED-EDG/29726/2017 and in the scope of the framework contract foreseen in the numbers 4, 5 and 6 of the article 23, of the Decree-Law 57/2016, of August 29, changed by Law 57/2017, of July 19.

REFERENCES

Bergdahl, N. et al, 2020. Disengagement, Engagement and Digital Skills in Technology-Enhanced Learning. *In Education and Information Technologies*, Vol. 25, No. 2, pp. 957–983.

Broadband Commission Working Group on Education, 2017. *Digital Skills for Life and Work*. UNESCO, Paris, France. Available at: https://unesdoc.unesco.org/ark:/48223/pf0000259013.

Carretero, S. et al, 2017. *DigComp 2.1: The Digital Competence Framework for Citizens with Eight Proficiency Levels and Examples of Use* (EUR 28558 EN). Publications Office of the European Union, Luxembourg. Available at: https://publications.jrc.ec.europa.eu/repository/handle/JRC106281

Cedefop, 2018. *Insights into Skill Shortages and Skill Mismatch: Learning from Cedefop's European Skills and Jobs Survey*. Publications Office of the European Union, Luxembourg. Available at: https://www.cedefop.europa.eu/en/publications/3075

Corrin, L. et al, 2018. The Myth of the Digital Native and What it Means for Higher Education. *In* Attrill-Smith, A. et al. (eds) *The Oxford Handbook of Cyberpsychology*. Oxford Handbooks Online, New York, USA, pp. 97–114.

Crawford, J. et al, 2020. Covid-19: 20 Countries' Higher Education Intra-Period Digital Pedagogy Responses. *In Journal of Applied Learning & Teaching*, Vol. 3, No. 1, pp. 1–20.

Ehlers, U.-D. (2020) *Future Skills: Future Learning and Future Higher Education*. Ulf-Daniel Ehlers, Karlsruhe, Germany.

Eshet, Y., 2012. Thinking in the Digital Era: A Revised Model for Digital Literacy. *In Issues in Informing Science and Information Technology*, Vol. 9, pp. 267–276.

European Commission, 2019. *Key Competences for Lifelong Learning*. Publications Office of the European Union, Luxembourg. Available at: https://op.europa.eu/en/publication-detail/-/publication/297a33c8-a1f3-11e9-9d01-01aa75ed71a1.

European Commission, 2020. *Digital Education Action Plan 2021-2027: Resetting Education and Training for the Digital Age*. Publications Office of the European Union, Luxembourg. Available at: https://education.ec.europa.eu/sites/default/files/document-library-docs/deap-communication-sept2020_en.pdf

European Political Strategy Centre, 2016. *Global Trends to 2030: The Future of Work and Workplaces, Individualism and Inequality*. European Strategy and Policy Analysis System. Available at: https://espas.secure.europarl.europa.eu/orbis/sites/default/files/generated/document/en/Ideas%20Paper%20Future%20of%20work%20V02.pdf

Ferrari, A., 2012. *Digital Competence in Practice: An Analysis of Frameworks* (EUR 25351 EN). Publications Office of the European Union, Luxembourg. Available at: https://ifap.ru/library/book522.pdf

Ferrari, A., 2013. *DIGCOMP: A Framework for Developing and Understanding Digital Competence in Europe* (EUR 26035). Publications Office of the European Union, Luxembourg. Available at: https://publications.jrc.ec.europa.eu/repository/handle/JRC83167

Gallego Arrufat, M.-J. et al, 2019. Competence of Future Teachers in the Digital Security Area. *In Comunicar*, Vol. 27, No. 61, pp. 53–62.

Gonzalez-Vazquez, I. et al, 2019. *The Changing Nature of Work and Skills in the Digital Age* (EUR 29823 EN). Publications Office of the European Union, Luxembourg. Available at: https://op.europa.eu/en/publication-detail/-/publication/508a476f-dc75-11e9-9c4e-01aa75ed71a1/language-en

Heidari, E. et al, 2021. The Role of Digital Informal Learning in the Relationship Between Students' Digital Competence and Academic Engagement During the COVID-19 Pandemic. *In Journal of Computer Assisted Learning*, Vol. 37, No. 4, pp. 1154–1166.

Ilomäki, L. et al, 2016. Digital Competence: An Emergent Boundary Concept for Policy and Educational Research. *In Education and Information Technologies*, Vol. 21, pp. 655–679.

Jørgensen, T., 2019. *Digital Skills: Where Universities Matter*. European University Association, Switzerland. Available at: https://eua.eu/downloads/publications/digital skills where universities matter.pdf.

Llorent-Vaquero, M. et al, 2020. Use of Information and Communication Technologies (ICTs) in Communication and Collaboration: A Comparative Study Between University Students from Spain and Italy. *In Sustainability*, Vol. 12, No. 10, 3969.

López-Meneses, E. et al, 2020. University Students' Digital Competence in Three Areas of the DigCom 2.1 Model: A Comparative Study at Three European Universities. *In Australasian Journal of Educational Technology*, Vol. 36, No. 3, pp. 69–88.

Lucas, M. et al, 2022. Digital Proficiency: Sorting Real Gaps From Myths Among Higher Education Students. *In British Journal of Educational Technology*, advance online publication, https://doi.org/10.1111/bjet.13220

Lupton, M. et al, 2019. Students' Professional Digital Identities. *In* Tippett, N. and Bridgstock, R. (eds) *Higher Education and the Future of Graduate Employability: A Connectedness Learning Approach*. Edward Elgar Publishing, United Kingdom, pp. 30–49.

Margaryan, A. et al, 2011. Are Digital Natives a Myth or Reality? University Students' Use of Digital Technologies. *In Computers and Education*, Vol. 56, No. 2, pp. 429–440.

Martzoukou, K. et al, 2020. A Study of Higher Education Students' Self-Perceived Digital Competences for Learning and Everyday Life Online Participation. *In Journal of Documentation*, Vol. 76, No. 6, pp. 1413–1458.

OECD, 2019. *OECD Skills Outlook 2019: Thriving in a Digital World*. OECD Publishing, Paris, France.

Santos, R. et al, 2013. Digital Divide in Higher Education Students' Digital Literacy'. *In* Kurbanoğlu, S. et al. (eds) *Worldwide Commonalities and Challenges in Information Literacy Research and Practice*. Springer International Publishing, Cham, Switzerland, pp. 178–183.

Strømsø, H. I. and Bråten, I., 2014. Students' Sourcing while Reading and Writing from Multiple Web Documents. *In Nordic Journal of Digital Literacy*, Vol. 9, No. 2, pp. 92–111.

Wild, S. and Schulze Heuling, L., 2020. How Do the Digital Competences of Students in Vocational Schools Differ From Those of Students in Cooperative Higher Education Institutions in Germany?. *In Empirical Research in Vocational Education and Training*, Vol. 12, No. 5, no pages.

Zhao, Y. et al, 2021. Digital Competence in Higher Education: Students' Perception and Personal Factors. *In Sustainability*, Vol. 13, No. 21, pp. 1–17.

VOTING TECHNOLOGIES – FROM OSTRACON TO E-VOTING

Elizabeta Trajanovska Srbinoska[1], Smilka Janeska Sarkanjac[2] and Branislav Sarkanjac[3]
[1]*Assembly of the Republic of Macedonia, 11Oktomvri 10, Skopje, Macedonia*
[2]*Ss Cyril and Methodius University in Skopje, Faculty of Computer Science and Engineering, Rugjer Boskovic 16, Skopje, Macedonia*
[3]*Ss Cyril and Methodius University in Skopje, Faculty of Philosophy, Goce Delcev 9a, Skopje, Macedonia*

ABSTRACT

One of the allocation methods of scarce resources, especially in the public sector, is based on majority rule. Modern societies use majority rule to elect representative governments that make some of the biggest decisions. Voting is a main vehicle of majority rule. There are four main factors that influence and are affected by voting – technology, law, politics and society. This paper provides an insight into the relationship of the voting technology and the level of development of democracy in a given state.

Voting technologies developed from ostracon in ancient Greece, to Australian paper ballot, to telephone, fax, various forms of electronic voting, internet voting, mobile voting, blockchain and AI supported voting.

E-voting could be considered as a form of display of the level of development of democracy in a given state. From ostracon to e-voting the main question is the same: how to have good life in a good state with good laws. From the answers offered by the Pythagoreans and Plato to today advocates of modern governance it has always been about how to organise a state so that we can live together the best way we can.

KEYWORDS

Voting Technology, Internet, Blockchain, AI voting, Governance

1. INTRODUCTION

Democracy is about choosing government that will make decisions that relate to wellbeing or prosperity of citizens. In most cases it is about determining the priorities of development. And that has to do with allocation of the state budget. One of the allocation methods of scarce resources (besides market price, command, contest, first-come, first-served etc. as in Parkin, 2012), especially in the public sector, is based on majority rule. This method allocates resources in the way that a majority of elected members of governing bodies in a state chooses. Being elected by citizens they indirectly choose what (majority) voters would choose. In general, modern societies use majority rule to elect representative governments that make some of the biggest decisions.

Majority rule is a decision rule that selects alternatives which have a majority, that is, more than half the votes. It is the binary decision rule used most often in influential decision-making bodies, including all the legislatures of democratic nations.

Voting is a main vehicle of majority rule. It is a formal expression of opinion or choice, either positive or negative, made by an individual or a group of individuals.

In modern societies, in a conventional paper voting system, voters visit the nearest polling station to cast their ballots. After the polling deadline, all ballots are counted manually by some trusted entity, such as the nation's electoral commission. Eventually, the casted ballots and voting results are securely stored and managed in some archival venue for a predetermined period of time.

More than thirty years ago, when WWW was invented and became a platform of mass media over the years, numerous and often far-reaching claims about the new media's transformative potential were made. Many authors enthusiastically argued that the Internet will fundamentally change democracy and politics by providing easy and universal access to information, undermining established structures of political power, democratizing the processes of agenda-setting, increasing the rates of political participation, improving the quality of deliberation and making plebiscitary forms of decision-making feasible (Lindner and Jennen, 2016; Weare, 2002; Anderson and Rainie, 2020).

Today, due to a large number of practical experiences, the debates on the Internet's effects on democracy are considerably less enthusiastic.

However, e-voting is indispensable in modern governance. There are many discussions on relationship of e-voting and e-governance. We would like to stress the importance of the wider context of e-voting that is governance. Here we refer to the framework given by socio-political governance of Jan Kooimann (1993). The theory of socio-political govermance focuses on the plurality and interactions of social and political actors in a social-political systems that are charcterazed by complexity, dynamics and diversity.

It is needless to say that governance includes democracy. If governance includes as many as possible political actors that means that democratic organized society is a prerequisite for good governance. Sophisticated, well thought out voting (electoral) system defines developed democracy (Evan, 2004; Robertson, 2006).

We argue that e-voting could be regarded as a "reality testing" of the progress of democracy in a state. Well organized e-voting is a demonstration of many accomplishments which are necessary for advancement of a democratic state, especially the so-called new democracies of the former Eastern Block. We single out two extremely important goals to be achieved: mature culture of lawfulness and low corruption society.

E-voting could be considered as a form of display of the level of development of democracy in a given state. If e-voting provides significantly more transparent process of election it certainly contributes to development of democracy. If this is the case, by accepting e-voting, the practice of buying votes, intimidation, bribery and manipulation during the elections are more difficult to organize.

Generally, there are four main factors that influence and are affected by voting - technology, law, politics and society (Krimmer, 2012).

2. FROM OSTRACON TO BALLOT

Voting as a process is presumed to begin somewhere in prehistoric times, when open voting, for example, elected a tribal leader or made a decision.

In the antiquity, with the introduction of democracy, the secret ballot begins to be used for voting. For example, the use of pebbles that were secretly placed in the candidate's vessels, pebbles of a different colour for the different candidate in a vessel, or the use of broken pottery 'ostrakoni' or other print media, where the name of the candidate-option being voted for, was written (Roisman, 2011).

The secret ballot enabled the further development of the democratic voting, i.e., the voter to express his voice without fear of the public, and at the same time the possibility of the so-called vote buying – i.e., rewarding the voter for the given vote for the candidate.

Paper voting as a form of secret ballot on a prescribed voting medium was first introduced in the Roman Republic by the Law on Voting (leges tabellariae) in 139 BC (Yakobson, 1995). This law also started the transition from public voting to secret ballot, for all types of voting. The reason for prescribing the ballot arose from the need for greater codification of the most important democratic process - the decision-making process, in the then largest country in the Mediterranean, Europe and (one of the largest in) the world, due to difficulties in maintaining the democracy of elections in it. Namely, the prescribed ostracons and other types of voting media are subject to connection, through some marking, handwriting, violation or other type of marking that could connect the voter with his vote and thus discredit the secrecy of the ballot. The ballot was actually the Egyptian papyrus, suitable for writing, but at the same time, under the strict supervision of the state. The law was valid everywhere in the Roman state and for all Roman colonies and protectorates (Yakobson,1995).

The Australian paper ballot, or also called a secret ballot, is the most widely used voting technology for elections in liberal democracies around the world. Victoria and South Australia were the first states to introduce secrecy of the ballot (1856), and for that reason the secret ballot is referred to as the Australian ballot. The system spread to Europe and the United States to meet the growing public and parliamentary demand for protection of voters. It is a system of voting in which voters mark their choices in controlled privacy in public places, on uniform ballots printed and distributed by the government or designate their choices by some other secret means.

Some 150 year after the introduction of Australian paper ballot there is nothing radically new on a large scale, despite many radical changes in our lives. We can travel with the speed of sound and exchange information with the speed of light. Admittedly, there were some technological breakthroughs that were included in voting, such as telephone, fax machines, computers, to internet, but most of them were not implemented in most of the countries.

Telephone voting allows remote voting, i.e., the voter can vote from home, without going to the polling station, but due to the difficulties in achieving the security and secrecy needed to vote in state and local elections, has very limited use. Instead, it is very much used in voting for TV shows (Eurovision song contest for example).

Fax voting (or voting by fax machine) is allowed in Alaska with prior registration. It is a way of remote voting with the conveniences it brings, but the problem with the security and secrecy of the ballot is even more pronounced than in a very similar telephone voting.

Electronic voting (e-voting), is a form of computer mediation voting in which voters make their selections using a computer. The voter usually makes his choice with the help of stand-alone electronic voting machines (EVMs), or with computers connected to the Internet (Gibson *et al.*, 2016; Hao and Ryan, 2016; Kersting and Baldersheim, 2004; Katz *et al.*, 2011).

To understand e-voting, it is convenient to consider three basic steps in the election process: 1- composition of ballots, in which voters make a choice, 2- registration of ballots, in which the system records the submitted ballots; and 3-summing, in which the votes are counted. Ballot casting, recording and summarizing are routinely done with computers even in non-electronic voting systems. Electronic voting is strictly a system where the first step, the composition of the ballots (and/or the selection), is done with the help of a computer.

Electronic voting technology may include drilling cards, mechanical machines withs wheels and buttons, membrane buttons, optical scanning voting machines, and specialized voting kiosks (such as touch screens) that include a direct recording electronic voting system (DRE) (Herrnson *et al.*, 2008; Herrnson *et al.*, 2009; Dill *et al.*, 2003).

Voting technology with optical scanning machines is not a direct electronic voting technology (Card and Moretti, 2007). The voter votes on a ballot paper which is then scanned to obtain an electronic record which is easily transmitted from the polling stations to the election officials. This can be done in two ways: by scanning and manual counting or by scanning and optical recognition and electronic counting. Because there is a paper ballot, the possibilities for electronic manipulation are reduced because in case of doubt, ballots can be manually reviewed and counted. The electronic part enables fast transfer of electronic data, faster summaries and publication of results. The disadvantage is that the technology consists of 2 systems - electronic and paper, which need to be organized separately and eventually integrated.

A typical DRE machine is composed of a touch screen connected to a computer. Ballots are presented to voters on the touch screen, where they make their choices and vote. The touch screen can be used to assist the voter in a variety of ways, including displaying large fonts and high contrast for the visually impaired, warning the voter to vote by choice, and preventing re-runs.

The DRE machine directly records the ballots and stores the data in its memory. Such a single machine is used for composing, voting and recording votes. The third step, writing the ballot to a memory device, is invisible to the voter (Kumar and Begum, 2012).

Introduction of new technology almost always has some shortcomings. What are the shortcomings of the use of a technology in voting?

Ensuring that voting is recorded, as voting relies on testing the hardware and software of the machine before the election, is the belief that the software running during the election is the same software as the one tested before the election. This is the subject of much controversy.

While testing hardware errors or unintentional software errors can be reliable, the same is not true for malware. Most security experts believe that an insider attack in the software development phase could reach the final product without being detected (although there is disagreement about the likelihood of such an attack). This problem is compounded by the fact that the source code is usually not available for public scrutiny (Dunn and Merkle, 2018).

Cryptographic techniques can partially solve the problem of software authentication. When software is evaluated and certified, a cryptographic hash (a short string of bits that serves as a kind of "signature" for computer code - for example, code length or code number of bits) can be calculated and stored. Just before the election, the hash is calculated. Any change in the certified software will cause the two hashes to be different. However, this technique may not prevent all attacks on the integrity of the software.

Computer viruses can infect a machine during elections. For this to happen, the machine must somehow communicate with another electronic device. Thus, connection to the Internet or wireless devices is usually not allowed. However, the voting session usually begins with the use of an activation card. An employee in the poll, after confirming the eligibility, sets the card to allow one voting session. After the session, the voter returns the card to the voting worker for reuse. At least one DRE system has been shown to be vulnerable to infection by the activation card. An infected machine can be made to register votes differently from those voted (Oo and Aung, 2014).

The threat posed by the DRE not to record votes as voters has led some individuals and organizations to argue that a paper-review report must be prepared for each ballot. DRE manufacturers have responded by adding a printer feature to their DRE. As a result, the systems produce both electronic records and paper records. However, problems with document handling and monitoring, both by voters and election officials, have led to much criticism of these hybrid systems. Many jurisdictions have already rejected them in favour of optical scanning technology (Stewart, 2011).

3. INTERNET VOTING AND MOBILE VOTING

Internet voting is remote electronic voting over the Internet where voters submit their ballots electronically to election authorities from any location. With the rapid use of the Internet, it seemed that the voting process would naturally migrate there. In this scenario, voters would choose from any computer connected to the Internet - including their home one. Beyond voting in regularly scheduled elections, many saw the emergence of these new technologies as an opportunity to transform democracy, enabling citizens to participate directly in the decision-making process. However, many countries have decided that the Internet is not secure enough for voting purposes.

As a first concern, denial-of-service attacks may block the system and to call into question the electoral process. Security experts are also concerned that many PCs are vulnerable to various types of malware. Such attacks can be used to block or replace legitimate votes, undermining the electoral process in an undisclosed manner (Jefferson et al., 2004).

The third concern about e-voting tackles the possibility of voter coercion and vote-selling when voting does not take place in a controlled environment. However, there is no consensus on the seriousness of this problem in stable democracies, as it is generally acknowledged that voting in general is less problematic in stable democracies. Taking the benefits of e-voting into account, it should be introduced more decisively in less stable democracies. Furthermore, this complaint also applies to absentee ballots, which have been widely used in the past, as well as by mail.

Electronic and online voting also provide some advanced opportunities in terms of increasing democracy (the ability to express the opinion of the individual and his influence on joint decisions), which would be very difficult to do with paper voting techniques. The first example is the open lists, i.e., the candidate lists where the parties, instead of a certain number of candidates for fixed positions, could propose extended lists and / or lists with variable order. Voters in this way have the opportunity not only to vote for a particular political party, but also to influence the order in that list, vote for people they trust and are nominated by different parties or independent lists and so on. In this case, the voting process might be preceded by a step of compiling the election-electoral list from the proposed ones, or even adding an unmentioned candidate (Tarasov and Tewari, 2017; Stewart III, 2011).

According to the survey by International IDEA from November 2020, only eight countries in the world allow their voters to cast their ballots to their national parliament elections, to the elections of the local government councils, or to the parliament of the EU online.

In countries such as France (piloting in 2003 and full access to online voting for all voters abroad in 2012), Panama or Pakistan, for example, the option to vote online is reserved for voters who live abroad. In other countries, such as Armenia, this possibility is offered only to diplomatic and military staff posted abroad. Several States in Australia implemented internet voting for the voters with disabilities, by using their computer screen reader tools when accessing a web-based platform.

In Switzerland, 15 cantons have offered internet voting to voters abroad and to a certain number of voters within their borders.

Some countries are taking advantage of online voting at the local level, such as Canada, where municipalities in the provinces of Ontario and Nova Scotia have been using internet voting since 2003 and 2008, respectively.[1]

Estonia remains the only country in the world in which any citizen can cast a remote online or mobile ballot during elections to their national parliament, to local government councils, or to the parliament of the EU – on all the election levels. Estonia became an online voting pioneer in 2005, is now a reference for the use of Internet voting technology. During the 2019, 47% of the votes to the European parliament were cast by online voters. Mainly because of denial-of-service attacks threats, Estonia maintains its traditional voting infrastructure along with the e-voting option.

4. FUTURE VOTING TECHNOLOGIES

E-voting is one of the sectors that can be advanced by blockchain technology. The idea of blockchain-enabled e-voting (BEV) is derived by analogy with the use of the digital assets like Bitcoin with which this technology was first introduced (and is the successor to the distributed P2P technologies that are used for a longer time). The BEV stipulates that every voter has a wallet - virtual access point that contains user credentials. Each voter, analogous to Bitcoin and other virtual currencies, receives one coin representing one voting opportunity (Susskind, 2017; Huang *et al.*, 2021).

The essential benefit of the introduction of the blockchain is that unlike the centralized management and verification of voting by the election authorities, voting councils or similar bodies, blockchain technology allows decentralization, i.e., the ability to check/verify from multiple places, even from each participant in the elections. With the blockchain, each participant actually has an insight into all the votes, the times when they occurred, these records cannot be changed, and everyone has access to them.

The blockchain technique used in the voting process provides increased voting security, and allows for greater immediacy: voting takes place on a perhaps more unreliable device - for example a mobile phone instead of a polling station, but with similar security.

5. AI-VOTING (ARTIFICIAL INTELLIGENCE SUPPORTED VOTING)

Voting assisted by AI-Artificial Intelligence implies 3 directions through which AI would improve the voting process:

- The first direction is to increase the security of voting through algorithms such as face recognition, voice, biometric information, handwriting for signature and the like.
- The second direction implies that in the voting process, a preliminary proposal of a vote is included, which the voter would then confirm, based on, for example, analysis of a questionnaire, analysis of programs of parties and candidates, previous voting of candidates, etc. The proposed questionnaire, for example, would guide the voter through a sequential question-based algorithm to the proposed vote. In this way, the voter is helped to cast his or her vote for the candidate who best meets his or her requirements.
- The third direction implies assistance to AI on the part of the candidates, politicians, parties, with better review and adjustment of the demands of the voters, i.e. their programs based on the proposals-analysis of the AI are adjusted, and at the same time better target the voters.

The use of AI in the voting process brings many benefits, improvements and increases the effectiveness and democracy of voting (Polonski, 2017), but in addition to the basic technological problems - ensuring secrecy and validity, which are similar to electronic and online voting, carries its own risks and ethical dilemmas, which should be further analysed and addressed.

---

[1] https://medium.com/edge-elections/which-countries-use-online-voting-3f7300ce2f0

6. WEIGHTED VOTING (W-VOTING)

Weighted voting is a concept-idea proposed by the first author of this paper, and represents an opportunity for a step forward in the voting process, supported by technology, towards greater democracy and immediacy of this process.

Weighted voting is based on advanced voting and / or voting with the help of AI. Basically, this technique will be based on electronic / internet / blockchain voting with adding weight to each group of voters, i.e., voters receive a different weight of their vote depending on the proximity to the voting topic or other criteria. In this way, for example, the citizens of the capital can vote for the development of the capital, as well as those who do not live in it, but the first should get more weight for their votes because they have a common capital with other citizens in the country, but additionally they live in it.

From the information technology side, in addition to the stated problems and solutions brought by electronic, internet and blockchain supported voting in this most advanced type of voting, the voter database should be complemented with the database of coefficients that would add or deduct weight to these votes. Coefficient databases can experience a rapid increase from 1 to thousands, similar to the growth of, for example, Google PageRank technology.

7. PUTING E-VOTING IN WIDER CONTEXT

E-voting is considered as a tool for fostering democracy and governance. However, it alone is not sufficient. And in reality, it is far from easy to implement it, despite expertise and political will.

There is no simple recipe. Many factors should be put in play. We will briefly mention two – probably the most important.

First, it is the education. Long-term investing in education is probably the first thing to do. Finland is probably the best example in Europe for how education changes a society – it raises the living standard; it eradicates unemployment and it improves competition. Better education doesn't come automatically with buying computers for every student. The process of education is a complex engagement. For illustration, The OECD's Programme for International Assessment (PISA) evaluates both digital and print reading performance. From education in general it is much is easier to go to political education.

Education is a prerequisite of democracy. As we mentioned before, corruption is linked with less stable democracies.

Second it is fighting the corruption. The fight against corruption must be accompanied with building a culture of lawfulness. E-voting without visible accomplishment in these areas is a weak factor of democracy. This implies only one thing. E-voting must be a part of a comprehensive development framework. E-voting is a concept that should be regarded as a test for digital literacy, culture of lawfulness and success in fight against corruption. The advocates of e-voting must be aware of the social and political context in a given country, triggering the discussions on corruption and lawfulness. So, even opening a discussion in a parliamentary committee, let alone a public debate, is a good sign. It is a sign of a new level of relationship of trust between state and citizens. And to put it succinctly, e-voting is all about trust.

Studies confirms that lower corruption is associated with an increase in the PISA scores across countries. Other indicators like access to education, enrolment, and schooling years show that there is significant relationship between education quality and corruption (De La Cruz Aquino, 2017). Access to education is indispensable for the implementation and sustainability of democracy (Climent, 2006).

Oelkers (2000), referring to Condorcet, argues that modern society needs active critical citizenship, citizens able to exchange arguments in the public arena of politics. The most important virtue in such society is civic courage, and this requires education.

8. CONCLUSION

The voting process is as old as civilization. It is not stationary, but a process that is constantly evolving and improving with the development of technology, social consciousness and democracy. The use of ICT technology in the voting process not only finds increasing application, but also becomes crucial and opens many new opportunities for comprehensive development of society. The benefits it brings are obvious. So, it is necessary to avoid the potential risks, in order to come up with solutions that will enable its smooth and safe use.

Potential risks are not only malware, hacking, attacks on databases of voters and many others security flaws.

Underdevelopment in key areas of a democratic society should be considered as risk, as well.

To put it in other words, e-voting enables voting in controlled privacy in private places. That means that you vote in privacy of your home on your personal computer. You are voting in privacy and you are not threatened or pressurized by another person. Control means that you cannot cast multiple votes, you cannot doodle on the e-paper, you cannot vote for another person etc. E-voting must provide software and security systems to prevent misuse or abuse of privacy. In our opinion that is only one side of the story. Control is not only surveillance and arresting evildoers. There are many democratic risks of vertical, bureaucratic, non-transparent and non-accountable state. E-voting works better in horizontal state with mature culture of lawfulness. Culture of lawfulness together with rule of law provide personal freedom (condition that guards a person from threats and coercion).

E-voting should be a test for advanced culture of lawfulness and measure of the success of fight against corruption. This claim is backed up with knowledge that countries that apply e-voting have lesser level of corruption. As lawfulness is indispensable in fighting corruption, it is indispensable in e-voting systems.

The main thesis in this study is that there is no e-voting in high corrupt states. And that there is a connection between low corruption and developed culture of lawfulness.

From ostracon to e-voting the main question is the same: how to have good life in a good state with good laws. From the answers offered by the Pythagoreans and Plato to today advocates of modern governance it has always been about how to organise a state so that we can live together the best way we can.

ACKNOWLEDGEMENT

This work was partially financed by the Faculty of Computer Science and Engineering at the Ss. Cyril and Methodius University in Skopje.

REFERENCES

Anderson, J. and Rainie, L., 2020. Many experts say digital disruption will hurt democracy. *Pew Research Center. Online: https://www. pewresearch. org/internet/2020/02/21/many-tech-experts-say-digital-disruption-will-hurt-democracy.*

Card, D. and Moretti, E., 2007. Does voting technology affect election outcomes? Touch-screen voting and the 2004 presidential election. *The Review of Economics and Statistics*, *89*(4), pp. 660-673.

Climent, A.C., 2006. *On the Distribution of Education and Democracy* (No. 0602).

De La Cruz Aquino, N., 2017. Correlation Between Corruption and Education in Developing Countries.

Dill, D.L., Schneier, B. and Simons, B., 2003. Voting and technology: Who gets to count your vote?. *Communications of the ACM*, *46*(8), pp. 29-31.

Dunn, M. and Merkle, L., 2018, March. Overview of Software Security Issues in Direct-Recording Electronic Voting Machines. In *Proceedings of the ICCWS 2018 13th International Conference on Cyber Warfare and Security, Washington, DC, USA* (pp. 8-9).

Evan, W.M., 2004. Voting technology, political institutions, legal institutions and civil society: a study of the hypothesis of cultural lag in reverse. *History and technology*, *20*(2), pp.165-183.

Gibson, J.P., Krimmer, R., Teague, V. and Pomares, J., 2016. A review of e-voting: the past, present and future. *Annals of Telecommunications*, *71*(7), pp.279-286.

Hao, F. and Ryan, P.Y. eds., 2016. *Real-world electronic voting: Design, analysis and deployment*. CRC Press.

Herrnson, P.S., Niemi, R.G., Hanmer, M.J., Bederson, B.B., Conrad, F.G. and Traugott, M.W., 2009. *Voting technology: The not-so-simple act of casting a ballot*. Brookings Institution Press.

Herrnson, P.S., Niemi, R.G., Hanmer, M.J., Francia, P.L., Bederson, B.B., Conrad, F.G. and Traugott, M.W., 2008. Voters' evaluations of electronic voting systems: Results from a usability field study. *American Politics Research*, *36*(4), pp. 580-611.

Huang, J., He, D., Obaidat, M.S., Vijayakumar, P., Luo, M. and Choo, K.K.R., 2021. The application of the blockchain technology in voting systems: A review. *ACM Computing Surveys (CSUR)*, *54*(3), pp.1-28.

Jefferson, D., Rubin, A.D., Simons, B. and Wagner, D., 2004. Analyzing internet voting security. *Communications of the ACM*, *47*(10), pp. 59-64.

Katz, G., Alvarez, R.M., Calvo, E., Escolar, M. and Pomares, J., 2011. Assessing the impact of alternative voting technologies on multi-party elections: Design features, heuristic processing and voter choice. *Political Behavior*, *33*(2), pp. 247-270.

Kersting, N. and Baldersheim, H. eds., 2004. *Electronic voting and democracy: a comparative analysis*. Springer.

Kooiman, J. ed., 1993. *Modern governance: new government-society interactions*. Sage.

Krimmer, R., 2012. The evolution of e-voting: why voting technology is used and how it affects democracy. *Tallinn University of Technology Doctoral Theses Series I: Social Sciences*, *19*.

Krimmer, R., 2019. A structure for new voting technologies: what they are, how they are used and why. In *The Art of Structuring* (pp. 421-426). Springer, Cham.

Kumar, D.A. and Begum, T.U.S., 2012, March. Electronic voting machine—a review. In *International Conference on Pattern Recognition, Informatics and Medical Engineering (PRIME-2012)* (pp. 41-48). IEEE.

Lindner, R., Aichholzer, G. and Jennen, L., 2016. Electronic democracy in Europe. *Prospects and challenges of e-publics, e-participation and e-voting. Cham: Springer*.

Oelkers, J., 2000. Demokratie und Bildung: über die Zukunft eines Problems. *Zeitschrift für Pädagogik*, *46*(3), pp. 333-347.

Oo, H.N. and Aung, A.M., 2014. A survey of different electronic voting systems. *International Journal of Scientific Engineering and Technology Research*, *3*(16), pp.3460-3464.

Polonski, V., 2017. The Good, the Bad and the Ugly Uses of Machine Learning in Election Campaigns. *Centre for Public Impact*.

Parkin, M., 2008. *Microeconomics*. Pearson Education.

Robertson, L., 2006. One Man One Vote: Trust between the Electorate, the Establishment, and Voting Technology. *Journal of Technology Studies*, *32*(2), pp.85-89.

Roisman, J., 2011. *Ancient Greece from Homer to Alexander: the evidence* (Vol. 10). John Wiley & Sons.

Stewart III, C., 2011. Voting technologies. *Annual Review of Political Science*, *14*, pp. 353-378.

Susskind, J., 2017. Decrypting democracy: Incentivizing blockchain voting technology for an improved election system. *San Diego L. Rev.*, *54*, p.785.

Tarasov, P. and Tewari, H., 2017. The future of e-voting. *IADIS International Journal on Computer Science & Information Systems*, *12*(2).

Weare, C., 2002. The Internet and democracy: The causal links between technology and politics. *International Journal of Public Administration*, *25*(5), pp. 659-691.

Yakobson, A., 1995. Secret ballot and its effects in the late Roman Republic. *Hermes*, *123*(H. 4), pp.426-442.

ARTIFICIAL INTELLIGENCE AND GENDER EQUALITY: A SYSTEMATIC MAPPING STUDY

J. David Patón-Romero[1], Ricardo Vinuesa[2], Letizia Jaccheri[1] and Maria Teresa Baldassarre[3]
[1]Department of Computer Science, Norwegian University of Science and Technology (NTNU), Sem Sælands Vei 7, 7034 Trondheim, Norway
[2]FLOW, Engineering Mechanics, KTH Royal Institute of Technology, SE-100 44 Stockholm, Sweden
[3]Department of Informatics, University of Bari "Aldo Moro" (UniBa), Via Edoardo Orabona, 4, 70126 Bari, Italy

ABSTRACT

Sustainability is not only understood as a manner to safeguard the environment, but also to fight against injustices and inequalities that exist on social and economic level. One of the biggest challenges that exists in social sustainability is to achieve gender equality, as defended by the Sustainable Development Goal (SDG) 5 of the 2030 Agenda. But this is a complex challenge and must be addressed from different spheres and fields of knowledge. Artificial Intelligence (AI) has proven to be an essential asset in the development of new and innovative technologies. Its development, adoption, and constant use by a growing part of the world's population demonstrates the social impact it entails and the importance of also becoming an asset for social sustainability and, in this case especially, for gender equality.
That is why this study aims to collect the current knowledge about the fields of AI and gender equality, through the development of a Systematic Mapping Study (SMS) that identifies the most significant advances in this regard, as well as the main gaps that must be covered. The results and findings obtained in this work show the novelty of joint analysis of both areas, as well as increasing attention they have received in recent years. Likewise, they also demonstrate the need to address specific and urgent issues within gender equality, both in the field of AI and caused by its development.

KEYWORDS

Artificial Intelligence, Gender Equality, Social Sustainability, Systematic Mapping Study

1. INTRODUCTION

Sustainability has become a key actor for the development and advancement of civilization. However, many times it is only interpreted as an environmental characteristic and other perspectives such as social and economic sustainability are ignored (Purvis et al., 2019). These three perspectives go hand in hand and must be addressed together, since, for example, it is not possible to aim to achieve sustainable development by and for the environment if changes are not conducted in society itself to support it.

According to the findings identified by (Harari, 2018), sustainability and Artificial Intelligence (AI) are two of the biggest challenges faced by humanity. From Information Technology (IT), AI has become one of the most relevant and innovative fields (Carleton et al., 2020; Menzies, 2019). The unstoppable progress that AI is experiencing demonstrates the importance of pursuing AI applications that can help to achieve a sustainable development and use in this regard (Nishant et al., 2020; Vinuesa et al., 2020). Thus, it is essential to relate both fields and work to achieve sustainability in and by AI. However, the focus has always been on the relationship between AI and technology in general with energy/climate neutrality (i.e., environmental sustainability) and marginally on its relationship with social and economic sustainability (Becker, 2015). Some of these aspects are discussed by (Vinuesa & Sirmacek, 2021).

Thus, this study aims to shed light into the issue and show the current relationship between AI and social sustainability, focusing on gender equality, which is one of the Sustainable Development Goals (SDGs) established by the 2030 Agenda (United Nation, 2015) which is having more focus and relevance regarding social sustainability (Rosa, 2017).

Current research shows that women are underrepresented in technology research, practice, and education (Albusays et al., 2021). Likewise, gender imbalance in technology has been seen as harming the economy, as highlighted by the *European Commission* when identifying an annual productivity loss of 16 billion Euro for the European economy (European Commission, 2018). In the same way, the OECD (*Organization for*

Economic Co-operation and Development) states that *"greater inclusion of women in the digital economy and increased diversity bring value, both social and economic"* (OECD, 2018).

Therefore, the present literature analysis through a Systematic Mapping Study in the areas of AI and gender equality will be useful, since it will allow knowing the latest knowledge and establishing the pillars that will guide the development of new and innovative research and ideas in this regard. Thanks to a greater understanding of the interplay between AI and gender equality, it will it be possible to understand the changes and challenges that exist towards achieving a sustainable development through the SDGs.

The rest of this study is organized as follows: Section 2 contains the background about 2030 Agenda, gender equality, and AI; Section 3 presents the research methodology followed to analyze the state of the art in the fields of gender equality and AI; Section 4 shows the results obtained from the analysis performed; Section 5 discusses the findings, limitations, and implications that have been reached; finally, Section 6 contains the conclusions and lines for future work in this regard. Likewise, Appendix A includes the list of primary studies selected during the analysis of the state of the art; and Appendix B shows the answers to the established research questions from each of these primary studies.

2. BACKGROUND

2.1 2030 Agenda & Sustainable Development Goal 5

The 2030 Agenda (United Nation, 2015) is an initiative promoted and agreed upon by the 193 Member States of the United Nations (UN) with the aim to achieve the so-called Sustainable Development Goals (SDGs). This includes a total of 17 Goals and 169 Targets that address the three pillars of sustainability (environmental, social, and economic) (Purvis et al., 2019), including areas such as climate change, economic inequality, innovation, natural resources consumption, peace, and justice, among other priorities. Likewise, for each of the Targets there is also a set of indicators that make it possible to measure the progress made in this regard (United Nations, 2017).

Among these SDGs, this study aims to focus on Goal 5 (Gender Equality), one of the Goals belonging to the field of social sustainability. The main purpose of this Goal is to *"achieve gender equality and empower all women and girls"*, for which it establishes 9 Targets (United Nation, 2015).

It is important to highlight that the 2030 Agenda identifies two types of Targets within the SDGs: 1) "outcome" Targets (i.e., circumstances to be attained), labeled by numbers; and 2) "means of implementation" Targets, labeled by lower case letters.

2.2 Gender Equality & Artificial Intelligence

In order to achieve gender equality, as defended by SDG 5 of the 2030 Agenda, actions must be conducted in all areas of knowledge. AI has proven to be an increasingly important actor in the development of new and innovative systems used by all levels of society (Lu et al., 2018). That is why it is vital that the entire life cycle of these systems is committed to achieving a better society and, therefore, gender equality must play an important role in this regard.

In general terms, it could be said that the main objective of AI in social sustainability is "the study and practice of design, build and use of AI systems with a positive impact on the society". However, when it comes to relating the terms of gender equality and AI, there is no clear criterion or definitions per se. To establish this relationship, the definitions and same logic as that used for the terms *Green by IT* and *Green in IT* (idea proposed in (Erdélyi, 2013)) will be followed, which defend sustainability in and by IT. Thus, in gender equality and AI we are faced with two perspectives:

- ***Gender by AI***: in which AI provides the necessary tools to achieve gender equality through different contexts (i.e., AI as an enabler).
- ***Gender in AI***: in which AI itself produces a negative impact on gender equality (e.g., lack of balance during the development of a system) and, therefore, said impact must be reduced (i.e., AI as a producer).

3. RESEARCH METHODOLOGY

A Systematic Mapping Study (SMS) is a research method used to collect, analyze, and categorize existing information from a specific context. In the specific case of this study, the guidelines established by (Kitchenham, 2007) have been followed, adopting also the lessons learned for the data extraction and analysis identified by (Brereton et al., 2007), and considering examples of application of SMSs in Software Engineering such as (Petersen et al., 2008). Thus, the characteristics established during the planning stage are shown below, as well as how the execution stage was conducted.

3.1 Planning Stage

3.1.1 Research Questions

The main goal of this study is to inspect the current state and existing relationship between the fields of artificial intelligence and gender equality. In this way, it is intended to collect and categorize all the information in this regard and identify the gaps that exist in order to develop new research proposals. To do this, the research questions (RQs) shown in Table 1 have been established.

Table 1. Research questions

Research question	Motivation
RQ1. What kind of studies exist on AI and gender equality?	Determine the type, number of publications, and trend over recent years in relation to AI and gender equality.
RQ2. What gender equality Targets are addressed in and by AI?	Determine what gender equality Targets are addressed in/by AI to identify possible opportunities and threats.
RQ3. What kind of AI proposals exist to address gender equality?	Determine the AI proposals that exist to address gender equality to identify trends and possible gaps in or by AI.

3.1.2 Search Strategy

As a strategy for search the relevant studies and information, the *Scopus* database will be used. To this end, we decided to conduct a general search and a search for each of the Targets identified by the SDG 5 of the 2030 Agenda (United Nation, 2015) (i.e., 10 searches). In this way, specific terms of each Target can be addressed in more detail and the identification of studies in this regard is facilitated. Thus, Table 2 shows the search strings that will be used. As can be seen, these search strings are divided into two main parts (the two contexts within the scope of this study).

Table 2. Search strings

Scope	Search string
General	("Artificial Intelligence" OR AI) AND (Gender OR "Women rights" OR "Social sustainability" OR "SDG 5")
Target 5.1	("Artificial Intelligence" OR AI) AND ((Women OR Girls OR Gender) AND Discrimination)
Target 5.2	("Artificial Intelligence" OR AI) AND ((Women OR Girls OR Gender) AND (Violence OR Exploitation OR Trafficking))
Target 5.3	("Artificial Intelligence" OR AI) AND (((Women OR Girls OR Gender) AND "Harmful practices") OR ((Child OR Early OR Forced) AND Marriage) OR "Genital mutilation")
Target 5.4	("Artificial Intelligence" OR AI) AND ("Care work" OR "Domestic work" OR "Social protection policies" OR "Shared responsibility")
Target 5.5	("Artificial Intelligence" OR AI) AND ((Women OR Girls OR Gender) AND ("Equal opportunities" OR Participation OR Leadership))
Target 5.6	("Artificial Intelligence" OR AI) AND ((Sexual OR Reproductive) AND (Health OR Rights))
Target 5.a	("Artificial Intelligence" OR AI) AND ((Women OR Girls OR Gender) AND Equal* AND Rights)
Target 5.b	("Artificial Intelligence" OR AI) AND ((Women OR Girls) AND Technology)
Target 5.c	("Artificial Intelligence" OR AI) AND ((Women OR Girls OR Gender) AND (Equal* OR Empower*))

These search strings will be applied to the title, abstract and keywords of the studies. Likewise, publications from 2010 and onwards will be considered, since it has been during the last decade when, mainly, the area of gender equality has had its momentum.

3.1.3 Selection Criteria

All the documents and information collected through the searches will be analyzed considering the title, abstract, and keywords of each one. This will determine which studies will be included for a more detailed analysis. To do this, on the one hand, those studies that meet the following inclusion criteria will be considered for further analysis:

- **I1.** Studies in English dealing with AI and gender equality.
- **I2.** Studies published between 2010 and 2021 in journals, conferences, and/or workshops, with peer review process.

On the other hand, the studies that meet any of the following exclusion criteria will be automatically discarded:

- **E1.** Discussion or opinion studies, as well as those that are only available as abstract or presentation.
- **E2.** Duplicate studies (in which case will be considered the most complete and recent).
- **E3.** Studies whose main contribution is not related to AI and gender equality, or where AI and gender equality are not related to each other.

In the same way, the snowballing effect (Wohlin, 2014) will be followed, so the documents referenced in the considered studies will also be evaluated for their possible inclusion.

3.1.4 Quality Assessment Criteria

One of the most critical points to obtain representative and relevant results and references for future research is the quality assessment of the studies. To do this, the following issues have been established that will be analyzed following a scoring system of three values (-1, 0, +1), generating a quality result for each study between -4 and +4:

a. The study presents a detailed description and guidance on how AI can contribute to gender equality.
   Yes (+1); Partially (0); No (-1).
b. The study validates the proposal or idea that it defends.
   Empirically validated (+1); Theoretically validated (0); Not validated (-1).
c. The study has been published in a relevant journal[1]/conference[2].
   High ranking (+1); Medium ranking (0); Low ranking or not indexed (-1).
d. The study has been cited by other authors in publications.
   More than five cites (+1); Between one and four cites or recently published in 2021 (0); Not cited (-1).

3.1.5 Data Extraction

A series of answers have been established for each of the research questions (as shown in Table 3). In this way, the same data extraction criteria will be applied to all studies, facilitating their analysis and categorization.

Table 3. Classification schema

Research question	Answers		
RQ1. What kind of studies exist on AI and gender equality?*	**a.** State of the art analysis **b.** Proposal	**c.** Validation **d.** Others	
RQ2. What gender equality Targets are addressed in and by AI?**	**a.** Target 5.1 **b.** Target 5.2 **c.** Target 5.3	**d.** Target 5.4 **e.** Target 5.5 **f.** Target 5.6	**g.** Target 5.a **h.** Target 5.b **i.** Target 5.c
RQ3. What kind of AI proposals exist to address gender equality?	**a.** *Gender by AI* **b.** *Gender in AI*		

*The answers to RQ1 follow the idea of the example of (Petersen et al., 2008).
**The answers to RQ2 have their origin in the Targets of the SDG 5 from the 2030 Agenda (United Nation, 2015).

3.1.6 Synthesis Methods

A both quantitative and qualitative synthesis of data will be conducted related to the answers to the research questions and the quality evaluations performed, respectively. These syntheses will be represented by tables and/or graphs with the results in a matter of numbers and/or percentages, as well as bubble plots to analyze how the research questions are related through their answers.

[1] Following the Journal Citation Reports (JCR): https://jcr.clarivate.com/
[2] Following the GII-GRIN-SCIE Conference Rating: https://scie.lcc.uma.es/

3.2 Execution Stage

In order to apply the protocol established during the planning stage, three main phases have been followed during the execution stage:

- **First phase.** Based on the identification of potential studies. To do this, first, after performing the 10 searches applying the search strings (cf. Table 2) on the *Scopus* database, 3,558 studies were obtained. Then, the selection criteria were applied to these studies, considering the abstract of each one, and 169 potential studies were obtained.
- **Second phase.** Oriented to the identification of primary studies, by means of which the selection criteria were applied again, but this time on the complete content of each of the 169 potential studies. As a result, 29 primary studies were obtained.
- **Third phase.** This last phase is dedicated to the compilation of results, for which the characterization of the primary studies was performed through the answers to the research questions and the obtaining of the main findings, as well as the quality assessment of said studies.

4. RESULTS

The general results obtained after the execution of the SMS are shown below, answering each of the established research questions. It is important to highlight that, related to these results, Appendix A includes the list of references of primary studies, while Appendix B contains a summary table with the mapping of the answers to the research questions of each of these primary studies.

4.1 RQ1. What Kind of Studies Exist on AI and Gender Equality?

The main objective of this RQ is based on identifying the type of studies that currently exist in AI and gender equality. In this way, it is possible to determine which are the most relevant studies when, e.g., conducting a new proposal in this regard or considering a specific proposal that is validated for the application of a case.

Our results (represented in Figure 1) show that about 24% of the studies found (S05, S08, S09, S11, S17, S18, and S21) are based on or contain some analysis of the state of the art on the field that concerns us.

Likewise, 69% of the studies (20 in total) deal with specific proposals to address gender equality from some point in and by AI. However, of all these proposals only 11 are validated by some empirical case (S01, S02, S04, S10, S13, S16, S22, S23, S27, S28, and S29).

It is also important to highlight 2 studies (S03 and S06), which are based on experiments aimed at assessing the impact of gender in and by AI.

Finally, related to this research question, it is also important to analyze the evolution of the studies over the last few years. Figure 2 shows how this progression has been, through which it can be seen that mainly in the last 2 years there has been a boom in publications in this regard. This is because it is in recent years when more efforts are beginning to be made to achieve the objectives of the 2030 Agenda (United Nation, 2015).

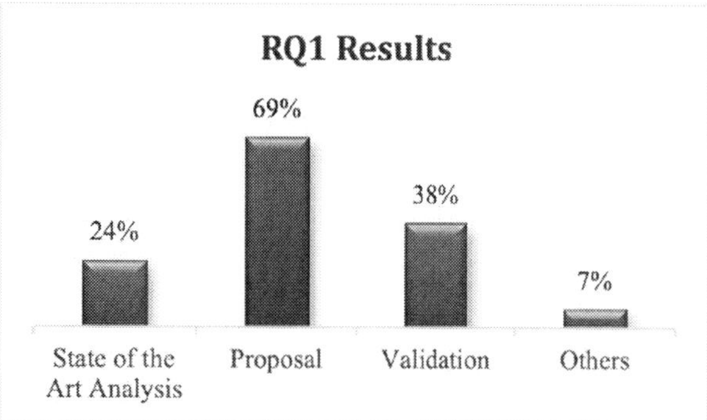

Figure 1. Results for the RQ1 (percentage of studies in each of the four categories)

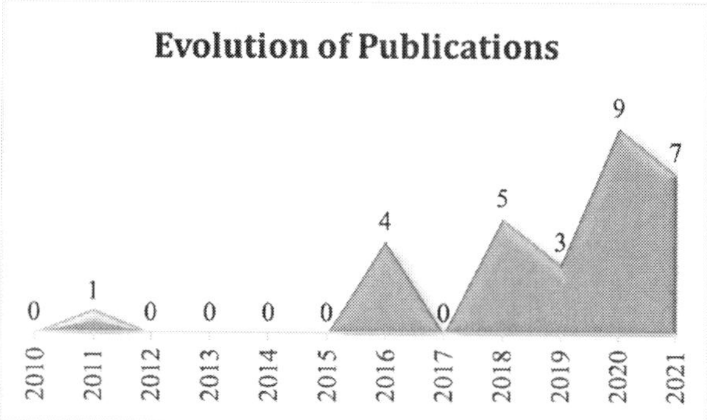

Figure 2. Evolution of the number of publications on gender equality and AI in recent years

4.2 RQ2. What Gender Equality Targets are Addressed in and by AI?

This research question is the main one of the present study, since its objective is to analyze and map the studies according to the Target(s) of the SDG 5 from the 2030 Agenda (United Nation, 2015) that address. In this way, it is possible to identify which are the Targets that are usually dealt with in and by AI, as well as the possible Targets that are not yet covered and need development in this regard.

From the results obtained (represented in Figure 3), there are only studies that address 5 of the 9 Targets established. Targets 5.3, 5.4, 5.5, and 5.a are not covered in any of the studies, so there is no evidence on possible developments in AI that address gender equality in the specific contexts of these Targets (it will be discussed later in detail).

Regarding the Targets that are covered, Target 5.1 is the one with the most development in the AI area, since 52% of the studies (15 in total) base their objective on addressing the context of this one. Likewise, following a decreasing order, Target 5.2 is found in 34% of the studies (10), Target 5.6 in 17% (5), Target 5.b in 17% (5), and Target 5.c in 7% (2).

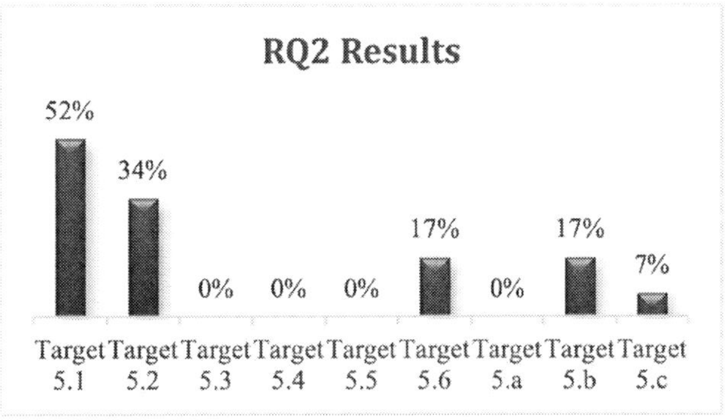

Figure 3. Results for the RQ2 (percentage of studies addressing each of the Targets within SDG 5)

4.3 RQ3. What Kind of AI Proposals Exist to Address Gender Equality?

The last of the research questions is focused on knowing the studies that are oriented both to achieve gender equality in different areas through the use of AI (*Gender by AI*) and to implement measures that help achieve better gender equality issues in AI itself (*Gender in AI*). In this regard, the results show equality in both perspectives, since 18 studies deal with *Gender by AI* and 13 with *Gender in AI*.

First, the studies S01, S02, S03, S04, S05, S06, S07, S10, S12, S13, S14, S15, S16, S23, S24, S25, S26, and S29 include some proposal or characteristics related to the context of *Gender by AI*. Second, *Gender in AI* is addressed in the studies S03, S05, S08, S09, S11, S17, S18, S19, S20, S21, S22, S27, and S28. And, finally, it should be noted that the studies S03 and S05 consider both perspectives.

Thus, 62% of the studies are framed in the perspective of *Green by AI*, while 45% deal with *Green in AI*.

5. DISCUSSION

5.1 Principal Findings

The main objective of this study is based on knowing the state of the art and the current relationship between gender equality and AI. In this way, it is intended to identify important aspects and gaps that help to develop new innovative ideas in this regard. After analyzing the results, the following observations can be made:

- **Focus on *Gender by AI*.** Although there is a high level of equality in the number of studies that deal with *Gender by AI* and *Gender in AI*, in recent years the main focus has been the development of studies related to *Gender by AI*. This could be due to the greater ease of developing a specific technology to address an aspect related to gender equality (such as, e.g., an AI device that detects dangerous situations for women) than, e.g., change the business/management processes that guide organizations when developing new AI proposals so that they follow a set of best practices that respect gender equality. Undoubtedly, the latter is more complicated, because it is not only necessary to understand both fields to develop useful and applicable best practices for most contexts, but also a high number of practical cases and the involvement of external actors who allow validation of these practices are needed. Therefore, the fact that the focus is currently on *Gender by AI* is an issue that can generate a lot of controversy, since the "cart is being put before the horse" and the question arises as to whether AI proposals that help gender equality can actually be developed, when in AI itself and in all the processes that surround it (i.e., the basis) there is no such equality.

- **Inequality when dealing with the Targets.** It is a normal result that there is a difference in the number of studies when addressing different contexts, but, in this case, the difference is quite large. To improve understanding, we can talk about three groups:
  - *Advanced development:* the Target 5.1 is the only one in this group, since it is the most addressed by the studies found. This is because it is a fairly general Target whose objective is *"end all forms of discrimination against all women and girls everywhere"* (United Nation, 2015). In general terms, the main purpose of SDG 5 of the 2030 Agenda is the one that defines this Target and that is why most studies tend to focus on it. However, we must not forget that there are other Targets with more specific purposes and that they must be addressed urgently.
  - *Medium development:* the Targets 5.2, 5.6, 5.b, and 5.c are found in this group. It is always relevant to find evidence that supports, in this case, the specific context of each of the Targets. However, the evidence is quite scarce, and it is necessary to continue developing new ideas, as well as improving the current ones. From a practical point of view, following the evidence found, these Targets can be addressed in a simple way in and by AI. For example, Target 5.2 aims to *"eliminate all forms of violence against all women and girls in the public and private spheres, including trafficking and sexual and other types of exploitation"* (United Nation, 2015), which has been shown to be easily addressed through specific AI proposals that analyze various situations to identify potential dangers affecting the integrity of women and girls. We are on the right path, but we must not get lost and continue dedicating efforts to develop proposals in and by AI in these Targets.
  - *Null development:* the Targets 5.3, 5.4, 5.5, and 5.a are not addressed by any study. This may be due to the difficulty of identifying an idea in and/or by the AI that effectively and efficiently addresses the specific contexts of these Targets. For example, regarding the Target 5.3 (*"eliminate all harmful practices, such as child, early and forced marriage and female genital mutilation"* (United Nation, 2015)), it is difficult to understand that AI can do something about it, since it deals with behaviors rooted in certain cultures and the application field of AI may not be sufficient to adequately address such a cultural change. However, the Target 5.a (*"undertake reforms to give women equal rights to economic resources, as well as access to ownership and control over land and other forms of property, financial services, inheritance and natural resources, in accordance with national laws"* (United Nation, 2015)) can be applied in the field of AI through, e.g., systems that analyze the profiles of certain candidates for obtaining economic resources of different kinds, without any type of bias related to the gender of each person. For this reason, all the Targets of the SDG 5 from the 2030 Agenda are equally important, but it is necessary and transcendental to guide and begin to dedicate efforts to develop these Targets that have not yet been explored, in order to propose ideas in and by AI that allow progress in gender equality in this regard.
- **Low number of practical cases.** When developing a proposal, it is essential to conduct practical cases that validate it and demonstrate its applicability, quality, effectiveness, and efficiency when addressing the objective for which it was developed. However, of the 20 proposals identified through the SMS, only 11 (i.e., 55%) have been validated. This supposes a too low number of validated proposals, since all or the vast majority should have been applied in some practical case, showing that they are complete and serious proposals. For this reason, it is extremely important that any development conducted in and/or by AI to, in this case, address some Target(s) of SDG 5 from the 2030 Agenda, be accompanied by a practical application and validation.
- **Lack of joint development of both fields.** Although there is evidence on the development of ideas in and by AI to address the specific context of gender equality, there is very little. This, together with the analysis of the progression of publications over the last few years, demonstrates the novelty of this field and the growing interest in conducting new research in this regard. It is very important to closely follow this progression, as well as analyze the new studies that arise and develop new ideas that contribute to this research field so important and necessary to achieve gender equality, improving the field of social sustainability and complying with the provisions of the 2030 Agenda.

5.2 Limitations

Although an attempt has been made to design this study to avoid or mitigate the possible limitations (such as the case of performing a general search and 9 searches for each of the Targets, with the aim to find studies with very specific terminology for certain contexts), there are always limitations that can affect when identifying and analyzing results and findings.

It should not be forgotten that the present analysis comes from the perspective of the authors and may not be interpreted in the same way by other researchers or professionals in the field. Likewise, it is possible that certain literature on the field has been overlooked, or even that some more recent evidence on the studies found has not yet been published at the time of the SMS execution. Consequently, to mitigate the risks in this regard, several authors analyzed and interpreted the data and results obtained here, contributed to the final consensus, thus reducing the bias among each other.

5.3 Implications

The development of this study has a high implication and significance for the fields in which it is found. As has been shown, there are few studies that put the fields of AI and gender equality in common. Thus, thanks to this study, not only the state of the art in this regard is identified, but also the gaps and possible lines of research that improve existing studies or that address new and innovative ideas not considered until now.

In section "0 5.1 Principal FINDINGS" a discussion has been conducted in which different lines of future work/research are identified. These can be used by researchers who are in the fields of AI, IT in general, gender equality and social sustainability. Therefore, this study is a necessary starting point and the demonstration of the importance of the fields that concern us, which will attract new researchers and professionals to the development of new proposals with the goal of achieving gender equality in and by AI.

6. CONCLUSIONS AND FUTURE WORK

The increasing use of technology and AI by a wide range of people around the world shows that they must be driven by and for the whole of society, avoiding gender, culture, religion, and other kinds of discrimination. However, women and other vulnerable and discriminated minorities are underrepresented in this regard and the progress to get around this situation is slow and scant (Adams & Khomh, 2020; Albusays et al., 2021).

That is why this study is focused on analyzing the state of the art in the fields of AI and gender equality. On the one hand, AI is becoming a fundamental field for the development of new and innovative technologies, so it is vital that it represents a positive asset for sustainability (Harari, 2018; Nishant, 2020). On the other hand, gender equality must be addressed in all fields without exception, and, above all, it needs a boost with new ideas and proposals in the field of technology and AI (European Commission, 2018; OECD, 2018).

Through the results presented here, not only the current status in this regard has been identified, but also a series of problems and gaps that must be addressed. The novelty of this work has been demonstrated, due to the small number of studies in this area, as well as the large increase in studies and the growing importance that these fields are taking in recent years. Therefore, it is necessary to continue with this momentum and address the gaps that exist by developing proposals and empirical validations that cover the different specific contexts of the Targets identified by the SDG 5 of the 2030 Agenda (United Nation, 2015) both in and by AI.

Thus, as future work, we are conducting new studies on gender equality in different areas related to technology, such as IT processes and entrepreneurship in the IT sector, in order to identify different points of view and links that help together to develop new and better proposals to address gender equality in this regard. Likewise, we also intend to develop a framework of best practices that establish the bases for the development, validation, evaluation, and improvement of proposals for both *Gender by AI* and *Gender in AI*. In this way, we want to facilitate and promote these fields both at the research and professional level in organizations.

A society unable to change will not generate any progress. Let us be the change our society needs, promoting new and inclusive ideas for all humankind.

ACKNOWLEDGEMENT

This work is result of a postdoc from the ERCIM "Alain Bensoussan" Fellowship Program conducted at the Norwegian University of Science and Technology (NTNU). This research work is also part of the COST Action - European Network for Gender Balance in Informatics project (CA19122), funded by the Horizon 2020 Framework Programme of the European Union. Likewise, RV acknowledges the financial support of the Swedish Research Council (VR).

REFERENCES

Adams, B. & Khomh, F., 2020. The Diversity Crisis of Software Engineering for Artificial Intelligence. *IEEE Software*, 37(5), pp. 104-108.

Albusays, K., Bjorn, P., Dabbish, L., Ford, D., Murphy-Hill, E., Serebrenik, A., & Storey, M., 2021. The Diversity Crisis in Software Development. *IEEE Software*, 38(2), pp. 19-25.

Becker, C., Betz, S., Chitchyan, R., Duboc, L., Easterbrook, S. M., Penzenstadler, B., Seyff, N., & Venters, C. C., 2015. Requirements: The Key to Sustainability. *IEEE Software*, 33(1), pp. 56-65.

Brereton, P., Kitchenham, B. A., Budgen, D., Turner, M., & Khalil, M., 2007. Lessons from applying the systematic literature review process within the software engineering domain. *Journal of Systems and Software*, 80(4), pp. 571-583.

Carleton, A. D., Harper, E., Menzies, T., Xie, T., Eldh, S., & Lyu, M. R., 2020. The AI Effect: Working at the Intersection of AI and SE. *IEEE Software*, 37(4), pp. 26-35.

Erdélyi, K., 2013. Special factors of development of green software supporting eco sustainability. *Proc. of IEEE 11th International Symposium on Intelligent Systems and Informatics (SISY 2013)*. Subotica, Serbia, pp. 337-340.

European Commission, 2018. *Women in the Digital Age*. SMART 2016/0025. European Commission, Brussels, Belgium.

Harari, Y. N., 2018. *21 Lessons for the 21st Century*. Random House, New York, NY, USA.

Kitchenham, B., 2007. *Guidelines for Performing Systematic Literature Reviews in Software Engineering* (Version 2.3). EBSE Technical Report. Keele University, Keele, UK.

Lu, H., Li, Y., Chen, M., Kim, H., & Serikawa, S., 2018. Brain Intelligence: Go Beyond Artificial Intelligence. *Mobile Networks and Applications*, 23(2), pp. 368-375.

Menzies, T., 2019. The Five Laws of SE for AI. *IEEE Software*, 37(1), pp. 81-85.

Nishant, R., Kennedy, M., & Corbett, J., 2020. Artificial intelligence for sustainability: Challenges, opportunities, and a research agenda. *International Journal of Information Management*, 53, pp. 102104.

OECD, 2018. *Bridging the Digital Gender Divide. Include, Upskill, Innovate*. Organization for Economic Co-operation and Development, Paris, France.

Petersen, K., Feldt, R., Mujtaba, S., & Mattsson, M., 2008. Systematic Mapping Studies in Software Engineering. *Proc. of 12th Int. Conference on Evaluation and Assessment in Software Engineering (EASE 2008)*. Bari, Italy, pp. 68-77.

Purvis, B., Mao, Y., & Robinson, D., 2019. Three pillars of sustainability: in search of conceptual origins. *Sustainability Science*, 14(3), pp. 681-695.

Rosa, W., 2017. Goal 5. Achieve Gender Equality and Empower All Women and Girls. *In Rosa, W. (eds.), 2017. A New Era in Global Health: Nursing and the United Nations 2030 Agenda for Sustainable Development. Springer Publishing Company, New York, NY, USA*, pp. 301-307.

United Nations, 2015. Transforming Our World: The 2030 Agenda for Sustainable Development. *Seventieth Session of the United Nations General Assembly*. Resolution A/RES/70/1. United Nations, New York, NY, USA.

United Nations, 2017. Global indicator framework for the Sustainable Development Goals and targets of the 2030 Agenda for Sustainable Development. *Seventy-first Session of the United Nations General Assembly*. Resolution A/RES/71/313. United Nations, New York, NY, USA.

Vinuesa, R., Azizpour, H., Leite, I., Balaam, M., Dignum, V., Domisch, S., Felländer, A., Langhans, S. D., Tegmark, M., & Nerini, F. F., 2020. The Role of Artificial Intelligence in Achieving the Sustainable Development Goals. *Nature Communications*, 11(1), pp. 1-10.

Vinuesa, R. & Sirmacek, B., 2021. Interpretable deep-learning models to help achieve the Sustainable Development Goals. *Nature Machine Intelligence*, 3(11), pp. 926.

Wohlin, C., 2014. Guidelines for Snowballing in Systematic Literature Studies and a Replication in Software Engineering. *Proc. of 18th Int. Conf. on Evaluation and Assessment in Software Engineering (EASE 2014)*. London, UK, pp. 1-10.

APPENDIX A. PRIMARY STUDIES SELECTED

Table 4. Primary studies selected

ID	Reference
S01	Hossain, N., Ovi, J. H., Tasnim, S., Islam, N., & Zishan, S. R., 2021. Design and development of Wearable multisensory smart device for human safety. *Proc. of 2021 IEEE International IOT, Electronics and Mechatronics Conference (IEMTRONICS 2021)*. Toronto, Canada, pp. 1-7.
S02	Islam, R., Keya, K. N., Zeng, Z., Pan, S., & Foulds, J., 2021. Debiasing Career Recommendations with Neural Fair Collaborative Filtering. *Proc. of Web Conference 2021 (WWW '21)*. Ljubljana, Slovenia, pp. 3779-3790.
S03	Winkle, K., Melsión, G. I., McMillan, D., & Leite, I., 2021. Boosting Robot Credibility and Challenging Gender Norms in Responding to Abusive Behaviour: A Case for Feminist Robots. *Proc. of Companion of the 2021 ACM/IEEE International Conference on Human-Robot Interaction (HRI '21 Companion)*. Boulder, CO, USA, pp. 29-37.
S04	Miranda, J. A., Canabal, M. F., Gutiérrez-Martín, L., Lanza-Gutierrez, J. M., Portela-García, M., & López-Ongil, C., 2021. Fear Recognition for Women Using a Reduced Set of Physiological Signals. *Sensors*, 21(5), pp. 1587.
S05	Guevara-Gómez, A., de Zárate-Alcarazo, L. O., & Criado, J. I., 2021. Feminist perspectives to artificial intelligence: Comparing the policy frames of the European Union and Spain. *Information Polity*, 26(2), pp. 173-192.
S06	Tomalin, M., Byrne, B., Concannon, S., Saunders, D., & Ullmann, S., 2021. The practical ethics of bias reduction in machine translation: Why domain adaptation is better than data debiasing. *Ethics and Information Technology*, 23, pp. 1-15.
S07	Siristatidis, C., Stavros, S., Drakeley, A., Bettocchi, S., Pouliakis, A., Drakakis, P., Papapanou, M., & Vlahos, N., 2021. Omics and Artificial Intelligence to Improve In Vitro Fertilization (IVF) Success: A Proposed Protocol. *Diagnostics*, 11(5), pp. 743.
S08	Cirillo, D., Catuara-Solarz, S., Morey, C., Guney, E., Subirats, L., Mellino, S., Gigante, A., Valencia, A., Rementeria, M. J., Chadha, A. S., & Mavridis, N., 2020. Sex and gender differences and biases in artificial intelligence for biomedicine and healthcare. *NPJ Digital Medicine*, 3(81), pp. 1-11.
S09	Cernadas, E. & Calvo-Iglesias, E., 2020. Gender perspective in Artificial Intelligence (AI). *Proc. of Eighth International Conference on Technological Ecosystems for Enhancing Multiculturality (TEEM'20)*. Salamanca, Spain, pp. 173-176.
S10	Peña, A., Serna, I., Morales, A., & Fierrez, J., 2020. Bias in Multimodal AI: Testbed for Fair Automatic Recruitment. *Proc. of 2020 IEEE/CVF Conference on Computer Vision and Pattern Recognition (CVPR 2020)*. Virtual, pp. 129-137.
S11	Wellner, G. & Rothman, T., 2020. Feminist AI: Can We Expect Our AI Systems to Become Feminist?. *Philosophy & Technology*, 33(2), pp. 191-205.
S12	Hernández-Álvarez, M. & Granizo, S. L., 2020. Detection of Human Trafficking Ads in Twitter Using Natural Language Processing and Image Processing. *Proc. of Advances in Artificial Intelligence, Software and Systems Engineering at the 11th Int. Conference on Applied Human Factors and Ergonomics (AHFE 2020)*. New York, NJ, USA, pp. 77-83.
S13	Mensa, E., Colla, D., Dalmasso, M., Giustini, M., Mamo, C., Pitidis, A., & Radicioni, D. P., 2020. Violence detection explanation via semantic roles embeddings. *BMC Medical Informatics and Decision Making*, 20(1), pp. 1-13.
S14	Khatri, H. & Abdellatif, I., 2020. A Multi-Modal Approach for Gender-Based Violence Detection. *Proc. of 2020 IEEE Cloud Summit*. Fairfax, VA, USA, pp. 144-149.
S15	Montiel Fernandez, Z. A., Torres Cruz, M. A., Peñaloza, C., & Hidalgo Morgan, J., 2020. Challenges of Smart Cities: How Smartphone Apps Can Improve the Safety of Women. *Proc. of 2020 4th International Conference on Smart Grid and Smart Cities (ICSGSC 2020)*. Sichuan, China, pp. 145-148.
S16	Bhagat, P., Prajapati, S. K., & Seth, A., 2020. Initial Lessons from Building an IVR-based Automated Question-Answering System. *Proc. of 2020 International Conference on Information and Communication Technologies and Development (ICTD2020)*. Guayaquil, Ecuador, pp. 1-5.
S17	Sun, T., Gaut, A., Tang, S., Huang, Y., ElSherief, M., Zhao, J., Mirza, D., Belding, E., Chang, K., & Wang, W. Y., 2019. Mitigating Gender Bias in Natural Language Processing: Literature Review. *Proc. of 57th Annual Meeting of the Association for Computational Linguistics (ACL 2019)*. Florence, Italy, pp. 1630-1640.
S18	Adams, R. & Loideáin, N. N., 2019. Addressing Indirect Discrimination and Gender Stereotypes in AI Virtual Personal Assistants: The Role of International Human Rights Law. *Cambridge International Law Journal*, 8(2), pp. 241-257.
S19	Johnson, K. N., 2019. Automating the Risk of Bias. *George Washington Law Review*, 87(6), pp. 1214-1271.
S20	Parsheera, S., 2018. A Gendered Perspective on Artificial Intelligence. *Proc. of 2018 ITU Kaleidoscope: Machine Learning for a 5G Future (ITU K 2018)*. Santa Fe, Argentina, pp. 1-7.
S21	Leavy, S., 2018. Gender Bias in Artificial Intelligence: The Need for Diversity and Gender Theory in Machine Learning. *Proc. of 1st Int. Workshop on Gender Equality in Software Engineering (GE '18)*. Gothenburg, Sweden, pp. 14-16.
S22	Sutton, A., Lansdall-Welfare, T., & Cristianini, N., 2018. Biased Embeddings from Wild Data: Measuring, Understanding and Removing. *Proc. of Int. Symposium on Intelligent Data Analysis (IDA 2018)*. Hertogenbosch, Netherlands, pp. 328-339.
S23	Kudva, V., Prasad, K., & Guruvare, S., 2018. Andriod Device-Based Cervical Cancer Screening for Resource-Poor Settings. *Journal of Digital Imaging*, 31(5), pp. 646-654.
S24	Rabbany, R., Bayani, D., & Dubrawski, A., 2018. Active Search of Connections for Case Building and Combating Human Trafficking. *Proc. of 24th Int. Conf. on Knowledge Discovery & Data Mining (KDD'18)*. London, UK, pp. 2120-2129.
S25	Berk, R. A., Sorenson, S. B., & Barnes, G., 2016. Forecasting Domestic Violence: A Machine Learning Approach to Help Inform Arraignment Decisions. *Journal of Empirical Legal Studies*, 13(1), pp. 94-115.
S26	Alvari, H., Shakarian, P., & Snyder, J. E. K., 2016. A Non-Parametric Learning Approach to Identify Online Human Trafficking. *Proc. of 2016 IEEE Conf. on Intelligence and Security Informatics (ISI 2016)*. Tucson, AZ, USA, pp. 133-138.

ID	Reference
S27	Bolukbasi, T., Chang, K., Zou, J., Saligrama, V., & Kalai, A., 2016. Man is to Computer Programmer as Woman is to Homemaker? Debiasing Word Embeddings. *Proc. of 30th International Conference on Neural Information Processing Systems (NIPS'16)*. Barcelona, Spain, pp. 4356-4364.
S28	Vachovsky, M. E., Wu, G., Chaturapruek, S., Russakovsky, O., Sommer, R., & Fei-Fei, L., 2016. Toward More Gender Diversity in CS through an Artificial Intelligence Summer Program for High School Girls. *Proc. of 47th ACM Technical Symposium on Computing Science Education (SIGCSE '16)*. Memphis, TN, USA, pp. 303-308.
S29	Crutzen, R., Peters, G. Y., Portugal, S. D., Fisser, E. M., & Grolleman, J. J., 2011. An Artificially Intelligent Chat Agent That Answers Adolescents' Questions Related to Sex, Drugs, and Alcohol: An Exploratory Study. *Journal of Adolescent Health*, 48(5), pp. 514-519.

APPENDIX B. PRIMARY STUDIES MAPPING

Table 5. Primary studies mapping

ID	RQ1				RQ2									RQ3		Quality
	a	b	c	d	a	b	c	d	e	f	g	h	i	a	b	score
S01		X	X			X								X		+1
S02		X	X		X							X		X		+3
S03				X	X	X						X		X	X	+4
S04		X	X			X								X		+2
S05	X				X							X	X	X	X	+1
S06				X	X									X		+3
S07		X								X				X		+2
S08	X				X					X					X	+3
S09	X				X							X			X	+1
S10		X	X		X									X		+3
S11	X				X										X	+2
S12		X				X								X		0
S13		X	X			X								X		+3
S14		X				X								X		0
S15		X				X								X		+1
S16		X	X							X				X		+2
S17	X				X										X	+3
S18	X				X								X		X	+3
S19		X			X										X	+3
S20		X			X										X	+1
S21	X				X										X	+1
S22		X	X		X										X	+3
S23		X	X							X				X		+3
S24		X				X								X		+3
S25		X				X								X		+3
S26		X				X								X		+3
S27		X	X		X										X	+4
S28		X	X									X			X	+4
S29		X	X							X				X		+4

IMPROVING PHISHING DETECTION VIA PSYCHOLOGICAL TRAIT SCORING

Sadat Shahriar, Arjun Mukherjee and Omprakash Gnawali
University of Houston, 4800 Calhoun Rd, Houston, TX 77004, USA

ABSTRACT

Phishing emails exhibit some unique psychological traits which are not present in legitimate emails. From empirical analysis and previous research, we find three psychological traits most dominant in Phishing emails – **A Sense of Urgency**, **Inducing Fear by Threatening**, and **Enticement with Desire**. We manually label 10% of all phishing emails in our training dataset for these three traits. We leverage that knowledge by training BERT, Sentence-BERT (SBERT), and Character-level-CNN models and capturing the nuances via the last layers that form the Phishing **Psychological Trait (PPT)** scores. For the phishing email detection task, we use the pretrained BERT and SBERT model, and concatenate the PPT scores to feed into a fully-connected neural network model. Our results show that the addition of PPT scores improves the model performance significantly, thus indicating the effectiveness of PPT scores in capturing the psychological nuances. Furthermore, to mitigate the effect of the imbalanced training dataset, we use the GPT-2 model to generate phishing emails (Radford et al., 2019. Our best model outperforms the current State-of-the-Art (SOTA) model's F1-score by 4.54%. Additionally, our analysis of individual PPTs suggests that Fear provides the strongest cue in detecting phishing emails.

KEYWORDS

Phishing, Email, BERT, Psychology

1. INTRODUCTION

Phishing is a technique used in electronic messaging to deceive the reader, where the phisher camouflages the message with a legitimate facade to access sensitive information or monetary gain (Vishwanath et al., 2011; Bose and Leung 2009). As the phishers prey on the vulnerability of the users, they often persuade people to take on some actions which may lead to undesirable consequences. Phishing attacks increased significantly in recent years and although researchers exploited several Natural Language Processing and Machine Learning techniques to detect phishing emails, the phishers evolved over time, making it harder to detect phishing emails (Almomani et al., 2013; Khonji, Iraqi, and Jones, 2013; FBI, 2021). Hence, phishing email detection systems must be smart enough to cope with the evolving nature of phishing techniques. In this work, we propose that all phishing emails exhibit some unique psychological traits, and detection of these traits can play a significant role in improving phishing vs. legitimate email classification.

Researchers analyzed the phishing attack from psychological perspectives, such as how persuasion is conducted (Akbar, 2014; Cialdini, 2001), the human factors in phishing attack (Stajano and Wilson, 2011; Jakobsson, 2007), and psychological mechanism in the effectiveness of phishing attacks (Luo et al., 2013). Research suggests that phishing messages often exhibit psychological cues, which can be crucial for their successful detection (Jones et al., 2019; Jakobsson, 2007). However, the current research is not adequate to quantify psychological traits expressed through the body of text. Consequently, how these traits play into detecting phishing emails is still an unexplored area of research. Nevertheless, for a smart detection of phishing emails, it is of immense importance to incorporate psychological attributes of the email's text, along with the linguistic model. We define *Phishing Psychological Traits (PPT)* as the psychological attributes evident in phishing emails. We claim that three major psychological attributes are evident in the phishing emails–based on whether the email sounds rushed (*a Sense of Urgency*), if the email induces fear (*Inducing Fear by Threatening*), and if there is an enticement through that email (*Enticement with Desire*). These traits can appear standalone or with a combination of any two and even three.

In this research, we capture the psychological traits by modeling a BERT, Sentence-BERT (SBERT) and Char-CNN network (Devlin et al., 2019; Reimers and Gurevych, 2019; Zhang, Zhao, and LeCun, 2015). We use these models to compute the softmax probability score (PPT score) for every phishing and legitimate emails. Next, we use pretrained BERT and SBERT model to find the feature-embedding (768-D) from text and concatenate the PPT scores with these embeddings. The concatenated features are fed to a fully-connected neural network to predict the email being phishing or legitimate. Our best performing model achieves the F1-score of 88.04%, which outperforms the current SOTA by 4.54%. We also observe a significant improvement of F1- score by up to 2.62% for the PPT-based model over the PPT-less model. The key to the consistent performance improvement is the PPT scores which provide reliable and unique cues by capturing the subtlety of psychological aspect expressed in the emails.

The novelty of this research is that our work is the first one to quantify the underlying psychological cues and leverage them for phishing email detection. We further investigate how the PPT scores help boosting the classification performance by providing t-SNE-based visualization (van der Maaten and Hinton, 2008). Furthermore, we analyze the contribution of individual PPT score and effectiveness of PPT scores in low-training-data situations. Our research provides important insights into the unique psychological attributes of phishing emails, which can create a new research direction in the phishing email detection paradigm.

2. RELATED WORKS

Phishing emails have been adversely affecting the internet world since 1996 (Salloum et al., 2021). Different NLP techniques were used to extract semantic, syntactic and contextual features which played important role for phishing detection along with classical machine learning techniques (Cui et al., 2020; Verma and Hossain, 2013; Park and Taylor, 2015; Blanzieri and Bryl, 2009; Gansterer and Pölz, 2009; Feng et al., 2016). However, the evolving nature of phishing emails entails more sophisticated techniques as often they do not contain the malicious code or word choices of known attacks (Lee, Saxe, and Harang, 2020a). The transformer-based models and their variants proved to have superiority over the traditional deep learning models mostly due to their transfer learning capabilities and they were successfully used in many deceptive text detection tasks (Vaswani et al., 2017; Shahriar et al., 2021). Lee et al. proposed a BERT-based phishing email detection model where they pruned half of the transformer blocks to better capture the semantics (Lee, Saxe, and Harang, 2020b). However, none of these works consider the unique psychological aspect of phishing emails, that can provide useful features to detect them. Naidoo suggested the *urgency* to be a dominant psychological feature in phishing email, but an ML-based automatic detection strategy is not present in their work (Naidoo, 2015). We address these research gaps by leveraging the context embedding of the BERT and SBERT model and using the psychological traits to detect phishing emails.

3. METHODOLOGY

Figure 1 explains the whole process in brief. We start by selecting 10% of the phishing emails and manually label them as 1 or 0 based on the presence of each PPTs. Next, we train a BERT, SBERT and Char-CNN model for each PPTs and use the trained model to compute the PPT scores of all emails. We concatenate these PPT scores with the fine-tuned BERT model and pretrained SBERT model and feed them to a Deep Neural Net model to detect the phishing email.

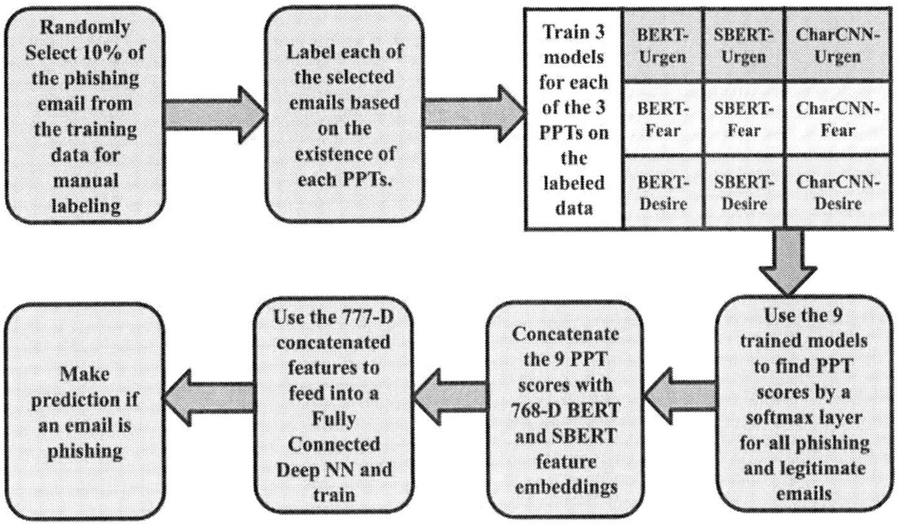

Figure 1. The complete flow diagram for phishing email detection using Phishing Psychological Traits (PPT)

3.1 Phishing Psychological Traits (PPT)

We claim that one or more persuasion strategies are implemented with the psychological traits expressed in the text. PPTs work as a broader umbrella and all phishing emails exhibit one or more of these PPTs.

3.1.1 A Sense of Urgency

One of the crucial characteristics of many phishing messages is the expression of urgency. Having a time constraint can induce stress in the readers' mind even if they are capable to do so within the stated time (Ordonez and Benson III, 1997). The phishing emails often urge the reader to take some action with promptness, thus reducing the time for the reader to reason or report (Aggarwal, Kumar, and Sudarsan, 2014). Additionally, it can induce impulsive behavior in the recipient that can lead to an error of judgment (Cui et al., 2020). Therefore, it is imperative to detect urgency from an email. We observe that urgency expressed in the phishing messages are direct and expressive, as the attackers want to maximize the possibility of a reader's response.

3.1.2 Inducing Fear by Threatening

Research shows that Fear by threatening is one of the most frequently exploited emotional trigger by the attackers (Sharma and Bashir, 2020, Halevi, Lewis, and Memon, 2013; Ferreira and Lenzini, 2015). The threat can be of many forms, for example, being locked out or blocked from one's account, losing access to information, getting hacked, stealing information, stealing currency, and individually targeted attack (Wang et al., 2012). Notably, Bitaab et al. stated that during the COVID-19 pandemic, the readers' fear is exploited by the attackers, which led to a high increase of phishing attacks (Bitaab et al., 2021). Hence it is evident that a direct or indirect threat can be a significant cue of phishing emails.

3.1.3 Enticement with Desire

Phishing email often lures by enticing the readers' personality trait of openness that can make them to be greedy and curious which can result in getting phished (Ding et al., 2015; Halevi, Lewis, and Memon, 2013). Phishing emails often contain a lucrative financial reward in exchange for clicking on some links, providing personal details or credit card information, and so on. Stajano and Wilson maintained that "Need and Greed" is one of the seven basic principles of scams, where people can be a victim of a lottery scam or a sexy swindler (Stajano and Wilson, 2011). Hence, the enticement with greed or curiosity can be an important signal of phishing emails.

3.2 Experimental Setup

To obtain the PPT scores of all emails, we train three models for each of the PPTs – BERT, SBERT and Char-CNN. We split all the manually-labeled-PPT emails in 80% for training and 20% for validation and repeat the experiments for three different splits. For the training of phishing email detection task, we use the BERT and SBERT model and apply the same train-validation split. We find the best set of hyperparameters by observing the performance on the validation set.

4. DATASET

The dataset we used was provided in the Anti-Phishing Pilot at ACM IWSPA 2018 (Verma, Zeng, and Faridi, 2019). Our test data size is 4300, from the first shared task, where emails are provided without the headers (*IWSPA_NH*). The IWSPA_NH training set has 5092 legitimate and 629 phishing emails. We also added 4082 legit, 503 phish emails from the IWSPA header-added dataset (*IWSPA_H*). Additionally, to examine how our trained model performs on other datasets, we curated a new small dataset called *UNIV_Phish*, containing 326 emails. We collected 163 phishing email from three different university websites: 72 emails from Stanford University, 68 from Lehigh University and 23 from University of Washington. We also added 163 emails from Enron "ham" emails. Notably, we made sure, none of these emails appeared in the IWSPA training set.

5. RESULT AND DISCUSSION

We hypothesize that when we add the PPTs along with the language model, the performance of phishing email detection improves. Therefore, we first need to find the PPT scores for all emails and then use these scores to detect phishing emails.

5.1 Detection of Phishing Psychological Traits

The randomly selected phishing emails are labeled by one of the authors as 1 or 0 for each PPTs. We find 82.54% emails are labeled as Urgent, where 71.42% emails are labeled as both Urgent and Fear. Only 4.76% emails exhibit all the three traits. We use BERT, SBERT, and Char-CNN to train on the manually labeled emails. Then, we make prediction on rest of the email using these models' last softmax layer and use the softmax output as PPT score.

Figure 2 depicts the distribution of BERT-based PPT scores in phishing vs legitimate emails. We observe that the PPT scores create distinguishable clouds for phishing and legitimate emails in Figure 2(a). Figure 2(b) shows the kernel density plot of individual PPTs, which represents the continuous probability density curve for these traits. For *Sense of Urgency* and *Fear by Threatening*, we observe a high density of Phishing emails on the right side of the curve which indicates the high probability of *urgency* and *fear* in the phishing emails. However, contrary to our intuition, phishing emails have a lower probability in the *Desire* score than the legitimate emails. One reason could be the lack of phishing emails in our dataset with this trait, which may have caused the failure to capture the *Desire* trait in the emails.

We explore how different words are related to the three PPTs. We observe "immediately", and "soon" are two most recurring adverbs in the *Urgent*-labeled emails. The *Desire*-labeled emails are mostly the offers of a free upgrade of some subscriptions. However, since the main goal of phishing emails is to persuade the user to take some action by clicking on a phishing link, we observe the words "link" and "click" frequently in all the phishing emails.

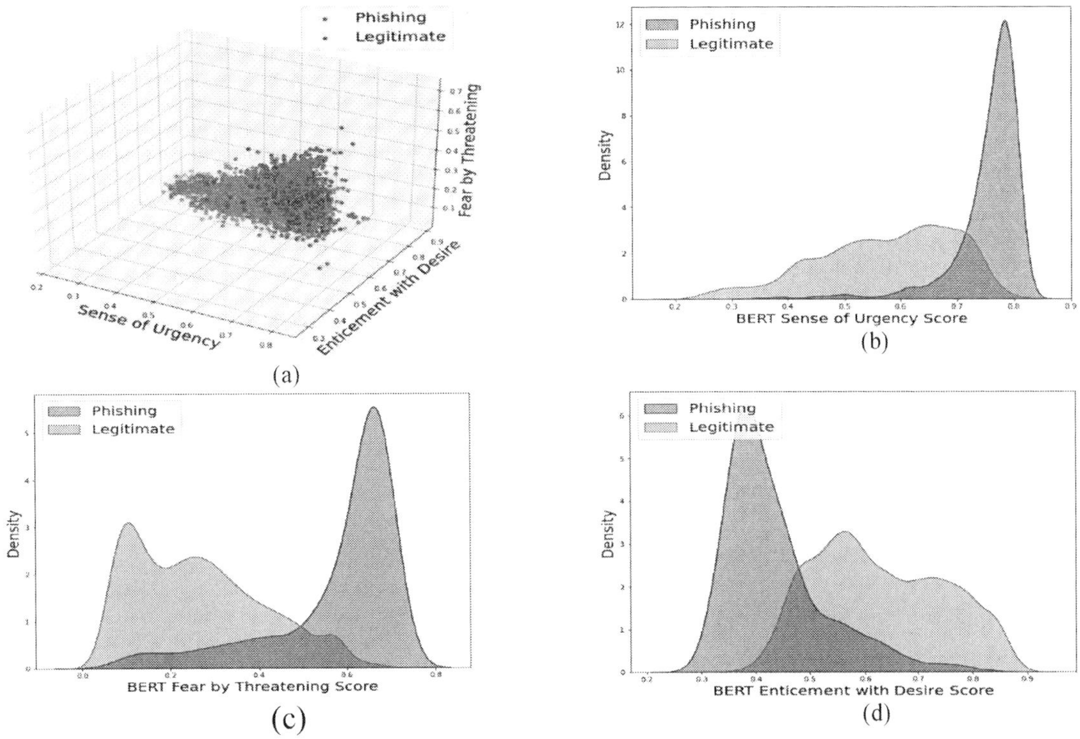

Figure 2. (a) 3-dimensional plot of PPT scores for BERT. (b) Kernel Density Estimation (KDE) plot of BERT-based PPT Score for Sense of Urgency (c) KDE plot Fear by Threatening (d) KDE plot for Enticement with Desire

5.2 Phishing Email Detection

From Table 1, we observe that, for every case, the performance improves when PPT scores are added to the training process. While trained on IWSPA_NH training data for the BERT model, we found that adding the PPTs improves the accuracy by 0.70% and F1-score significantly by 2.62% (p-value=0.04). We find the same trend for SBERT model as well. Adding the psychological traits improves the accuracy by 1.31% and F1-score by 1.34%.

Next, we added the additional training data for IWSPA with header information (IWSPA_H). The additional training data helps improve the performance. However, adding the PPTs again improves the performance. For BERT model, the improvement is by 0.63% in accuracy and 2.19% (p-value=0.03) in F1-score. Similarly, for the SBERT model, F1-score has a significant improvement of 2.50% (P-value=0.02). It may be noted that the current SOTA F1-score for this test set is 83.5%. Adding the PPTs with IWSPA_NH and IWSPA_H training data, we outperform the current SOTA (85.16% for BERT and 83.88% for SBERT).

A major challenge in our task is the lack of training data in the phishing email category due to the corporations being reluctant and individuals being ashamed to share such sensitive data (Aassal et al., 2018). In order to balance the training dataset, we tried different approaches like SMOTE (Chawla et al., 2002), cost-sensitive learning methods (Thai-Nghe, Gantner, and Schmidt-Thieme, 2010), and the addition of GPT- 2-generated phishing emails (Radford et al., 2019). However, we did not find any performance improvement for the first two. For GPT-2, we observe the performance boost for BERT models by 1.02% in accuracy and 3.59% in F1-score. When we added the psychological trait features with GPT- 2-generated emails, it also improved the accuracy by 0.45%, and F1-score by 1.38% .

Table 1. Performance of the IWSPA test set and UNIV_Phish dataset while trained on different training data. We observe the best performance is found when we use IWSPA header-less, header-added data, GPT-2-generated phishing emails, and PPT features added with the BERT model

		BERT		BERT + PPT		SBERT		SBERT + PPT	
Tested On	***Training Data***	**Acc (%)**	**F1 (%)**	**Acc (%)**	**F1 (%)**	**Acc (%)**	**F1 (%)**	**Acc (%)**	**F1 (%)**
IWSPA Test set	*IWSPA_NH*	95.11	79.34	95.81	81.96	95.39	79.29	95.70	80.63
	IWSPA_NH + IWSPA_H	96.00	83.07	96.63	85.61	96.23	81.38	96.54	83.88
	IWSPA_NH+IWSPA_H+GPT2	97.02	86.66	97.47	**88.04**	94.32	77.19	95.04	79.41
UNIV_ Phish	*IWSPA_NH*	85.27	84.71	86.81	85.61	81.29	79.38	82.82	80.70
	IWSPA_NH + IWSPA_H	86.50	85.23	87.11	86.17	82.21	80.01	83.74	82.15
	IWSPA_NH+IWSPA_H+GPT2	87.42	86.98	88.03	**87.77**	80.06	76.63	82.82	80.55

However, contrary to the BERT model, the addition of GPT-2 generated data created a significant decline in the SBERT performance. The reason could be the poor coherence of some of the generated data. While for the BERT model, we use the output from the [*cls*] token, SBERT uses a pooling strategy, leading to poor sentence embedding of non-coherent texts. Additionally, as suggested in Reimers and Gurveych (Reimers and Gurevych, 2019), since SBERT cannot be used to update all the internal layers of BERT architecture, it may not be well suited for transfer learning.

We further test our model performance on UNIV_Phish dataset. From Table 1, we observe that added PPT improves the performance up to 4.02% in F1-score. We also observe the similar performance boost with added IWSPA_H set (up to 1.45%) and added GPT-2-generated email set (up to 1.75%). Hence, the performance of UNIV_Phish dataset further strengthens our model validity.

Next, we analyze the embedding representation of the emails using t-SNE plot (Maaten and Hinton, 2008). Figure 3 shows that the cloud of misclassified samples is mainly in the overlapped region of phishing and legitimate emails, which indicates the lack of better embedding representation of these emails. However, we observe that the distance between the center of phishing email and the legitimate email cloud increased by adding the PPTs by 9.56%, and 15.29% using Euclidean and Manhattan distance, respectively. Thus, the addition of psychological traits seems to improve the embedding representation.

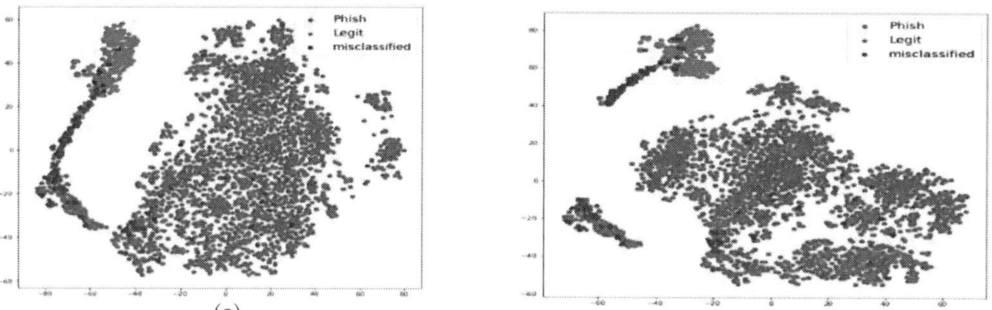

(a) (b)

Figure 3. t-SNE representation of Phishing and Legitimate email with (a) BERT-based feature only (b) BERT-based + Phishing Psychological Trait features. The figure shows the misclassification zones in blue

We further examine the effectiveness of the Phishing Psychological Traits model by ablation experiments (Meyes et al. 2019). From Figure 4 (a), we observe that the performance decreases the most, when we remove the *Urgency* trait (0.91% in F1-score, 1.01% in accuracy), followed by the *Fear* trait (0.88% in F1-score, 0.99% in accuracy). *Desire* had the least effect on performance (0.60% in F1-score and 0.71% in accuracy), which is consistent with our previous analysis.

Finally, we vary the training data proportion to examine the effect of PPT when we have a small amount of training data. Figure 4(b) shows that while with 100% training data, PPT scores improve the F1-score by 2.62%, with only 20% training data, the F1-score improvement is by 3.94% indicating the effectiveness of PPTs even with insufficient training data. Figure 4(b) also demonstrates the effect of adding PPTs individually. We observe that as a standalone PPT, Fear by Threatening has a better impact on performance than the others. Nevertheless, the three PPTs combined provide the best cue for detecting phishing emails.

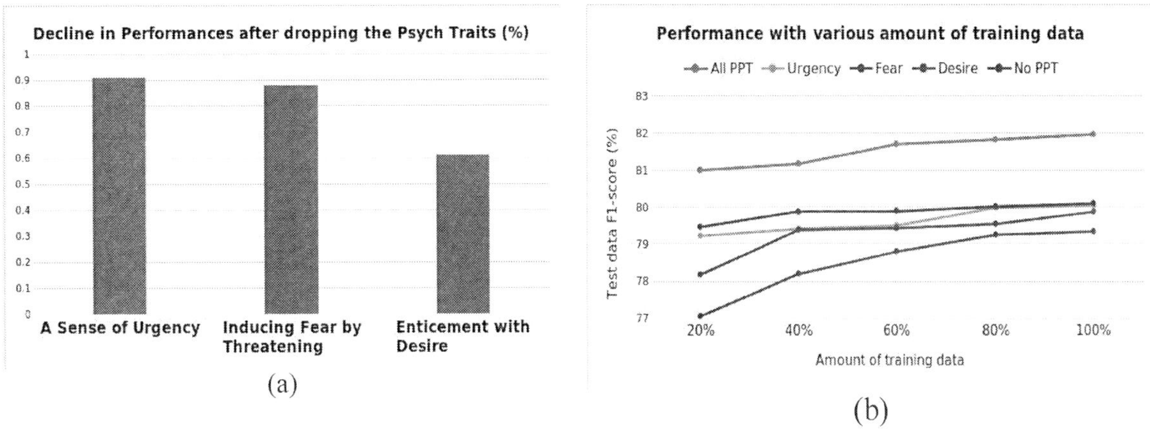

Figure 4. (a) Decline in performance after dropping the PPTs one at a time (b) Performance in test data at varying proportion of training data

6. CONCLUSION

Quantifying the psychological traits of an email can provide key signals which help improve the phishing email detection performance. In this paper, we define, analyze, and quantify the PPTs that can successfully capture the nuances of an email's intent and show promising results. Hence, our work may provide potential research direction to win the battle against evolving nature of phishing. However, we still have limitations and room for further improvement. First, we will obtain ground truth for the PPTs by labeling them with multiple human raters enabling us to measure the kappa statistics for testing inter-rater reliability, which in turn can provide a more accurate estimation of PPTs. Second, further research might be required to understand the flow of psychological traits in conversational turns to detect more organized phishing than single email-based phishing. Finally, an investigation of how the individual PPTs contribute to forming a phishing email can provide valuable insight that can be utilized for more efficient phishing email detection.

ACKNOWLEDGMENT

Research was supported in part by grants NSF 1838147, ARO W911NF-20-1-0254. The views and conclusions contained in this document are those of the authors and not of the sponsors. The U.S. Government is authorized to reproduce and distribute reprints for Government purposes notwithstanding any copyright notation herein.

REFERENCES

Aassal, A. E., Moraes, L. F., Baki, S., Moraes, L., & Verma, R. (2018). Anti-Phishing Pilot at ACM IWSPA 2018 Evaluating Performance with New Metrics for Unbalanced Datasets. *1st Anti-Phishing Shared Task at 4th ACM IWSPA (IWSPA-AP).* Retrieved from http://ceur-ws.org/Vol-2124/invited_paper_1.pdf

Aggarwal, S., Kumar, V., & Sudarsan, S. D. (2014). Identification and Detection of Phishing Emails Using Natural Language Processing Techniques. *Proceedings of the 7th International Conference on Security of Information and Networks* (pp. 217–222). New York, NY, USA: Association for Computing Machinery. doi:10.1145/2659651.2659691

Agrawal, A., Venkatraman, J., Leonard, S., & Paepcke, A. (2015). YouEDU: Addressing Confusion in MOOC Discussion Forums by Recommending Instructional Video Clips. *EDM.*

Akbar, N. (2014, October). Analysing Persuasion Principles in Phishing Emails. *Analysing Persuasion Principles in Phishing Emails.* Retrieved from http://essay.utwente.nl/66177/

Akoury, N., Wang, S., Whiting, J., Hood, S., Peng, N., & Iyyer, M. (2020). STORIUM: A Dataset and Evaluation Platform for Machine-in-the-Loop Story Generation. *arXiv preprint arXiv:2010.01717.*

Aljofey, A., Jiang, Q., Qu, Q., Huang, M., & Niyigena, J.-P. (2020). An Effective Phishing Detection Model Based on Character Level Convolutional Neural Network from URL. *Electronics, 9.* doi:10.3390/electronics9091514

Almomani, A., Gupta, B. B., Atawneh, S., Meulenberg, A., & Almomani, E. (2013). A Survey of Phishing Email Filtering Techniques. *IEEE Communications Surveys Tutorials, 15*, 2070-2090. doi:10.1109/SURV.2013.030713.00020

Bitaab, M., Cho, H., Oest, A., Zhang, P., Sun, Z., Pourmohamad, R., . . . others. (2021). Scam Pandemic: How Attackers Exploit Public Fear through Phishing. *arXiv preprint arXiv:2103.12843*.

Blanzieri, E., & Bryl, A. (2009). A survey of learning-based techniques of email spam filtering. *Artificial Intelligence Review, 29*, 63-92.

Bose, I., & Leung, A. C. (2009). Technical opinion What drives the adoption of antiphishing measures by Hong Kong banks? *Communications of the ACM, 52*, 141–143.

Chawla, N. V., Bowyer, K. W., Hall, L. O., & Kegelmeyer, W. P. (2002, June). SMOTE: Synthetic Minority over-Sampling Technique. *J. Artif. Int. Res., 16*, 321–357.

Cialdini, R. B. (2001). The science of persuasion. *Scientific American, 284*, 76–81.

Cui, X., Ge, Y., Qu, W., & Zhang, K. (2020). Effects of Recipient Information and Urgency Cues on Phishing Detection. In C. Stephanidis, & M. Antona (Ed.), *HCI International 2020 - Posters* (pp. 520–525). Cham: Springer International Publishing.

Devlin, J., Chang, M.-W., Lee, K., & Toutanova, K. (2019, June). BERT: Pre-training of Deep Bidirectional Transformers for Language Understanding. *Proceedings of the 2019 Conference of the North American Chapter of the Association for Computational Linguistics: Human Language Technologies, Volume 1 (Long and Short Papers)* (pp. 4171–4186). Minneapolis: Association for Computational Linguistics. doi:10.18653/v1/N19-1423

Ding, K., Pantic, N., Lu, Y., Manna, S., & Husain, M. I. (2015). Towards building a word similarity dictionary for personality bias classification of phishing email contents. *Proceedings of the 2015 IEEE 9th International Conference on Semantic Computing (IEEE ICSC 2015)*, (pp. 252-259). doi:10.1109/ICOSC.2015.7050815

Fang, Y., Zhang, C., Huang, C., Liu, L., & Yang, Y. (2019). Phishing Email Detection Using Improved RCNN Model With Multilevel Vectors and Attention Mechanism. *IEEE Access, 7*, 56329-56340. doi:10.1109/ACCESS.2019.2913705

FBI. (2021). *Internet Crime Report 2020.* Retrieved from https://www.ic3.gov/Media/PDF/AnnualReport/2020_IC3Report.pdf

Feng, W., Sun, J., Zhang, L., Cao, C., & Yang, Q. (2016). A support vector machine based naive Bayes algorithm for spam filtering. *2016 IEEE 35th International Performance Computing and Communications Conference (IPCCC)*, (pp. 1–8).

Ferreira, A., & Lenzini, G. (2015). An analysis of social engineering principles in effective phishing. *2015 Workshop on Socio-Technical Aspects in Security and Trust*, (pp. 9-16). doi:10.1109/STAST.2015.10

Gansterer, W., & Pölz, D. (2009, April). E-Mail Classification for Phishing Defense., (pp. 449-460). doi:10.1007/978-3-642-00958-7_40

Goldberg, L. R. (1982). From Ace to Zombie: Some explorations in the language of personality. *Advances in personality assessment, 1*, 203–234.

Halevi, T., Lewis, J., & Memon, N. (2013). A Pilot Study of Cyber Security and Privacy Related Behavior and Personality Traits. *Proceedings of the 22nd International Conference on World Wide Web* (pp. 737–744). New York, NY, USA: Association for Computing Machinery. doi:10.1145/2487788.2488034

Jakobsson, M. (2007). The human factor in phishing. *Privacy & Security of Consumer Information, 7*, 1–19.

Jones, H. S., Towse, J. N., Race, N., & Harrison, T. (2019). Email fraud: The search for psychological predictors of susceptibility. *PloS one, 14*, e0209684.

Kejriwal, M., & Zhou, P. (2019). Low-Supervision Urgency Detection and Transfer in Short Crisis Messages. *Proceedings of the 2019 IEEE/ACM International Conference on Advances in Social Networks Analysis and Mining* (pp. 353–356). New York, NY, USA: Association for Computing Machinery. doi:10.1145/3341161.3342936

Khonji, M., Iraqi, Y., & Jones, A. (2013). Phishing Detection: A Literature Survey. *IEEE Communications Surveys Tutorials, 15*, 2091-2121. doi:10.1109/SURV.2013.032213.00009

Lee, Y., Saxe, J., & Harang, R. (2020). CATBERT: Context-Aware Tiny BERT for Detecting Social Engineering Emails. *CATBERT: Context-Aware Tiny BERT for Detecting Social Engineering Emails*.

Li, Q., Cheng, M., Wang, J., & Sun, B. (2020). LSTM based Phishing Detection for Big Email Data. *IEEE Transactions on Big Data*, 1-1. doi:10.1109/TBDATA.2020.2978915

Luo, X. (., Zhang, W., Burd, S., & Seazzu, A. (2013). Investigating phishing victimization with the Heuristic–Systematic Model: A theoretical framework and an exploration. *Computers & Security, 38*, 28-38. doi:https://doi.org/10.1016/j.cose.2012.12.003

Maaten, L. V., & Hinton, G. E. (2008). Visualizing Data using t-SNE. *Journal of Machine Learning Research, 9*, 2579-2605.

Meyes, R., Lu, M., Puiseau, C. W., & Meisen, T. (2019). Ablation Studies in Artificial Neural Networks. *ArXiv, abs/1901.08644*.

Microsoft (2021) *The quiet evolution of phishing*. Available at: https://www.microsoft.com/security/blog/2019/12/11/ the-quiet-evolution-of-phishing (accessed August 20, 2021).

Naidoo, R. (2015, February). Analysing urgency and trust cues exploited in phishing scam designs *in 10th International Conference on Cyber Warfare and Security (p. 216).*

Ordonez, L., & Benson III, L. (1997). Decisions under time pressure: How time constraint affects risky decision making. *Organizational Behavior and Human Decision Processes, 71*, 121–140.

Park, G., & Taylor, J. M. (2015). Using Syntactic Features for Phishing Detection. *CoRR, abs/1506.00037*. Retrieved from http://arxiv.org/abs/1506.00037

Parrish Jr, J. L., Bailey, J. L., & Courtney, J. F. (2009). A personality based model for determining susceptibility to phishing attacks. *Little Rock: University of Arkansas*, 285–296.

Radford, A., Wu, J., Child, R., Luan, D., Amodei, D., Sutskever, I., & others. (2019). Language models are unsupervised multitask learners. *OpenAI blog, 1*, 9.

Reimers, N., & Gurevych, I. (2019, November). Sentence-BERT: Sentence Embeddings using Siamese BERT-Networks. *Proceedings of the 2019 Conference on Empirical Methods in Natural Language Processing.* Association for Computational Linguistics. Retrieved from https://arxiv.org/abs/1908.10084

Salloum, S., Gaber, T., Vadera, S., & Shaalan, K. (2021). Phishing Email Detection Using Natural Language Processing Techniques: A Literature Survey. *Procedia Computer Science, 189*, 19-28. doi:https://doi.org/10.1016/j.procs.2021.05.077

SBERT (2021) *SBERT Pretrained Models.* Available at: https: //www.sbert.net/docs/pretrained models.html (accessed July 29, 2021).

Shahriar, S., Mukherjee, A., & Gnawali, O. (2021, September). A Domain-Independent Holistic Approach to Deception Detection. *Proceedings of the International Conference on Recent Advances in Natural Language Processing (RANLP 2021).* (pp. 1308-1317)

Sharma, T., & Bashir, M. (2020). An Analysis of Phishing Emails and How the Human Vulnerabilities are Exploited. In I. Corradini, E. Nardelli, & T. Ahram (Ed.), *Advances in Human Factors in Cybersecurity* (pp. 49–55). Cham: Springer International Publishing.

Sheng, S., Wardman, B., Warner, G., Cranor, L., Hong, J. I., & Zhang, C. (2009). An Empirical Analysis of Phishing Blacklists. *CEAS 2009.*

Stajano, F., & Wilson, P. (2011, March). Understanding Scam Victims: Seven Principles for Systems Security. *Commun. ACM, 54*, 70–75. doi:10.1145/1897852.1897872

Thai-Nghe, N., Gantner, Z., & Schmidt-Thieme, L. (2010). Cost-sensitive learning methods for imbalanced data. *The 2010 International Joint Conference on Neural Networks (IJCNN)*, (pp. 1-8). doi:10.1109/IJCNN.2010.5596486

van der Maaten, L., & Hinton, G. (2008). Visualizing Data using t-SNE. *Journal of Machine Learning Research, 9*, 2579–2605. Retrieved from http://www.jmlr.org/papers/v9/vandermaaten08a.html

Vaswani, A., Shazeer, N. M., Parmar, N., Uszkoreit, J., Jones, L., Gomez, A. N., . . . Polosukhin, I. (2017). Attention is All you Need. *ArXiv, abs/1706.03762.*

Verma, R. M., & Hossain, N. (2013). Semantic Feature Selection for Text with Application to Phishing Email Detection. *ICISC.*

Verma, R. M., Zeng, V., & Faridi, H. (2019). Data Quality for Security Challenges: Case Studies of Phishing, Malware and Intrusion Detection Datasets. *Proceedings of the 2019 ACM SIGSAC Conference on Computer and Communications Security* (pp. 2605–2607). New York, NY, USA: Association for Computing Machinery. doi:10.1145/3319535.3363267

Vishwanath, A., Herath, T., Chen, R., Wang, J., & Rao, H. R. (2011). Why do people get phished? Testing individual differences in phishing vulnerability within an integrated, information processing model. *Decision Support Systems, 51*, 576-586. doi:https://doi.org/10.1016/j.dss.2011.03.002

Wang, J., Herath, T., Chen, R., Vishwanath, A., & Rao, H. R. (2012). Research article phishing susceptibility: An investigation into the processing of a targeted spear phishing email. *IEEE transactions on professional communication, 55*, 345–362.

Williams, E. J., Beardmore, A., & Joinson, A. N. (2017). Individual differences in susceptibility to online influence: A theoretical review. *Computers in Human Behavior, 72*, 412-421. doi:https://doi.org/10.1016/j.chb.2017.03.002

Workman, M. (2008, February). Wisecrackers: A Theory-Grounded Investigation of Phishing and Pretext Social Engineering Threats to Information Security. *J. Am. Soc. Inf. Sci. Technol., 59*, 662–674.

Zhang, N., & Yuan, Y. (2012). Phishing Detection Using Neural Network.

Zhang, X., Zhao, J. J., & LeCun, Y. (2015). Character-level Convolutional Networks for Text Classification. *CoRR, abs/1509.01626.* Retrieved from http://arxiv.org/abs/1509.01626

CENTRALIZED OR DE-CENTRALIZED DATA AND ALGORITHMS IN THE FINNISH HEALTH CARE INFRASTRUCTURE

Jussi Salmi[1,2] and Lisse-Lotte Hermansson[1]
[1]*BCB Medical Ltd., Ruukinkatu 2, 20540 Turku, Finland*
[2]*Åbo Akademi University, Tuomiokirkontori 3, 20500 Turku, Finland*

ABSTRACT

The Secondary Use Act came into effect in Finland in 2019. It created a national authority, Findata, as the only authority in Finland which can authorize the use of clinical patient information for academic and other users. This has both advantages and disadvantages. When a researcher requests clinical information all the requests go through a single authority and data from more than one register can be requested with a single application. In this article we discuss the advantages and disadvantages of this health care research jurisdiction and compare the Finnish model to the systems in use in other Nordic countries. We also describe the possibilities and limitations this poses to modern data sharing paradigms.

KEYWORDS

Health Information Exchange, Electronic Health Record, Findata, Federated Learning, Common Data Model

1. INTRODUCTION

Comprehensive electronic health registers have been used in Finland for over 20 years. The health care system is funded and organized publicly through 20 hospital districts. Specialized health care is divided into 5 regions. These regions each have a comprehensive electronic health record system for tertiary care which includes almost the whole population.

Recently there have been many initiatives aiming at improving the use of clinical health data for research (Kuiper et al. 2015). The Finnish biobank act came into effect in 2013, after which 8 biobanks have been founded. There also exists a central biobank organization, FinBB, which can combine samples, data and even genetic information requests from several biobanks (FinBB 2022). For the past 10 years, biobanks have made it easier to share patient samples and associated data both to academic research groups and pharmaceutical companies. The benefit regarding the older system with specially recruited patients is that biobanks use a protocol where a patient signs a single consent and all of his/her present and future samples and data can be shared many times. The consent can be withdrawn at any time. A recent article reported a process of combining data from 3 biobanks in a pharmacogenomic retrospective study with 7000 individuals. Collecting the data took 16 months (Lähteenmäki et al. 2021).

A further development was the act on the Secondary Use of Health and Social Data which came into effect on 1.5.2019 (The Ministry of Health and Social Services 2019). The act aims at creating a unified system for granting access to researchers and pharmaceutical companies to patient data which has been recorded in hospitals for clinical use. An authority, Findata, was created. The researchers can apply for a dataset gathered from many sources, e.g. several hospital districts or other register owners. See Table 1 for data providers. Findata gathers the data and places it in a secure server. The individual-level data cannot be downloaded, but it can be analyzed using standard software and the aggregated results can be downloaded after an administrator has checked that the requested dataset follows the information security rules.

Table 1. Register owners in Finland accessible through Findata (Findata register owners 2022)

Finnish Centre for Pensions
Regional State Administrative Agency
Digital and Population State Services Agency
The Social Insurance Institution
Finnish Medicines Agency
National Supervisory Authority for Welfare and Health
Ministry of Social Affairs and Health
Finnish Institute for Health and Welfare
Statistics Finland
Finnish Institute of Occupational Health
All public health and social support service providers
All private health services providers

The act does not concern data that has been collected solely for research purposes or that is only from one register. A notable group of data which is only for research purposes is the omics data. The information security status of genetic data is not yet clear, but a Genome Act is being prepared and it should clarify that (The Ministry of Health and Social Services 2022). The Secondary Use Act requires Findata to provide a secure data handling environment, but it also allows other operators to provide such an environment if they follow the security requirements given by Findata. The detailed requirements were given in October 2020 (Findata, 2020).

2. FEDERATED AND CENTRALIZED DATA

Recently, federated learning has been gaining ground in health data processing (Kuiper et al. 2015, Sheller et al. 2020). The traditional model of developing algorithms in one place and transferring the data to the computing environment is facing challenges due to the stricter privacy laws that are being implemented all over Europe. It is becoming more and more difficult to gain access to the data and getting permission to move data out of the clinical systems. At the same time the amount of health data that can in principle be used for statistical or machine learning applications is increasing rapidly. A solution to this problem is federated learning. The algorithms can be run locally close to the data. The data is accessible often from a secure cloud environment which can provide virtual machines with considerable computational power and up-to-date libraries, thus making it easy to run the algorithms remotely without moving the data anywhere. The idea is to have less bureaucracy and more data security. See Figure 1 for a schematic on different approaches to data sharing.

With federated learning it is possible to combine several smaller datasets to get one larger set. This is especially useful with rare diseases when there is not one place where there is enough data to train an AI model or perform statistical analysis. One can do statistical analysis on the data locally and combine aggregate results from several sites elsewhere. The aggregate results are not sensitive if the number of patients is large enough.

Another way of sharing data in a federated way is to produce synthetic data from the original data. Synthetic data is created from the original data by creating a model of the original data and drawing new samples from that model. The model can be e.g. created by probabilistic Bayesian networks (Gogoshin 2021) or neural network methods. The synthetic data method creates a new data set where the synthetic patients resemble the original patients as closely as possible but not so closely that the original patients can be identified from the synthetic data. The synthetic data can then be used in place of the original data so that the same analyses can be carried out with it and the results are still useful. The synthetic data can either follow closely the original data, in which case the identification problem can persist, or it can be more generalized, in which case the statistical properties can be further from the properties of the real patient population. The solution to this trade-off between statistical power and privacy is a hot research topic. Different synthetic data generating methods have different requirements, regarding e.g. the number of patients needed to build a high-quality data set, the type of data that can be included in the data and the quality of modelling the statistical dependencies between variables.

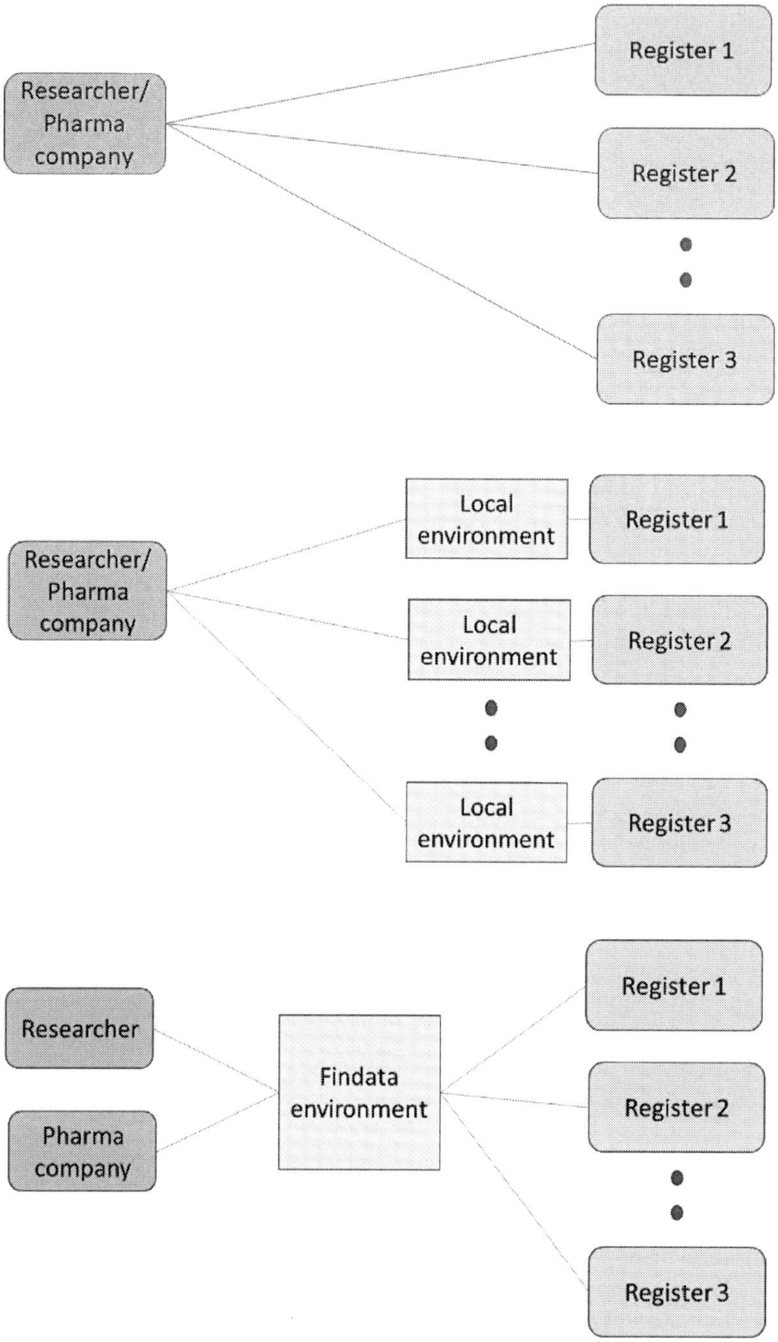

Figure 1. Three methods for sharing data. copying data to researcher's environment (above), federated data sharing (middle) and Findata-style centralized data sharing (below) between several data users and data providers

In the federated setting, the synthetic data can be created in the local computation environment as in the middle image in Figure 1. The synthetic data can be shared outside the local environment because the data security laws don't dictate its use: no person's personal information is revealed in the process. The different sets of synthetic data can then be combined to form a complete data set. For example, in the case or rare diseases, this opens new possibilities because the statistical analyses have more power with a larger data set gathered from a larger group of people.

Another issue is the harmonization of the data from many different sources. Not only the data that is collected can differ between hospitals but also the ways e.g. laboratory results are analyzed can be different and the treatment lines are different. This normally requires the recruitment of local clinicians to help in deciphering the data. There are projects with the aim of presenting standardized data structures. Most notable of these is the OMOP project of harmonization (OHDSI 2021). OMOP contains a Common Data Model (CDM) which defines the data structures and metadata elements which must be included in the harmonization. See Figure 2 for an overview of the transformation process.

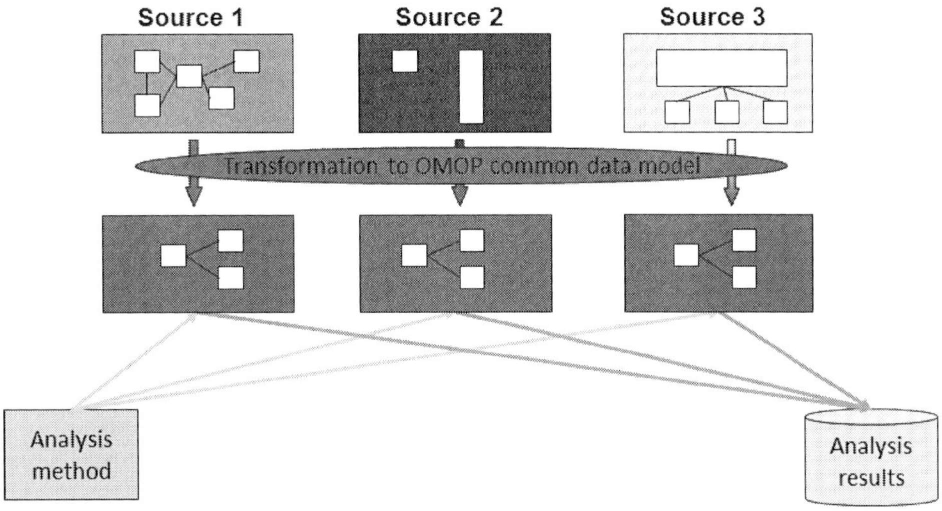

Figure 2. Data from different sources is transformed to a common data model which can be used by different analysis methods (OHDSI 2022)

3. COMPARISON TO OTHER COUNTRIES

In Nordic countries there generally are comprehensive and practical health care registries (Laugesen et al. 2021). This is due to the fact that there is an individual and unique social security number which makes following a person in different registers easy. The health sector is very centralized and tightly controlled. In each country a public authority controls the field and uses a single system of registers to control the quality and effectiveness of health care. Electronic health care registers were taken into use in the 90's or early 2000's and thus they allow following the health of a large share of the population retrospectively. Even in the Nordic countries there are local actors which have autonomy in selecting their register providers but there also exists many registers that are national. Also, the IT technology and research sector is developed which means that there is a history of using the registers for research and development purposes.

In Sweden there currently does not exist a single authority tasked with giving research permits and collecting the data from multiple sources. The National Board of Health and Welfare oversees the quality registers in Sweden and currently they are in a process to develop further the Swedish system. In a recent study they found the following problems in the registers in Sweden (Socialstyrelsen 2021). The data in the registers is not uniform and changes depending on the treatment that was given to a patient. The report tentatively found that a unified register could solve some of these problems. Some institutions proposed the Finnish model with Findata as an example.

In Denmark there exists a public organization, Danish Health Data Authority which runs the national health registries and grants access to the data. The data must be accessed in the Forskermaskinen (the Research Engine), a high-security computing environment. Foreign research groups should cooperate with a Danish institution to get access to the data. Denmark is also well known for the quality and availability of its health data (Schmidt 2015).

4. CRITICISM OF FINDATA

The Secondary Use Act and especially the strict data security requirements received a lot of criticism (Pajula et al. 2021). The act allowed also other actors in the health care area to develop their own secure environments. At first only the Windows-based Findata environment was available. The systems are very complex, and they must be audited before they can be taken into use. The modern data analysis methods require a lot of computing power, either CPU or GPU -based and they often run in Linux. Neither of these are available in the Findata environment as of March 2022, 1,5 years after the requirements were published. The users cannot install their own software unless it runs on the Findata Windows server's R and Python environments. The price of the Findata services is high for especially small projects, contains too much bureaucracy and the queue for the data permits is 8 months long currently. These factors, according to many researchers, make small clinical research projects and longer follow-up studies much more difficult and costly to implement. Artificial intelligence methods, big image or genetic data sets cannot be used at all currently. To ease the problems, the Finnish parliament granted a 1-year continuation to the deadline (until 1.5.2022) for fulfilling the data security requirements, so that more environments would be available.

Comparing it to the modern federated learning and synthetic data approaches it is clear that the central service provider (Findata) approach has positive and negative aspects. It is clearly based on the traditional paradigm of central data storage and computation platform where the researcher can run his/her algorithms on data from many Finnish sources. To the researcher there is only one layer of bureaucracy, but internally there are two: the Findata authority and the register owner, which must cooperate in fetching the data for Findata. Currently, there is only one Windows server without high performance AI capabilities, but it is likely that there will be more service providers with a wider selection of virtual machines. Whether they can answer every need remains to be seen. Currently, it is difficult to process image or genomic data together with individual-level clinical information in the same algorithm. Data harmonization is done by Findata according to their own practices which may differ from what the client has used elsewhere. So, the flexibility and possibility for customization is relatively low. The waiting time for Findata decisions and actions is very long. In 2021 on average the collection of data and getting the permit cost 6700 euros for data that was from the biggest hospital district in Finland, of which 5800 euros went to the hospital district and 900 euros to Findata (Seppänen 2021). On top of that there is a cost for using the computation environment. The smallest virtual machine costs 2790 euros per year.

It can be argued that the centralized Findata approach works as a national computation hub enabling the use of Finnish clinical data in research. It provides a single entry to the whole Finnish health care data system. This can be seen as a good thing if the service provided will in the future fulfil the requirements of data driven health care research. But it can also be seen as conflicting with the federated approach of not moving the data far away from the clinical systems. An extra layer of bureaucracy, delays and costs is inserted between the data provider and the user. A user cannot coordinate research directly with the register owner about the register because Findata is inevitably included in the process. The register owner and the user do not have control over the Findata processes. As everyone who has done clinical research with data knows, the intricacies in the data acquisition can be decisive in understanding the meaning of the data and selecting the correct way of analyzing it. In practice Findata does not negotiate about this, it is up to the researcher to recruit a clinician with inside knowledge of a specific register in a specific region.

So far there is not much first-hand experience from using the strict data security environments. The reception of the law was very critical in the research community. The status of the genetic information is unclear at the moment. In the future, more and more genomic information will be produced for clinical purposes in the context of personalized medicine. This will have great value for researchers, but the availability and the possibility of using that data is a matter of discussion now that the genomic law is under preparation. The biobank law states that the results of the analyses produced from biobank samples must be returned to the biobank and the citizen, whose data it is, can inquire about the use of the data. Genomic information is produced from the biobank samples, and it should be usable for anyone accessing the biobank data and samples. In case the clinically produced genomic data falls under the Secondary Use Act it will also be restricted to being analyzed in the Findata server or similar server with very tight security restrictions.

With the COVID-19 pandemic rapid sharing of information between epidemiologists has been crucial in finding and sharing the best treatments to the disease. This requires the ability to quickly build ad hoc teams and take into use data environments where knowledge of different patients and the outcomes of treatments could be collected (Pihlava 2020). This seems to be in contradiction to the Findata approach. Because the epidemiologists in different institutes have their data in different register owners' registers according to the Secondary Use act they should apply for a permission to combine their data in the Findata environment. The waiting time for permissions is currently 8 months which is clearly too much. In the case of pandemic, the pandemic research could be prioritized but still the inbuilt bureaucracy would hamper crucial research. It is unclear what kind of process would have been used in the case of COVID-19 if the Secondary Use act had already been in force at the time when the pandemic started. And during a milder but still threatening epidemic like a new strain of the enterovirus rapid information sharing for research purposes would be needed even if it was not recognized to be as threatening to the society as the COVID-19 pandemic.

5. CONCLUSION

The Finnish Secondary Use Act provides a way for researchers to use the clinical data that has accumulated in the health service providers' databases over several decades. This is a wealth of information that has great value to the research and pharmaceutical actors. The architecture chosen in the law both provides opportunities and presents problems. The opportunity is the fact that a single authority can provide access to all the health and social support data. This is valuable especially in the case of relatively rare diseases where a single hospital does not have enough patients to carry through research. The difficulty is in the centralized and rather bureaucratic architecture. It goes only halfway into providing federated access to a register owner's data. This problem can be mitigated if the registry owners provide powerful computing environments with enough resources to use modern AI methods and which fulfil the data security requirements of Findata.

In a comparison to other Nordic countries, Findata resembles the Danish Health Data Authority. Both provide access to all registers through a provided secure data server. Findata does not own a single register for a disease group like the Danish authority, but it collects and combines data from several regional health providers. In the Swedish model the registers are dispersed and owned by several actors which creates problems in acquiring the data but is perhaps more compatible with the federated approach.

REFERENCES

Findata register owners, 2022. https://findata.fi/en/data/

Findata, 2020. *Regulation by the Health and Social Data Permit Authority: Requirements for other service providers' secure operating environments*. https://findata.fi/wp-content/uploads/sites/3/2020/10/048ba8a0-findata-regulation-1-2020-requirements-for-other-service-providers-secure-operating-environments.pdf. (Fetched 13.1.2022)

Finnish Biobanks FinBB, 2022. https://finbb.fi/en/. Fetched 13.1.2022

Gogoshin, G., Branciamore, S.Branciamore, & Rodin, A.S, 2021. "Synthetic data generation with probabilistic Bayesian Networks". *Mathematical Biosciences and Engineering*, 18(6): 8603-8621. doi: 10.3934/mbe.2021426

Kuiper J, van den Heuvel EW and Swertz MA., 2015. "The Hybrid Synthetic Microdata Platform: A Method for Statistical Disclosure Control." *Biopreserv. Biobank* 2015 Jun. p.178-182.

Laugesen K, Ludvigsson JF, Schmidt M, Gissler M, Valdimarsdottir UA, Lunde A, Sørensen HT, 2021. "Nordic Health Registry-Based Research: A Review of Health Care Systems and Key Registries". *Clin Epidemiol.* Jul 19;13:533-554.

Lähteenmäki J, Vuorinen A-L, Pajula J, Harno K, Lehto M, Niemi M, Van Gils M, 2021. "Integrating data from multiple Finnish biobanks and national health-care registers for retrospective studies: Practical experiences." Scandinavian *Journal of Public Health*. 2021 Apr 12:14034948211004421. doi: 10.1177/14034948211004421. Epub ahead of print. PMID: 33845693.

The Ministry of Health and Social Services, 2019. *Secondary use of health and social data*. https://stm.fi/en/secondary-use-of-health-and-social-data. Fetched 13.1.2022.

The Ministry of Health and Social Services, 2022. *National Genome Centre*. https://stm.fi/en/genome-center. Fetched 13.1.2022

OHDSI, 2021. *The Book of Ohdsi.*. https://ohdsi.github.io/TheBookOfOhdsi/ (fetched 29.3.2022)

OHDSI, 2022. *OMOP Common Data Model*. https://www.ohdsi.org/data-standardization/the-common-data-model/ (fetched 29.3.2022)

Pajula J., Viiri S., Similä H., Lähteenmäki J., & Tuomi-Nikula A, 2021. *Toisiolain vaikutukset tutkimukseen ja data-analytiikan sovelluksiin: Hyteairon analytiikkatyöryhmän selvitys*. VTT Technical Research Centre of Finland. VTT reports No. VTT-R-00118-21. Fetched 13.1.2022 (in Finnish)

Pihlava, M., 2020 "Toisiolaki torppasi tutkimusta". *Lääkärilehti* 48/2020, pp. 2574 – 2578 (in Finnish)

Seppänen, J., 2022. "Mitä rekisteritutkimuksen aineisto maksaa?". *Lääkärilehti* 28.3.2022.

Sheller MJ, Edwards B, Reina GA, Martin J, Pati S, Kotrotsou A, Milchenko M, Xu W, Marcus D, Colen RR, Bakas S, 2020. "Federated learning in medicine: facilitating multi-institutional collaborations without sharing patient data". *Sci Rep.* 2020 Jul 28;10(1):12598.

Schmidt, M., Schmidt, S. A., Sandegaard, J. L., Ehrenstein, V., Pedersen, L., & Sørensen, H. T, 2015. "The Danish National Patient Registry: a review of content, data quality, and research potential". *Clinical epidemiology*, 7, 449–490. https://doi.org/10.2147/CLEP.S91125

Socialstyrelsen, 2022. *Kartläggning av datamängder av nationellt intresse på hälsodataområdet (delrapport)*. https://www.socialstyrelsen.se/ globalassets/sharepoint-dokument/artikelkatalog/ovrigt/2022-3-7781.pdf (Fetched 28.3.2022).

HUMAN MOVEMENT VARIABILITY ANALYSIS IN OFFICE-WORKERS: A REVIEW

Maria Eduarda Oliosi[1], Catia Cepeda[2], Luís Silva[2], Daniel Zagalo[2], Phillip Probst[2], Ana Rita Pinheiro[3], João Paulo Vilas-Boas[1] and Hugo Gamboa[2]

[1]*LABIOMEP (Biomechanics Laboratory), Faculty of Sport of University of Porto, Porto, Portugal*

[2]*LIBPhys (Laboratory for Instrumentation, Biomedical Engineering and Radiation Physics), Faculdade de Ciências e Tecnologia, Universidade Nova de Lisboa, Caparica, Portugal*

[3]*IBIMED (Institute of Biomedicine), School of Health Sciences, University of Aveiro, Aveiro, Portugal*

ABSTRACT

Occupational disorders have not only an effect on mortality worldwide but also a significant impact on the workers' quality of life, in terms of organizational productivity and insurance costs. In specific, work-related musculoskeletal disorders are the most prevalent among office-workers whose occupational exposure presupposes the frequent adoption of maintained postures. Given that there is a close relationship between the state of health of the worker and the human movement system, the monitorization of risk factors associated with the human movement, i.e. of the motor variability in office-workers, has been suggested to prevent or mitigate musculoskeletal complaints. This paper presents a literature review of the existing methodologies and tools that explore the motor variability in office-workers in a sitting position based on linear and nonlinear measures and emphasizes their potential importance in understanding human movement to prevent work-related musculoskeletal disorders.

KEYWORDS

Human Movement, Occupational Health, Office-Work, Motor Variability, Nonlinear, Features Extraction

1. INTRODUCTION

Work-related disorders constitute a public health problem (Bevan, 2015). When focusing on workers within Europe, exposure to several ergonomic risk factors that induce work-related disorders is especially prevalent (Hulslof et al., 2021). These risk factors range from individual and physical to psychosocial, and organizational factors (Mingels et al., 2021; Hulslof et al., 2021). Office-working activities are performed repeatedly over long periods of time, often in a constrained or static sitting posture and capable of generating employees' discomfort, fatigue, perceived tension and/or feeling of pain (Srinivasan and Mathiassen, 2012a,b; Arippa et al., 2022). Long periods of time in a sitting position are associated with several conditions, such as musculoskeletal disorders (MSD), diabetes, cardiometabolic diseases and mortality (Madeleine et al., 2021; Arippa et al., 2022). MSDs have a high impact on society and individual workers as these have negative repercussions on their daily life and constrain them in the execution of tasks at work (Bevan, 2015; Arippa et al., 2022). With this in mind, it is essential to encourage workers to adopt more appropriate postures over time.

The variability analysis has been becoming more popular and demonstrates a high impact on quantifying normal variations that occur in motor performance over time (Stergiou and Decker, 2011; Harbourne and Stergiou, 2009). The motor control strategies are revealed in the structure of motor variability (Latash, Scholz, and Schöner, 2002). Motor variability can be observed and quantified using linear and nonlinear statistical tools in several extensions and tasks (Karwowski et al., 2019; Karimi et al., 2021; Emanuelsen et al., 2019; Giovanini et al., 2017).

The presented work is part of the Prevention of Occupational Disorders in Public Administrations based on Artificial Intelligence (PrevOccupAI) project, in which the main objective is the promotion of occupational health through the identification of occupational risks. Through non-invasive sensors an estimation of occupational risks for office-workers will be performed, with the goal of minimizing injuries and costs. For this purpose, we have already developed a cross-platform application to assess occupational health and it is prepared to acquire data from self-reports and sensors from a smartphone, a smartwatch and from a wearable device that acquires biosignals such as electromyography (EMG) and movements (inertial measurements) (Silva et al., 2022).

An important aspect of this project is to provide feedback to the workers regarding their health status and how their posture has an influence on it. To assess risk factors in human movement variability, meaningful linear and nonlinear measures should be extracted from these devices using libraries such as Time Series Feature Extraction Library (TSFEL) (Barandas et al., 2020). It is crucial to decide which measures are more relevant and how they should be combined (see e.g. in Rodrigues et al. (2021)).

This review is thus focused on the existing linear and nonlinear metrics, and their importance in understanding human movement. The document is structured as follows: Section 2 will make an introduction to the concept of variability and will explain the relevance of occupational variability; Section 3 describes the approach of using the linear and nonlinear tools; Section 4 will cover some research applications of these tools in the context of sedentary work tasks, and Section 5 will conclude with their potential in the context of the PrevOccupAI project.

2. VARIABILITY

Does motor variability matter? To answer this question it is relevant to understand the concept of "repetition without repetition" (Emanuelsen et al., 2019; Bibbo et al., 2020) and why there is valuable information in what was previously considered. Regarding the former, it is essential to understand that, no matter how much we perform a specific task, the way it is executed is never identical. Thus, there is a degree of predictability with high complexity. When it comes to noise, it is interesting to note that previously it was considered an unwanted phenomenon (Latash, Scholz and Schöner, 2002). However, the so-called noise that we can find among the events of repetitions over repetitions has physiological and motor meaning. The statistical properties that characterize these events over time can differentiate between natural and impaired systems. In other words, the optimal variability does not correspond to a set of repetitions with too much or less variability. There is a specific kind of variability inherent to health systems that can be characterized as optimal variability or complexity (Harbourne and Stergiou, 2009; Stergiou and Decker, 2011).

The model that explains the optimal variability is called the Theoretical Model of Complexity. It describes the relationship that could exist between complexity and predictability. In clinical settings, a healthy pattern presents statistical properties that correspond to a high level of complexity with some degree of predictability. A flawed system is characterized by a lower degree of complexity and can have either a high or low degree of predictability, that is, its events over time can be random or with equal time spacing between them. When a high degree of predictability is present, the dynamical system is more rigid and periodic. When predictability is low, the relationship among the events that make up the phenomenon is random and independent. That is, too much or too little variability is not desirable. The relevance of this model in terms of health is that in disease, this optimal variability is destroyed, losing its complexity, and becoming more or less predictable. With this in mind, optimal variability has a direct relationship with physiological and motor health. Therefore, variability reflects various options for movement, providing flexible, adaptive strategies that are not reliant on rigorousness for each task or each changing condition encountered (Harbourne and Stergiou, 2009; Guastello, 2017; Karwowski et al., 2019). This natural variability is destroyed with disability as seen during aging and disease (Stergiou and Decker, 2011; Busa and van Emmerik, 2016; Madeleine et al., 2021). In this situation, the events over time become more random, presenting more variability, or turn out close to invariant, that is, showing less variability.

2.1 Occupational Variability

It is reasonable to wonder that, if variability can be a biomarker of disability due to aging and disease, other similar conditions that contribute to poor health conditions will show alterations in variability. People spend a considerable time continuously working, namely when performing repetitive work (Srinivasan and Mathiassen, 2012a,b; Emanuelsen et al., 2019). Physiological and inertial measures provide information about changes over time in work conditions that can be used by algorithms for occupational risk estimation at work. Thus, we can define as *Occupational Variability* the quantification of physiological and inertial variability in the work context for risk stratification.

Interestingly, to avoid exposure to repetitive work or static postures for long periods, extrinsic methods such as breaks with rest periods, job rotations, and task variation have been suggested (Srinivasan and Mathiassen, 2012a,b; Arippa et al., 2022). However, the use of these measures usually lacks in adjusting individual characteristics of the worker (e.g personal factors, such as previous experience), increasing occupational risk. Therefore, intrinsic variation has been brought to the occupational context (Srinivasan and Mathiassen, 2012a,b; Gaudez, Gilles, and Savin, 2016).

Motor variability, that is the natural variation that occurs in movements and postures over time while performing a certain task, brings resources of analysis that can explain both general workload risk associated with different workstations and workers' motor strategies to overcome fatigue over time. Thus, refining the definition of occupational variability, we can state that it is the set of natural variations over time, by means of physiological and inertial measures, related to a certain task in a work context.

In this way, task and individual characteristics influence intrinsic motor variability, and its evaluation brings new insights to ergonomics decisions such as workplace design and real-time risk evaluation. Employees working at an unbefitting workstation may suffer from muscle fatigue and musculoskeletal discomfort that can result in work-related MSD (Gaudez, Gilles and Savin, 2016). The evolution of wearable devices brings new opportunities that can be explored. To achieve this purpose, a toolkit of options for processing and decision-making can be applied according to signal (time-series) to establish occupational risk metrics.

3. LINEAR AND NONLINEAR MEASURES OF VARIABILITY

The two main dimensions to measure variability, depending on the signal and problem under study, are known as linear and nonlinear (Saito et al., 2021; Harbourne and Stergiou, 2009). Both types of measures are valuable and should be used considering the information that is required.

Linear measures are more commonly used in the clinical field for prediction and problem-solving. Besides the relevance of these measures, they fail to characterize the variability structure because their focus is on only one dimension as a straight line (Harbourne and Stergiou, 2009). However, linear approaches do not appropriately handle physiological and psychological temporal structures (França et al., 2019). For instance, two temporal structures representing a system (time-series) could have the same mean and standard deviation (SD) when they evolve differently over time. Several linear measures are considered in the literature to quantify the amount of variability (e.g., Chen, 2021) such as SD, coefficient of variation, root mean square (RMS), range, interquartile range, and standard error. Such linear analysis is often complemented with nonlinear analysis, as it provides measurements of the pattern of movement over time (Harbourne and Stergiou, 2009; Emanuelsen et al., 2019).

Nonlinear measures allow us to understand the adaptability of a biological system to change conditions. These nonlinear tools quantify the variation in how a motor behavior emerges in time scales (Saito et al., 2021). Therefore, nonlinear measures provide further information and allow an understanding of complexity (Harbourne and Stergiou, 2009). Thus, the nonlinear model, often used with the term "dynamics", can be defined as a system with a nonlinear proportion between its in- and outputs, thus inherently having a higher complexity. In other words, the linear statistical tools reflect the amount or magnitude of variation in movement patterns. Conversely, the nonlinear tools described the temporal organization or structure of the variability (as opposed to amount) (Saito et al., 2021; Harbourne and Stergiou, 2009). Following, the most common nonlinear measures that can be applied to quantify variability are described.

Entropy is the loss of information in a time-series or signal (Yentes et al., 2013). Entropy metrics are commonly pursued to estimate the temporal order of variation in a time-series (Karimi et al., 2021). It is the natural logarithm of a conditional probability, interpreted as the rate of information generation and it estimates the complexity of the underlying system producing the dynamics in question (Søndergaard et al., 2010; Yentes et al., 2013; Costa, Goldberger and Peng, 2005). The approximate entropy (ApEn) quantifies the regularity or predictability. Increasing ApEn values reveal more significant irregularity, else ways, lower values reveal a more regular or periodic behavior (Harbourne and Stergiou, 2009; Stergiou and Decker, 2011). Sample entropy (SampEn) is a unitless, non-negative number describing a time-series' structural irregularity or complexity, which has been used to analyze kinetic or kinematic signals (Emanuelsen et al., 2019), more details about this algorithm in Yentes et al. (2013). The multiscale entropy (MSE) measure differs from SampEn and ApEn by including multiple measurement time scales. MSE applies the SampEn algorithm to calculate the entropy value at each time scale. The inclusion of these various measurements allows for distinct benefits such as the overall quantification of the complexity of a system in short and long scales, calculated as the sum of the entropy values of the individual time scales. Combining these features allows researchers to identify the time scales at which the breakdown in complexity occurs and the overall complexity that takes all of the time scales into account. The multiscale approach has been used to identify how physiological changes impact the overall complexity of physiological processes (Busa and van Emmerik, 2016; Karimi et al., 2021; Costa, Goldberger and Peng, 2005).

The *largest Lyapunov exponent* (LyE) measures the divergence of movement trajectories, quantifying chaos between two extremes by rating at which nearby orbits converge or diverge (Harbourne and Stergiou, 2009; Guo et al., 2015). The values between random and periodic define complexity or highly variable fluctuations in several physiological processes resembling mathematical chaos. A faster divergence of trajectories represents a higher instability of the system that is measured (e.g. postural stability) (Karimi et al., 2021; Karwowski et al., 2019). The ability to quantify the separation of trajectories makes the LyE also function to measure the average predictability of a system that exhibits nonlinear dynamics (Harboune & Stergiou, 2009). Two algorithmic options are commonly used: Wolf (Wolf, Swift, Swinney and Vastano, 1985) and Rosenstein (Rosenstein, Collins, and Luca, 1993).

The idea beyond *Fractal Analysis* is that self-similarity is present when an object is broken in smaller scales over and over again, independently of their size they look the same. This concept is applicable in time-series. If we have sequential events over time (e.g., electrocardiogram RR, a direction of postural sway), the time-series that is made up of those events can be divided in different window lengths, known as lags or scales. If their fluctuations at different scales are similar, then the structure of that time-series is fractal (Hausdorff, 2007; França, 2019; Ihlen, 2012; Stanley et al., 1999; Kantelhardt et al., 2002). Detrended Fluctuation Analysis (DFA) was introduced by Peng et al. (1995) to study the presence or not of long-range correlations in nucleotide sequences. The presence of long-range correlations or memory of the time-series means that the events of that time-series are dependent on each other but are not the same. DFA is based on the concept of a fractal time-series as statistical properties that can be described by a power law (Ihlen, 2012). MultiFractal Detrended Fluctuation Analysis (MFDFA) assumes that the phenomenon that is measured has a monofractal structure, requiring just a single power low (Ilhen, 2012). However, this assumption is overly restrictive for human behavior. Monofractal analysis considers that fractal scaling is independent of time and space. To avoid this assumption, MFDFA operates with different scaling orders revealing information about small and large fluctuations. These spatial and temporal variations can be defined by a multifractal spectrum of power law exponents. These characteristics were identified by Stanley et al. (1999) and the mathematical procedures were developed by Kantelhardt et al. (2002).

4. RESEARCH APPLICATION

In the following, a closer look will be taken at how linear and nonlinear measures were applied in research applications to measure the motor variability while being seated, with a focus on office-work scenarios and computer tasks. Measures have been extracted using a variety of systems and principally sensors to quantify the center of pressure (COP) and EMG signals.

4.1 Linear Measures

Considering linear measures, motor variability has been measured to assess new equipment and at the occupational level, to be associated with cognitive tasks and to evaluate pain.

In terms of equipment analysis, Sardini et al. (2015) developed and validated a T-shirt with integrated inductive sensors, running along the front and the back to assess the sitting posture in the lab and in real-world scenarios. The shirt is able to send data wirelessly to a computer and also contains a vibro-feedback system. The posture is estimated through the change of impedance in the inductive sensor that results from geometric changes occurring when the posture is altered. The output is expressed in the range of motion (ROM) that is based on a reference ROM to ensure comparability between results. When deploying the shirt in a home scenario, they were able to show that it is appropriate to assess posture variability.

Chen et al. (2021) analyzed posture variability among participants using three types of chairs (stool, computer chair and gaming chair), associated with the most comfortable sitting postures. The authors recorded 2D markers positions at the head, trunk and knee joint angles using a motion analysis system, which also determines the joint angles to further analysis. Unexpectedly, when the participants sat at a posture that they perceived to be the most comfortable, high variability was found within chairs type and gender. To evaluate how the posture is affected by cognitive engaging tasks in relation to dynamics and sway, Bibbo et al. (2020) developed a sensorized office chair, without a backrest, equipped with four load cells. From this setup, they calculated the COP and extracted the mean distance, RMS distance, mean velocity, sway area, 95% confidence circle and ellipse area. They were capable of supporting the hypothesis that tasks that have a higher cognitive demand lead to lower stability in the seated posture.

Studies had been conducted to associate musculoskeletal pain with linear measures extracted from sitting posture during computer work tasks. Recently, Arippa et al. (2022) aimed to analyze movement patterns during computer work and associate these patterns with musculoskeletal pain. Seat pan pressure and trunk sway parameters were quantified by using a pressure-sensitive mat of two groups: 14 breakers that stoop up at least once during the procedure, and 14 prolongers that remained sitting throughout the process. The linear measures analyzed were associated with COP: mean position, sway path, sway area, sway velocity, maximum displacement, and in-chair movements. The authors concluded that the trunk sway parameters tend to decrease over time, and breakers have more consistent movement during the procedure and present lower discomfort than prolongers.

The relationship between back pain and sitting behavior in call-centre employees was explored by Bontrup et al. (2019). Similar to the previously described study, they used a pressure mat, however in this case, a textile mat connected to a mobile phone application, where data was stored. Four groups of participants were considered to analyze the sitting behavior (no pain, no functional disability, with pain, with functional disability), regarding the features: mean number of movements per working hour, mean number of positional changes per working hour, mean time period of stable sitting, and percentage of transient periods during the whole working period. Although a high variability was found within the groups, chronic low back pain subjects tended to show a more static sitting behavior. A sitting position classification based on all pressure values and using a random forest algorithm and Leave-one-out cross-validation achieved an overall classification accuracy of 90%.

Kelson et al. (2019) also assessed pain in computer workers and its association with EMG sensors to record the trapezius muscle activity. To analyze the data, they used the tool exposure variation analysis that expresses continuous exposure-vs-time data in categories defined by exposure amplitude and accumulated time, thus giving way to an understanding of the differences in the temporal structure of exposures at different amplitude levels. They extracted five amplitude and five duration categories, to compare pain and control groups, concluding that neck-shoulder pain spent less time at low amplitudes and had longer continuous durations of muscle activation. Furthermore, the results of this study suggest that workers with pain have chronic motor control changes occurring in the adaptation of pain.

Mingels et al. (2021) used a camera system to make a comparative non-randomized study of the spinal postural variability (SPV) during a computer task of 18 people suffering cervicogenic headaches (CH) and a matched control group (CG). The spinal posture was acquired using 12 infrared Bonita T10 and two video cameras. From the gathered data, left sagittal angles were calculated and significant differences could be observed between both groups. For example, the SPV of the CH group was higher than the CG. Furthermore, they observed that for both groups, the upper spinal variability was generally lower than the lower spinal variability, except for the upper thoracic variability, which was higher for the CH group when compared to the variability of the lower lumbar.

4.2 Nonlinear Measures

Adding complexity to the analysis of motor behavior, some authors considered nonlinear variables to assess different equipment, pain and, at the occupational level, to be associated with cognitive tasks and aging.

Using optical motion capture, Lau et al., (2015) concluded that the movement performance is a multifractal behavior, which can be associated with the multiple strategies behind the motor control, and neural activities, by the analysis of the dynamic properties of spinal curvature during postural sway based on the MFDFA. Sondergaard et al. (2010) used both linear and nonlinear analysis techniques to measure perceived discomfort during prolonged sitting. Subjects were seated on a force platform with no armrests, nor back and foot support, leaving only a force platform for contact. The mean, SD and SampEn of the COP displacement in both anterior-posterior (AP) and medial-lateral (ML) directions were extracted. The resulting amount of variability for both lumbar curvature and COP displacement increased over time with an increasing perceived discomfort, suggesting a relationship. Further studies explored nonlinear measurements while subjects were performing computer tasks, related to occupational contexts. Regarding the dynamics of sitting during cognitive tasks, Madeleine et al. (2018) recorded mental fatigue ratings, overall performance and kinetic, the last using a force and torque 3D transducer on an office chair. Sitting dynamics were analyzed using the average displacement, SD and SampEn values and the displacement of the COP in APand ML) directions. They concluded that, for their procedure, computer work did not change the dynamics of sitting. However, they observed increased size and structure of variability in the ML than in the AP direction. Still associated with cognitive tasks, Caballero et al. (2021) analyzed differences in motor variability related to the learning rate. Motor variability was recorded via two-axis force sensor (ML and AP) to measure variability structure based on DFA, and motor synergy variability based on good and bad variability ratio (GV/BV). No results were found related to DFA, but higher initial GV/BV was related to greater performance improvements than those with lower GV/BV.

Also, with nonlinear measures, EMG was used to study the spatial distribution of upper trapezius muscle activity during computer work as an effect of biofeedback. Samani et al. (2010) extracted two-dimensional maps of RMS, relative rest time and permuted sample entropy (PeSampEn) from EMG. To quantify changes in the spatial distribution of muscle activity, the centre of gravity (CoG) and entropy of maps were also calculated. They used PeSampEn as a measure of temporal heterogeneity and the entropy of RMS as a measure of spatial heterogeneity. The findings of this study suggest that there is a potential benefit of superimposed muscle contraction in relation to the spatial organization of muscle activity during computer work.

Madeleine et al. (2021) set out to investigate the sitting dynamics of computer work, and how these differ between age groups. To this end, an instrumented office chair was designed with a 3D transducer using strain gauge technology. This allowed for recording applied forces and moments on the link between the seat pan and chair shaft. The range, velocity, area, SD and SampEn values were extracted using this data. The results contradicted the authors' initial hypothesis regarding changes across the task's duration, as no major changes were observed. Age, however, was shown to play a bigger role: for older subjects, the range and velocity of the COP displacement were larger. In contrast, complexity was lower among younger users.

The literature research explored in this paper focuses on the motor variability during sitting posture, more specifically in office-worker scenarios. There are several applications for innovative ergonomics analysis. The major instruments referred to in this review were the plate of force, such as instrumented chairs and also EMG sensors. To our surprise, a well-known nonlinear measure was not used in the office-work studies, the LyE. Although this paper has a focus on the quantification of COP and EMG signals, an interesting alternative approach would be the use of wearables to provide direct measurements for ergonomics research (Santos et al., 2020).

5. CONCLUSION

In this literature review, linear and nonlinear measures to assess motor variability while executing sedentary tasks (i.e. office scenario) were presented. Motor variability assessment remains a challenging task due to the complexity and versatility of human motion. Thus, the presented works used specialized systems like optical motion tracking systems, sensorized office chairs, and wearable fabrics with integrated inductive sensors. While these systems allow for a detailed analysis of human movement they are, in most cases, only deployable in lab scenarios (e.g. optical motion tracking), are associated with extra costs (e.g. sensorized office chair), or can, to this stage, not be easily produced (e.g. sensorized shirt). Furthermore, the shown systems were mostly tested in limited scenarios.

Therefore, an effort has to be made to develop systems that make use of easily accessible technology like smartphones and smartwatches. Subsequently, these systems need to be tested over long periods of time in real-world scenarios (i.e. offices) while subjects are working in their natural environment. Research has to be carried out on how to extract meaningful linear and nonlinear measures from these devices to comprehensively assess human movement variability in the mentioned scenario. Finally, these measures need to be presented in an easy-to-understand way together with recommendations on what users can do to improve their occupational health and mitigate work-related risk factors.

ACKNOWLEDGEMENT

This work was partly supported by Fundação para a Ciência e Tecnologia, under project PREVOCUPAI (DSAIPA/AI/0105/2019).

REFERENCES

Arippa, F., Nguyen, A., Pau, M. and Harris-Adamson, C., 2022. Postural strategies among office workers during a prolonged sitting bout. *Applied Ergonomics*, 102, p. 103723.

Barandas, M., Folgado, D., Fernandes, L., Santos, S., Abreu, M., Bota, P., Liu, H., Schultz, T. and Gamboa, H., 2020. TSFEL: Time series feature extraction library. *SoftwareX*, 11, p. 100456.

Bevan, S., 2015. Economic impact of musculoskeletal disorders (MSDs) on work in Europe. *Best Practice & Research Clinical Rheumatology*, 29(3), pp. 356-373.

Bibbo, D., Conforto, S., Schmid, M. and Battisti, F., 2020. The Influence of Different Levels of Cognitive Engagement on the Seated Postural Sway. *Electronics*, 9(4), p. 601.

Bontrup, Carolin, et al., 2019. Low back pain and its relationship with sitting behaviour among sedentary office workers. *Applied ergonomics*, 81, p. 102894.

Busa, M.A. and van Emmerik, R.E., 2016. Multiscale entropy: a tool for understanding the complexity of postural control. *Journal of Sport and Health Science*, 5(1), pp. 44-51.

Caballero, C., Moreno, F.J. and Barbado, D., 2021. Motor Synergies Measurement Reveals the Relevant Role of Variability in Reward-Based Learning. *Sensors*, 21(19), p. 6448.

Chen, Y.L., Chan, Y.C. and Zhang, L.P., 2021, December. Postural Variabilities Associated with the Most Comfortable Sitting Postures: A Preliminary Study. *In Healthcare* (Vol. 9, No. 12, p. 1685).

Costa, M., Goldberger, A.L. and Peng, C.K., 2005. Multiscale entropy analysis of biological signals. *Physical review E*, 71(2), p. 021906.

Emanuelsen, A., Madeleine, P., Voigt, M. and Hansen, E.A., 2019. Motor variability in elicited repeated bout rate enhancement is associated with higher sample entropy. *Human Movement Science*, 68, p. 102520.

França, L.G.S., Montoya, P. and Miranda, J.G.V., 2019. On multifractals: a non-linear study of actigraphy data. *Physica A: Statistical Mechanics and its Applications*, 514, pp. 612-619.

Gaudez, C., Gilles, M.A. and Savin, J., 2016. Intrinsic movement variability at work. How long is the path from motor control to design engineering?. *Applied ergonomics*, 53, pp. 71-78.

Giovanini, L.H., Silva, S.M., Manffra, E.F. and Nievola, J.C., 2017. Sampling and digital filtering effects when recognizing postural control with statistical tools and the decision tree classifier. *Procedia Computer Science*, 108, pp. 129-138.

Guastello, S.J., 2017. Nonlinear dynamical systems for theory and research in ergonomics. *Ergonomics*, 60(2), pp. 167-193.

Guo, Y., Naik, G.R., Huang, S., Abraham, A. and Nguyen, H.T., 2015. Nonlinear multiscale Maximal Lyapunov Exponent for accurate myoelectric signal classification. *Applied Soft Computing*, 36, pp. 633-640.

Harbourne, R.T. and Stergiou, N., 2009. Movement variability and the use of nonlinear tools: principles to guide physical therapist practice. *Physical therapy*, 89(3), pp. 267-282.

Hausdorff, J.M., 2007. Gait dynamics, fractals and falls: finding meaning in the stride-to-stride fluctuations of human walking. *Human movement science*, 26(4), pp. 555-589.

Hulshof, C.T., Pega, F., Neupane, S., van der Molen, H.F., Colosio, C., Daams, J.G., Descatha, A., Kc, P., Kuijer, P.P., Mandic-Rajcevic, S. and Masci, F., 2021. The prevalence of occupational exposure to ergonomic risk factors: A systematic review and meta-analysis from the WHO/ILO Joint Estimates of the Work-related Burden of Disease and Injury. *Environment international*, 146, p. 106157.

Ihlen, E.A.F.E., 2012. Introduction to multifractal detrended fluctuation analysis in Matlab. *Frontiers*, 3, p. 141.

Kantelhardt, J.W., Zschiegner, S.A., Koscielny-Bunde, E., Havlin, S., Bunde, A. and Stanley, H.E., 2002. Multifractal detrended fluctuation analysis of nonstationary time series. *Physica A: Statistical Mechanics and its Applications* 316(1-4), pp.87-114.

Karimi, Z., Mazloumi, A., Sharifnezhad, A., Jafari, A.H., Kazemi, Z., Keihani, A. and Mohebbi, I., 2021. Determining the interactions between postural variability structure and discomfort development using nonlinear analysis techniques during prolonged standing work. *Applied Ergonomics*, 96, p. 103489.

Karwowski, W., Kern, D., Murata, A., Ahram, T., Gutiérrez, E., Sapkota, N. and Marek, T., 2019. The complexity of human performance variability on watch standing task. *Applied ergonomics*, 79, pp. 169-177.

Kelson, D.M., Mathiassen, S.E. and Srinivasan, D., 2019. Trapezius muscle activity variation during computer work performed by individuals with and without neck-shoulder pain. *Applied ergonomics*, 81, p.102908.

Latash, M.L., Scholz, J.P. and Schöner, G., 2002. Motor control strategies revealed in the structure of motor variability. *Exercise and sport sciences reviews*, 30(1), pp.26-31.

Lau, N.M., Choy, C.S.T. and Chow, D.H.K., 2015. Identifying multifractality structure on postural sway. *Journal of Ergonomics*, 5(137), p.2.

Madeleine, P., Marandi, R.Z., Norheim, K.L., Andersen, J.B. and Samani, A., 2021. Sitting dynamics during computer work are age-dependent. *Applied Ergonomics*, 93, p.103391.

Madeleine, P., Marandi, R.Z., Norheim, K.L., Vuillerme, N. and Samani, A., 2018, August. Characterization of the dynamics of sitting during a sustained and mentally demanding computer task. *In Congress of the International Ergonomics Association* (pp. 338-344). Springer, Cham.

Mingels, S., Dankaerts, W., van Etten, L., Bruckers, L. and Granitzer, M., 2021. Lower spinal postural variability during laptop-work in subjects with cervicogenic headache compared to healthy controls. *Scientific reports*, 11(1), pp. 1-11.

Peng, C.K., Havlin, S., Hausdorff, J.M., Mietus, J.E., Stanley, H.E. and Goldberger, A., 1995. Fractal mechanisms and heart rate dynamics: long-range correlations and their breakdown with disease. Journal of electrocardiology, 28, pp.59-65.

Rodrigues, J., Probst, P., Cepeda, C., Guede-Fernández, F., Silva, S., Gamboa, P., Fujão, C., Quaresma, C.R. and Gamboa, H., 2021, March. microErgo: A Concept for an Ergonomic Self-Assessment Tool. *In 2021 Seventh International conference on Bio Signals, Images, and Instrumentation (ICBSII)* (pp. 1-6). IEEE.

Rosenstein, M.T., Collins, J.J. and De Luca, C.J., 1993. A practical method for calculating largest Lyapunov exponents from small data sets. *Physica D: Nonlinear Phenomena*, 65(1-2), pp.117-134.

Saito, H., Watanabe, Y., Kutsuna, T., Futohashi, T., Kusumoto, Y., Chiba, H., Kubo, M. and Takasaki, H., 2021. Spinal movement variability associated with low back pain: *A scoping review*. Plos one, 16(5), p.e0252141

Samani, A., Holtermann, A., Søgaard, K. and Madeleine, P., 2010. Active biofeedback changes the spatial distribution of upper trapezius muscle activity during computer work. *European journal of applied physiology,* 110(2), pp. 415-423.

Santos, S., Folgado, D., Rodrigues, J., Mollaei, N., Fujão, C. and Gamboa, H., 2020, February. Exploring Inertial Sensor Fusion Methods for Direct Ergonomic Assessments. *In International Joint Conference on Biomedical Engineering Systems and Technologies* (pp. 289-303). Springer, Cham.

Sardini, E., Serpelloni, M. and Pasqui, V., 2015, May. Daylong sitting posture measurement with a new wearable system for at home body movement monitoring. In 2015 *IEEE International Instrumentation and Measurement Technology Conference (I2MTC) Proceedings* (pp. 652-657). IEEE.

Silva, S., Cepeda, C., Rodrigues, J., Probst, P., & Gamboa, H. Assessing Occupational Health with a Cross-Platform Application based on Self-Reports and Biosignals. *International Joint Conference on Biomedical Engineering Systems and Technologies*. Springer, Cham, 2022.

Søndergaard, K.H., Olesen, C.G., Søndergaard, E.K., De Zee, M. and Madeleine, P., 2010. The variability and complexity of sitting postural control are associated with discomfort. *Journal of biomechanics*, 43(10), pp.1997-2001.

Stanley, H.E., Amaral, L.N., Goldberger, A.L., Havlin, S., Ivanov, P.C. and Peng, C.K., 1999. Statistical physics and physiology: monofractal and multifractal approaches. *Physica A*, 270(1-2), pp.309-324.

Stergiou, N. and Decker, L.M., 2011. Human movement variability, nonlinear dynamics, and pathology: is there a connection?. *Human movement science*, 30(5), pp.869-888.

Srinivasan, D. and Mathiassen, S.E., 2012. Motor variability in occupational health and performance. *Clinical biomechanics*, 27(10), pp.979-993.

Srinivasan, D. and Mathiassen, S.E., 2012. Motor variability–an important issue in occupational life. *Work*, 41, pp. 2527-2534.

Wolf, A., Swift, J.B., Swinney, H.L. and Vastano, J.A., 1985. Determining Lyapunov exponents from a time series. *Physica D: nonlinear phenomena*, 16(3), pp.285-317.

Yentes, J.M., Hunt, N., Schmid, K.K., Kaipust, J.P., McGrath, D. and Stergiou, N., 2013. The appropriate use of approximate entropy and sample entropy with short data sets. *Annals of biomedical engineering*, 41(2), pp.349-365.

CHARACTERIZING MEDICAL ANDROID APPS

Raina Samuel, Iulian Neamtiu, Sydur Rahaman and James Geller
New Jersey Institute of Technology, USA

ABSTRACT

There is a proliferation of medical mobile apps: Google Play alone has thousands of apps in the "Medical" category. Many such apps perform critical tasks (e.g., are used with a medical device or in lieu of a device); handle sensitive patient related information; perform diagnosis; or treat diseases. However, there are wide gaps between an app's claims and users' expectations as well as between app implementations and regulatory frameworks' mandates. We perform the first study, based on analyzing more than 4,000 Android apps, that characterizes medical apps. We begin by introducing an automated classification scheme that integrates textual information extracted from multiple sources to establish the purpose and target audience for an app, based on fine-grained traits and high-level categories; we found that the most common functionalities involved connecting to medical devices (e.g., hearing aids, glucometers), offering tele-health services, or patient management. We then dive deeper into app nature and characterize according to the function and domain of the app. We reveal actionable findings found in various facets of medical applications, regulatory frameworks and user privacy and safety.

KEYWORDS

Mobile Computing, Characterization, Mobile Health Applications, Android, Health Informatics

1. INTRODUCTION AND BACKGROUND

Over the past decade, the digital/mobile health area has grown substantially, as devices have become more advanced and more ubiquitous (Statista, 2021). On the Android platform alone, this pervasiveness has led to thousands of health-related apps. Furthermore, virtually all hospitals have enabled patients to access their health information via portal apps in both the outpatient and inpatient setting (Johnson, 2021).

Many users of these medical apps are unfamiliar with the app landscape and unsuspectingly trust that the apps are safe. Users should not be expected to question the legitimacy of a medical app or "dissect" an app to understand what it is doing with personal data. Medical apps can be valuable tools, but there is no universal standard that defines what is effective and does not put personal data at risk. A 2015 study of apps that evaluate symptoms for self-diagnosis and triage revealed that many deficits exist in both aspects (Semigran, 2015). Such lapses are potentially a public health issue, as apps are often used to make healthcare decisions.

Moreover, with the widening scope of medical apps, their capabilities and intended audience remain unclear. For example, the app **Instant Heart Rate: HR Monitor & Pulse Checker** has over 10,000,000 installs; the app's description states that it is the "most accurate" heart app and has been used in research. While its functionality is legitimate, the disclaimer states that "Instant Heart Rate should be used for entertainment purposes." In order to triage potential app abuses or misleading claims, a clear consistent classification scheme of apps and their functionalities is essential.

We present a characterization of medical apps in Section 2. We begin by categorizing apps using a multifaceted analysis employing three main sources from an app's metadata: app description, XML (extensible markup language) assets, and image assets. By this process we extract relevant medical keywords. The keywords are used to determine orthogonal fine-grained traits that describe app functionality (e.g., sending patient data, found in 775 apps, or handling insurance, found in 376 apps). Combining these traits leads to a higher-level categorization scheme that allows us to better understand the app's intended audience. We establish six categories: virtual visit, patient portal, medical device, professional, reference, and patient. Virtual visit apps allow users to interact with medical professionals remotely. Patient portal apps allow users to access information regarding visits or book appointments. Medical device apps interface with an external device, such as a glucometer. Professional apps are intended for medical professional uses in office management or patient care. Reference apps provide study and reference material. Finally, patient apps are intended for general personal health reminders. We find that the most popular category is *patient*, with 1,993 apps, followed by *reference* with 1,590 apps. Our classification scheme shows that the most common functionalities of medical apps involve connecting to medical devices, tele-health, and medical calculators.

Section 3 discusses actionable findings from our research. We investigate possible lapses found in the way regulatory agencies approve and determine medical apps and their functionalities. We discuss privacy implications of handling user data and how developers and marketplaces should be more transparent in how sensitive data is handled.

Our paper makes several contributions:

1. An automated approach and study that characterize medical apps into sub-categories to better understand their purposes and functionalities.
2. A discovery of the most common functionalities of medical apps, such as connecting to medical devices, providing telehealth management, or patient management.
3. A discussion of regulatory frameworks and user privacy practices and how these can be improved for the benefits of both developers and users.

Prior Work. Safety concerns regarding medical mobile apps have been a prominent subject of study. Magrabi et al. (Magrabi, 2019) (1) argue that it is difficult to regulate healthcare apps due to the fact anyone can develop an app and (2) confirm that there is little to no monitoring of use or formal evaluation of such apps, which is exceptionally concerning as more and more medical apps are being produced and recommended by physicians to patients to help track and monitor symptoms. Mobile mental health apps are a growing field aimed to help patients manage their mental health conveniently on their mobile devices. However, Terry et al. (Terry, 2018) revealed a lack of clear regulations of mobile mental health apps and created a typology of mobile mental health apps. Terry et al. also discuss that it is very difficult to judge the quality and efficacy of mental health apps, especially since many of the apps were developed outside of traditional healthcare spaces, revealing deficiencies in current regulatory frameworks. However, none of the aforementioned studies provide an overview of current regulatory frameworks globally found in our study.

Developing safe medical apps and understanding the risks that apps entail has been a topic of discussion and research. Yet there is a lack of discussion on the primary audience(s) apps are intended for. For example, Lewis et al. (Lewis,) evaluates and creates a risk framework of medical apps based on functionality. Additionally, Wicks et al. (Wicks, 2015) provide methods on how one can develop a medical app safely and securely. Akbar et al. (Akbar, 2020) performed a series of meta-analyses on 74 app studies; they exposed a variety of safety concerns in medical apps and grouped apps based on functionality. In our work, however, we determined app functionality on thousands of medical apps currently on Google Play.

Tangari et al. (Tangari, 2021) discovered severe privacy issues in 88% of medical apps used in their study, i.e., medical apps could potentially share user data with third parties, namely advertising and tracking services. Despite conducting a study with thousands of medical applications, the authors did not provide a characterization scheme as we have.

Tools assessing mobile medical app quality have been developed, namely Stoyanov et al.'s MARS (Medical App Rating Scale). However, they focus on iOS apps and do not rate apps based on overall intended usage and audience as well as potential risks (Stoyanov, 2015).

2. CHARACTERIZATION

App characterization -- understanding the nature, purpose, and target audience of an app -- is challenging, as detailed next. To address these challenges, we use multi-source information along with a multi-rater human approach. First, we use information retrieval to extract terms of interest, and then define first-order low-level *traits*. Building upon traits, we then establish high-level *categories*.

Challenges. Characterization is a major challenge for several reasons. First, apps may serve more than one purpose, e.g., an app may manage a patient's prescriptions, help locate the nearest emergency room, and support video chats with the provider. Second, app features are hard to detect automatically (e.g., video chat software can be home-made as opposed to using a video chat library); similarly, location/mapping services can serve several purposes, thus the presence of such a library simply indicates that the app provides location-relevant services. Third, the app description on Google Play is at the developers' discretion, and can be incomplete, inaccurate, or downright misleading. Fourth, actual app functionality can only be reconstituted from heterogeneous sources via a multi-faceted analysis of app description, embedded images, app bytecode, etc.

We started by retrieving all apps (APK files) from Google Play's *Medical* category along with their descriptions. We only retained those apps that had English descriptions and at least 1,000 installs, for a total of 2,215 apps. Our approach for extracting relevant text is shown in Figure 1 and described in detail next.

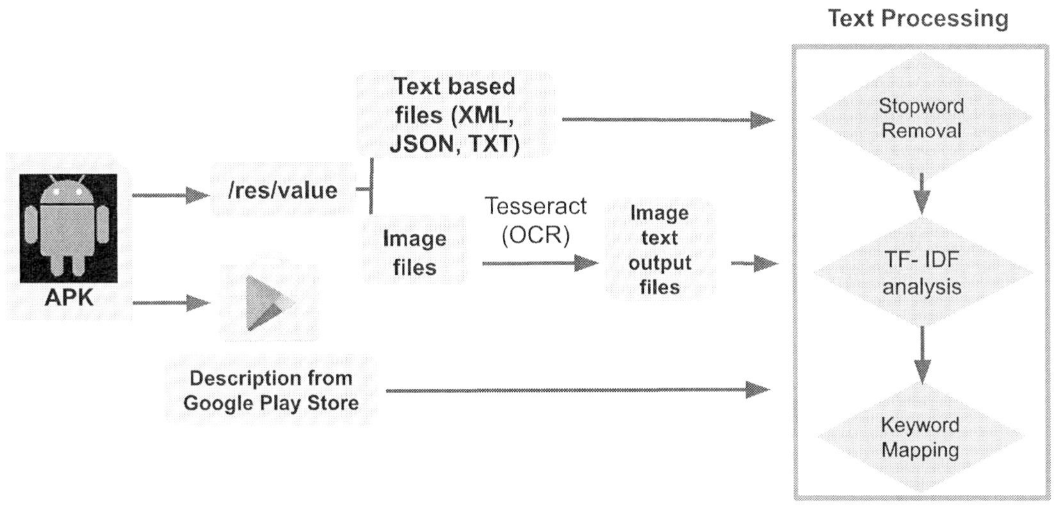

Figure 1. Trait extraction

2.1 Sources

An APK is essentially an archive of directories and files that include bytecode, resource files, assets, and libraries. Among all these files, we find that text relevant to app functionality and user interactions with the app appears in three main locations: XML assets, images, and app descriptions, as shown on the left of Figure 1. We describe each of these and provide evidence why all three sources are needed.

Table 1. Location and frequency of relevant keywords

App	Frequency (%)		
	XML Assets	Image Assets	Description
SimplePharmacology	31	69	-
MyNM by Northwestern	100	-	-
Medicine AnthroCalc	-	-	100

XML Assets. Layout XML files, stored in the app's *res/* directory, define the user interface by storing all text views, buttons, and other UI (user interface) elements. We are interested in these features as we can discover what information the app is requesting from the user and what kind of information the user provides to the app by interacting with it. String XML files store strings accessed by the application, which constitute another key location of medical terminology that can be extracted.

Image Assets. These assets, also stored in the app's res directory, are relevant as well. For example, some apps may use an image as a button, rather than defining the button's text string as an XML asset. In general, images may contain relevant text. We use the Tesseract OCR (optical character recognition) package (Tesseract, no date) on image files to extract the English text present in images.

Descriptions. App descriptions are found on Google Play and not within the app itself. As a result, the description of an application allows a user to understand what an app does prior to installation. As the description is the first impression a user has of the app, its functionalities should be clearly defined in a way that help users establish an app's purpose confidently and securely. However, as the description is usually written by the app developers themselves, the app is often portrayed in a flattering and overly positive way to attract users. Therefore, app descriptions can be inaccurate, misleading, or incomplete, which we found to be another essential aspect of an app that should be considered in our analysis.

Why all three sources are necessary. Table 1 illustrates why using only one of the three sources is insufficient: the table shows, for three apps, where the relevant keywords are located. For the app **SimplePharmacology**, 69% of the relevant keywords are in the image assets while the description contains no relevant keywords whatsoever. In contrast, for the app **MyNM**, all the keywords are found in the XML assets; images and the description contain no relevant terms. Finally, for the app **AnthroCalc**, all keywords are in the app description. Therefore, we need to analyze and integrate information from all three sources.

2.2 Methodology

In order to create a clear classification scheme based on app functionality, we referred to ICD'11 (International Classification of Diseases) and PHI (Protected Health Information) terms. ICD codes provide a reliable established standard of diseases and health conditions. PHI terms allow us to obtain a broad idea of what data is required from users in certain apps to determine their functionalities. We began our text processing with extracting the descriptions.

First, the descriptions had stop words removed to focus on conceptual information in the text, followed by a TF-IDF (term frequency – inverse document frequency) analysis (tf-idf, no date) based on ICD keywords and PHI terms. As a result, we could provide a preliminary classification of each application based on keyword matches and frequencies. The process was repeated for XML and image assets. With the resulting keywords extracted, we observed which resources provided the most relevant results. More keywords were found in the XML files of apps as opposed to image files (via Tesseract), or app descriptions. This evidences that descriptions do not paint a complete picture.

Defining Traits and Categories. We employed a multiple-rater approach (Green, 1997) to determine traits and categories: three human raters had to come to 100% agreement on what constituted and differentiated the various traits. Raters had to agree first on what should be considered a unique trait of a medical app and which keywords should be used in determining that trait (traits essentially define low-level orthogonal functionality "facets" for apps). As a result, 19 traits were determined. Subsequently, raters would then agree on what combination of traits would dictate the category of an app. Once the baseline was set, the app was classified using traits into categories, as illustrated in Figure 2. In this way, some apps may have multiple traits and belong to various categories. However, such categorization provides a more nuanced view on the general functionality of certain medical apps.

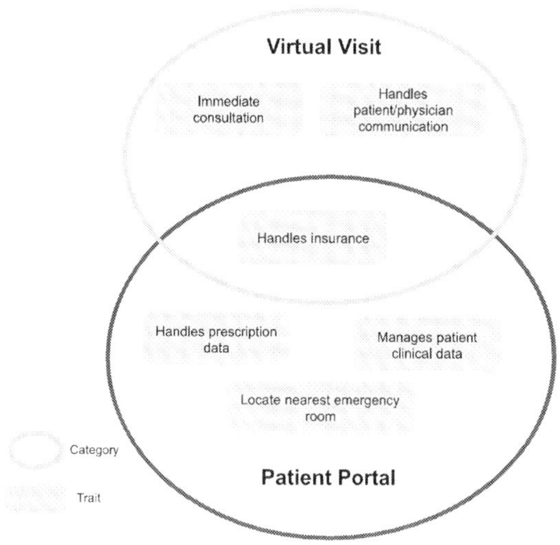

Figure 2. An example of category determination based on traits

2.3 Traits

Traits are defined as single aspects of app functionality, orthogonal to other aspects. Apps can exhibit multiple traits, as apps can provide several functionalities. We determined traits by finding common ICD terms and medical keywords. As a result, we defined 19 unique traits; their definitions and frequencies are shown in Table 2. Many of the traits are self-explanatory and commonly used, such as 'Anatomy' and 'Locate nearest emergency room.' There are certain traits that needed further refinement, specifically those dealing with patient data management. Table 2 reveals that many common traits involve reference material for medical professionals. For instance, Medical Student Study Aids are found in the top five traits in medical apps, as well as medical calculators, which are often used by professionals.

Table 2. Frequency of traits found in apps

Trait	Description	% Apps	#Apps
Anatomy	Anatomy reference material	61	1351
Well-being reminder	Keeps track of patient habits	51	1129
Medical student study aids	Medical student study material	38	841
Medical calculator	Calculates unsaved patient readings	37	819
Sends acquired patient data	Sends inputted patient data to a provider	35	775
Handles prescription data	Patient info regarding prescriptions	28	620
Manages patient clinical data	Stores medical history of a patient	27	598
Visual guide	Visual reference material used by a professional	26	575
Medical procedures	Procedural references for professionals	22	487
Patient journal/diary	Patient behavior or progress with disease or nutrition	20	443
Disease name	Disease reference material used by a professional	20	443
Handles patient/physician comm.	Stores and transmits patient info between patient and provider	20	443
Patient symptom tracker	Keeps track of various symptoms, may lead to diagnosis	19	420
Drug name	Drug name and pharmaceutical reference material	18	398
Handles insurance	Stores patient medical history related to insurance policy	17	376
Immediate consultation	Virtual consultations with a provider	16	354
Dose calculator	Calculations for patients and providers to administer medication	16	354
Locate nearest emergency room	Using current location to find an ER	14	310
Device measuring patient data	Using an external device to collect readings, e.g., blood pressure	14	310

2.4 Categories

Table 3. Category determination from traits

Trait \ Category	Reference	Patient Portal	Professional	Patient	Virtual Visit	Medical Device
Sends acquired patient data			●			●
Handles prescription data		●		●		
Handles patient/phys. comm.			●	●	●	
Handles insurance		●		●	●	
Manages patient clinical data		●	●			
Device measuring patient data						●
Patient symptom tracker				●		

Well-being reminder				●		
Dose calculator			●			●
Patient journal/diary				●		
Immediate consultation			●		●	
Anatomy	●		●			
Medical student study aids	●		●			
Medical calculator			●	●		
Disease name	●					
Medical procedures	●					
Visual guide	●					
Drug name	●		●			
Locate nearest ER		●	●	●		
Total number of apps	1,590	327	509	1,993	1,269	609
Total percentage of apps	72%	15%	23%	89%	57%	27%

The various combinations of certain traits allow us to determine specific categories of medical apps as is evident in Table 3. We established six unique categories that apps may fall into.

Reference. These apps serve either as general references regarding medical terms or first-aid procedures. Some apps are study aids or provide quizzes for medical professionals in training.

Patient Portal. Users can schedule and make appointments with their medical providers and view their lab results or test results and data from their visits. In addition, users can search for nearby providers.

Professional. These apps are directed towards medical professionals ranging from medical staff to office assistants. Many apps help medical clinics with scheduling and handling patient data in a professional setting.

Patient. Apps in this category are aimed at patients to help them log their daily progress or daily habits such as sleeping or pill reminders.

Virtual Visit. These apps provide for virtual visits, e.g., via a video call with a medical professional. In doing so, users often provide personal information and discuss their symptoms.

Medical Device. These apps are considered as medical devices or work in tandem with devices, such as hearing aids, glucometers, or sphygmomanometers for hypertension. Apps in this category can be used to store device readings and be maintained as a log or can be used as a remote control for the device.

3. ACTIONABLE FINDINGS

Our categorization has revealed that medical apps serve a broad audience and variety of purposes. However, because many such purposes are sensitive or even critical, and not intended for a general audience, there should be barriers for app access control. Theoretically, as these apps are free, and found on a public app distribution platform, anyone can download and use them, even though the apps are meant exclusively for professionals. Generally, apps that are meant for professionals in a hospital or clinical setting usually require credentials to access such systems. However, there are professional apps which can potentially result in a diagnosis or interface with a medical device for a procedure; if such apps are available for general use, it can lead to possible user harm. Hence there is a need for strong regulatory frameworks protecting end-users.

We now describe actionable findings covering various aspects of medical apps. We review current regulations and definitions regarding mobile health and medical apps established by various legal entities throughout the world, while also finding certain lapses and difficulties in implementing these guidelines. From these definitions, we discuss potential privacy implications and user safety concerns.

3.1 Regulatory Framework Enforcement

Actionable finding: Regulatory frameworks should be clearer defined and more accessible for developers when creating medical apps managing user data.

Medical apps can perform critical functions that involve patient data or other sensitive information. Overall, app users generally assume that apps are "certified" and trustworthy when making medical decisions. The question that arises is whether these apps are indeed approved by regulators and safe for use.

For example, in the United States, the FTC (Federal Trade Commission) provides definitions and guidelines for app developers. The guidelines indicate whether the app is a medical device, or a medical app; as well as whether the FTC will apply any regulatory oversight (FTC. 2019f). Additionally, the FDA (United States Food and Drug Administration) regulates functions of mobile devices that use device sensors (camera, light, vibrations) to perform medical device functions (e.g., measuring blood pressure), connecting a mobile device to a medical device and being able to manipulate it from the mobile device (e.g., alter settings of an implant), or active patient monitoring (e.g., acquiring signals from a cardiac monitor) (FTC, 2019d).

EU regulation of mobile medical apps focuses on potential privacy concerns that may arise. Mobile health apps must comply with data protection laws (Data Protection Directive) that were enacted, as well as ensuring that apps provide 'clear and unambiguous information about processing to end users before app installation' (Crossley, 2016).

Some Asian countries, such as China and Japan, regulate standalone medical software as medical devices, though depending on the overall software class, whose definition is based on functionality (Gross, 2017).

Overall, regulatory bodies have general guidelines on medical app behavior and functionality. However, there is no clear standard for app developers to easily refer to when developing a medical app. Having an accessible flowchart or a streamlined explanation of definitions would aid developers as well as app markets (Google Play, Apple's App Store) in managing the apps, especially apps handling users' medical data.

3.2 User Security and Safety

Actionable finding: Medical apps should be more transparent regarding user data management prior to installation.

App functionality plays a large role in determining whether the app falls under a regulatory framework. Medical apps often manage identifiable and private health information, that is, demographic information related to a user's health or condition that can be used to identify the user. For instance, in the US, if such apps work with health care providers or HIPAA entities, they are subject to HIPAA rules regarding security (FTC, 2019e) and privacy (FTC, 2019a) and what must be done when a breach has occurred (FTC, 2019b). However, not all data acquired by an app is considered identifiable health information. For example, an app measuring a user's weight and blood pressure is not considered a big security risk, compared to an app that tracks patient activity and prescriptions. Thus, certain apps pose lower risks to user privacy and would not need to be under scrutiny from regulatory bodies. An example would be apps that are general aids or of general purpose (e.g., magnifying glass); automate general office functions in healthcare and are not used for diagnosis; and educational apps (e.g., flashcards, encyclopedias, textbooks). These apps are neither regulated nor will have any discretionary enforcement exercised on them. However, as discussed previously, many medical apps handle patient data, and despite regulations and guidelines, users do not know how securely their data is managed or transmitted. App developers and markets must be more forthcoming and transparent about patient data management, by concisely explaining to users prior to installation what happens to their private health information. Potentially, these entities should be held accountable, should any leaks occur.

4. CONCLUSION

Medical apps across many categories have been implemented and publicized that serve millions of users and provide a multitude of functions. To better understand the app landscape, our study categorizes medical apps based on stated and observed functionality. Overall, our research makes several contributions. First, we provide an automated approach and study that characterize medical apps into sub-categories to better understand their purposes and functionalities. Second, observe the most common functionalities of medical apps. Third, we discuss regulatory frameworks and user privacy practices. By doing so, we are better equipped to undertake further studies into app behavior, app security, app claims, etc.; and ultimately improve the health and well-being of app users.

ACKNOWLEDGMENT

We thank the anonymous reviewers for their feedback. This material is based upon work supported by the National Science Foundation under Grant No. CCF-2106710.

REFERENCES

Akbar, S. et al. (2020) "Safety concerns with consumer-facing mobile health applications and their consequences: A scoping review", *J. Am. Med. Inform. Assoc.* Oxford University Press (OUP), 27(2), bll 330–340.

Crossley, S. (2016) "EU regulation of health information technology, software and mobile apps". Available at: https://uk.practicallaw.thomsonreuters.com/2-619-5533?contextData=(sc.Default).

FTC (2019a) "Breach Notification Rule". Available at: https://www.hhs.gov/hipaa/for-professionals/privacy/laws-regulations/index.html.

FTC (2019b) "Breach Notification Rule". Available at: https://www.hhs.gov/hipaa/for-professionals/breach-notification/index.html.

FTC (2019c) "Examples of Device Software Functions the FDA Regulates". Available at: https://www.fda.gov/medical-devices/device-software-functions-including-mobile-medical-applications/examples-device-software-functions-fda-regulates.

FTC (2019d) "Examples of Software Functions for Which the FDA Will Exercise Enforcement Discretion". Available at: https://www.fda.gov/medical-devices/device-software-functions-including-mobile-medical-applications/examples-software-functions-which-fda-will-exercise-enforcement-discretion.

FTC (2019e) "Health Information Laws". Available at: https://www.hhs.gov/hipaa/for-professionals/security/index.html.

FTC (2019f) "MOBILE HEALTH APPS INTERACTIVE TOOL". Available at: https://www.FTC.gov/tips-advice/business-center/guidance/mobile-health-apps-interactive-tool.

FTC (2019g) "MOBILE HEALTH APPS INTERACTIVE TOOL Glossary". Available at: https://www.FTC.gov/tips-advice/business-center/guidance/mobile-health-apps-interactive-tool#glossary.

Gross, A. (2017) "Column - Medical Software Regulations in Asia 2017", *MedTech Intelligence*. MedTech Intelligence. Available at: https://www.medtechintelligence.com/column/medical-software-regulations-asia-2017/.

Green, A. M. (1997) "Kappa Statistics for Multiple Raters Using Categorical Classifications", in *Proceedings of the 22nd annual SAS User Group International conference*, bl 4.

Johnson, C. et al. (2021) "Hospital Capabilities to Enable Patient Electronic Access to Health Information, 2019". ONC Data Brief.

Lewis, T. L.et al. (2014) "mHealth and Mobile Medical Apps: A Framework to Assess Risk and Promote Safer Use", *J Med Internet Res*. doi: 10.2196/jmir.3133.

Magrabi, F. *et al.* (2019) "Why is it so difficult to govern mobile apps in healthcare?", *BMJ Health & Care Informatics*. BMJ Specialist Journals, 26(1). doi: 10.1136/bmjhci-2019-100006.

Semigran, H. L. *et al.* (2015) "Evaluation of symptom checkers for self diagnosis and triage: audit study", *BMJ*. BMJ Publishing Group Ltd, 351. doi: 10.1136/bmj.h3480.

Shuren, J., Patel, B. en Gottlieb, S. (2018) "FDA regulation of Mobile Medical Apps", *JAMA*, 320(4), bl 337. doi: 10.1001/jama.2018.8832.

Statista Research Department (2021) "Healthcare apps available Google Play 2021", *Statista*. Available at: https://www.statista.com/statistics/779919/health-apps-available-google-play-worldwide/.

Stoyanov, S. R., Hides, L., Kavanagh, D. J., Zelenko, O., Tjondronegoro, D., & Mani, M. (2015). "Mobile app rating scale: a new tool for assessing the quality of health mobile apps." JMIR mHealth and uHealth, 3(1), e27. https://doi.org/10.2196/mhealth.3422

Tangari, G. *et al.* (2021) "Mobile Health and Privacy: Cross Sectional Study", *BMJ*. doi: 10.1136/bmj.n1248.

Tesseract-Ocr (no date) "tesseract-ocr/tesseract: Tesseract Open Source OCR Engine (main repository)", *GitHub*. Available at: https://github.com/tesseract-ocr/tesseract.

Terry, N. P.at al. (2018) "Regulating mobile mental health apps", *Behavioral Sciences & the Law*, 36(2), bll 136–144. doi: 10.1002/bsl.2339.

"tfidf: A Single-Page Tutorial - Information Retrieval and Text Mining" (no date). Available at: http://www.tfidf.com/.

Wicks, P. et al. (2015) "'Trust but verify' – five approaches to ensure safe medical apps", *BMC Medicine*, 13(1). doi: 10.1186/s12916-015-0451-z.

FELIX THE DIGIBUD: UNVEILING THE DESIGN OF AN ICT-SUPPORTED INTERVENTION FOR OCCUPATIONAL STRESS MANAGEMENT

Manoja Weerasekara[1,2] and Åsa Smedberg[1]

[1]*The Department of Computer and Systems Sciences, Stockholm University, Kista, Sweden*

[2]*The Department of Computer and Systems Sciences, NSBM Green University Town, Homagama, Sri Lanka*

ABSTRACT

Digital stress management is an evolving yet promising approach to the continuum of stress management programs. There are shreds of evidence on the benefits that could be achieved. However, numerous studies discuss the challenges that hinder the potential benefits of digital stress management interventions. Such challenges mainly include less engagement, insufficient theoretical underpinning, high attrition, and lack of personalisation. Thus, the current study aims to propose a design of a digital intervention to bridge the identified gaps. The proposed intervention, Felix the DigiBud, was co-designed and developed using a multidisciplinary team based on empirical and literature evidence. The mentioned empirical studies mainly focused on gathering user requirements from different stakeholder groups. These stakeholder groups involved software employees, counsellors and human resource managers working at software companies. All the empirical studies were carried out in Sri Lanka, applying quantitative and qualitative approaches. Seven design principles were built to govern the design process and fulfil the defined requirements. The first iteration of the design cycle resulted in a clickable mock-up with a web-based front end to visualise the user interfaces and the process flow of the intended features and functionalities. It has a setting module and eight functional modules that reflect different stress management activities. The designers made a comprehensive effort to embed gamification and digital micro intervention concepts in the intervention design to increase user engagement. In the next phase, the ICT-supported intervention will be demonstrated to the stakeholders to determine to what extent the artefact fulfils the identified problems.

KEYWORDS

eHealth, Health Intervention, Digital Micro Intervention, Occupational Stress Management, Gamification

1. INTRODUCTION

Occupational stress is a type of stress associated with jobs. When the work demands exceed the person's capacity and capability to cope, individuals tend to fall into stress conditions. Most occupations are subjected to stress; however, the software industry is vulnerable to yielding stressed employees due to its demanding and technology-driven nature. The software industries around the globe practice a software development phenomenon called Global Software Development (GSD) (Amin et al., 2011). Following this, software applications are developed through organisations, people, and technologies across geographical locations, cultures, languages, and work patterns (Herbsleb & Moitra, 2001). Though this has brought enormous benefits in knowledge and resource sharing, it has also created negative pressure on software employees. Such pressures include 1. global pressures in terms of market and competition, 2. technology pressures in updating and developing skills, and 3. local pressures in terms of working hours, changing work culture, changing teams, and changing peer group (Sonnentag et al., 1994). These pressures make the software occupations more demanding and contribute to the risk of increasing the stress levels of software employees.

The ubiquitous use of technology has touched almost all everyday objects and activities. Industries and researchers have made numerous efforts to bring technological interventions to various domains, including healthcare. With the growing number of demands and scarcity of medical resources, health industries sought the possibility of using the power of technology to render affordable and personalised services to patients beyond geographical and time boundaries (Burman & Goswami, 2018). Though there are many examples of how technology is applied in healthcare, its application in mental health care is still at its blooming stage.

Although digital mental health interventions are a growing phenomenon, considerable research on technological interventions in stress management has been conducted. Companies have recognised digital stress management interventions as a promising addition to the continuum of programs used to manage employees' stress and alleviate their well-being. However, the research shows mixed results on the efficacy of such programs due to challenges identified in the intervention. Such challenges include, but are not limited to, less engagement, poor adherence, high attrition, and lack of personalisation (Scholten & Granic, 2019). The literature claims that it is essential to identify the user requirements and involve a multidisciplinary team in the design and development process to gain the ultimate benefits from such interventions (Ariani et al., 2017). Moreover, the literature highlights several research gaps that have lowered the intervention's impact. The insufficient theoretical underpinnings and generic design without targeting the occupational category are among them (Oman et al., 2006; Tveito & Eriksen, 2009; Weerasekara & Smedberg, 2019).

Thus, the current study presents a stress-management intervention prototype designed explicitly for software employees. The intervention was co-designed by a multidisciplinary team based on rigorous empirical and literature evidence targeting software employees. The design process undertaken in the intervention is comprehensively discussed in the upcoming sections.

2. LITERATURE REVIEW

2.1 Transactional Model of Stress

The Lazarus transactional model of stress and coping is the most widely used framework in stress management research. This model elaborates stressful experiences as a person-environment transaction. According to the framework, a person's appraisal of the stressor and the availability of psychological and social resources could mediate the impact of the stressor or demand (Lazarus & Folkman, 1984). The current study follows this model and applies part of the transactional model of stress and coping. Particular emphasis was given to making available required resources and setting up supportive personal and social environments to practice behaviours that enable them to manage stressful encounters.

2.2 Digital Stress Management

Technological advancements have placed a greater interest in digital stress management interventions. Numerous efforts have been made toward designing and developing a continuum of interventions involving different devices and technologies. Since stress is a common phenomenon in both personal and work life, individuals and organisations have made significant efforts toward identifying suitable interventions to support the stress management process. However, research shows mixed results of digital interventions due to their adherence and engagement barriers (Howe et al., 2022). Thus, more focus is given to designing and developing interventions capturing the user needs and providing a personalised solution to its users. The existing literature on digital stress management interventions shows that many interventions are delivered as web-based internet interventions. Other modalities like mobiles, sensors, and smart devices have also been used to produce the interventions. Previous studies emphasise selecting the suitable modality to deliver the intervention. The modality establishes technological limitations to deliver diversified digital content. Selection of modality with convenient access could minimise the switching cost for its users (Howe et al., 2022). Support and guidance have also played a significant role in the intervention delivery and supported increased adherence and engagement.

The support or guidance has mainly been delivered in human- and machine-based support (Weerasekara & Smedberg, 2019). In the human-based approach, qualified and trained counsellors, therapists, or peers provide support and guidance. The machine support option renders the guidance and support as chat, email, or text-based communication. In both the options, it was evident that providing support varied from fully supported to moderately supported to less supported options. Research has also highlighted that intervention users appreciated peers' and experts' involvement with the intervention delivery and sought support only at a moderate level (Weerasekara & Smedberg, 2019). Moreover, literature shows that most interventions have not used a solid theoretical foundation to structure the intervention (Oman et al., 2006; Tveito & Eriksen, 2009).

Thus, such intervention outcomes were difficult to justify or explain due to the lack of a theoretical baseline (Oman et al., 2006; Tveito & Eriksen, 2009; Weerasekara & Smedberg, 2019).

The digital micro intervention is considered a novel approach to designing intervention content to lower the entry barriers and minimise user effort to engage with interventional content (Baumel et al., 2020). This has enabled users to engage more with the content while improving the usability of the intervention(Baumel et al., 2020). Recent studies extensively used gamification, and emerging technologies like virtual reality, social networking, and wearables sensors. These concepts have increased user engagement and adherence (de Witte et al., 2021). Moreover, studies argue that the design of the intervention, study context, and user characteristics directly influence the usage and engagement with digital behavioural change interventions (Perski et al., 2017). Thus, it is essential to identify the user needs and design a solution (Floryan et al., 2019) with a theoretical foundation and rigorous empirical evidence.

2.3 Gamification

Gamification has gained increased attention in technology-based behavioural and mental health intervention design and development (Floryan et al., 2019). It is considered a way of embedding entertainment and creative dimensions into the non-gaming context (Deterding et al., 2011). Evidence shows that gamification has contributed to increasing adherence, engagement, and motivation to behavioural interventions (Floryan Mark and Chow, 2020). Many existing interventions use gamification elements like badges, leader boards, points and levels, challenges and quests, social engagement loops, and onboarding (Zichermann & Cunningham, 2011). However, research highlights the importance of using a gamification framework to guide the implementation of these elements within a technology-based intervention (Floryan Mark and Chow, 2020). The consolidated gamification framework (Floryan et al., 2019) elaborates a model with five design principles mapped into four categories. The five main principles include 1. support player archetypes (traits and values the user brings to the intervention that helps to increase engagement); 2. meaningful/freedom of choice (close to the concept of personalisation, allows users to choose the desired path to reach the goal); 3. meaningful purpose/knowledge of benefit (the purpose of undertaking an activity and how to practice is visible to the user); 4. feedback (user receives feedback on ongoing activities) and 5. Visibility of progress and path to the destination (user can visualise/understand their progression). This framework provides comprehensive guidance for applying gamification principles to behavioural and mental-health-focused internet interventions. Unlike in the past, research studies show that companies also use gamification to increase employee productivity. The research shows evidence that not only males dominate in this area but that the use has spread across all ages without any gender disparity (Kiselicki et al., 2018).

2.4 Digital Micro Interventions

Numerous research studies show misalignment of user behaviour and intervention design. On one side, they have recognised the users' lack of motivation and a minimal investment of effort and time to engage in the intervention. At the same time, the interventions expected a considerable effort and time from the users' end to achieve the desired results. This mismatch has been acknowledged and addressed in the novel concept of digital micro intervention. Where digital micro-interventions tend to provide beneficial therapeutic support with a minimum burden to the user, such interventions offer several shorter and much more focused interventions (Baumel et al., 2020). Thus, it could lower the cost of entry to the intervention and the commitment and effort required for focused engagement (Baumel et al., 2020). Such digital micro-interventions are based on three main components: events (individual in-the-moment attempt that has an impact on the overall target), decision rules (decides which and when events should be deployed), and assessments (measure the impact of the individual event and the impact to the overall aim of the intervention) (Baumel et al., 2020). An event can be informational or interventional. From the stress management perspective, an event may provide educational material at the required time (informational) or encourage practising some form of action like breathing exercises (interventional). A digital micro intervention can be based on an individual event (Ayers et al., 2015; Strauman et al., 2015) or multiple events combined (Elefant et al., 2017) and sequenced to match a context and user requirements. Then, the decision rules decide when and which event to be deployed. The assessment targets to assess the impact of the intervention, which can be seen as two-fold. On the one hand, it may target evaluating the intervention's overall outcome. On the other hand, it may aim at determining the more proximal effects of deployed events.

3. METHODOLOGY

The proposed study reveals a prototype of an ICT-supported intervention for occupational stress management. A series of empirical studies were carried out involving multiple stakeholder groups to assist the intervention design process. The stakeholder groups comprised software employees and human resource managers in Sri Lankan-based software companies and counsellors engaged in stress management practices. Quantitative and qualitative data collection and analysis techniques complemented each approach and supported yielding valuable insights into the user needs. Seven core user requirements were identified based on the empirical studies and literature evidence (see Table 2). Next, with extensive literature support and careful mapping of the user requirements, the researchers derived seven design principles to govern the design process. The following table (see Table 1) shows the design principles and the operationalised descriptions.

Table 1. Derived Design Principles to Govern the Design Process

Derived Design Principle [Focus on ....]	**Description**
Modality and Focus	The intervention delivery platform and the target level (individual, group, hybrid, organisational, etc.) must be selected at the initial stage.
Content Provision	The intervention carefully selects the techniques and technologies to deliver the right content to the user efficiently.
Personalisation	The system renders personalised content based on the user requirements acknowledging their familiarity and experience.
Social Connectedness	The intervention enables the user to connect with selected social circles (peers, experts, etc.)
Use of Gamification	Possibility of embedding gamified content (both gaming elements and dynamics) into the intervention to increase user engagement
Adaptive Learning	The intervention can learn from the user and provide content and support based on user behaviour.
Aesthetics and Simplicity	The intervention must select visual elements to offer pleasing aesthetics while assuring the simplicity of the intervention.

These seven principles governed the design process and covered the intervention's functional and non-functional aspects. The researchers worked collaboratively with a multidisciplinary team of software engineers, UI/UX engineers, and counsellors to establish the design features. When designing elements of the functional units, the design team heavily relied on the digital micro intervention components like forming events, decision rules and setting up assessment strategies and gamification techniques. After establishing design features, a clickable prototype was designed by the team. The following table (Table 2) shows the overview of the design requirements and proposed design features in the prototype supported by the literature.

When designing each core functional module in the Felix prototype designers, besides building Felix's micro-interventional unit, the design team made a rigorous effort to apply gamification elements to the intervention. Gamification elements like points, badges, visualised dashboards, progress bars, leader boards, status, and individual task lists were used to render gamification dynamics in the intervention. Table 3 (see Table 3) shows examples of how the intervention implemented the gamification principles.

Table 2. Mapping of User Requirements into Design Features with Literature Evidence

User Requirement	**Literature Evidence**	**Associated Design Principles**	**Design Features presented in the proposed intervention**
Ability to perform stress assessment and monitoring	(Hänsel, 2016; Maclean et al., n.d.; Rodrigues et al., 2015; Sharmin et al., 2015)	Content provision on assessment and monitoring. Use of gamification to support interactive visualisation.	*Assessment:* The system can capture stress measures through user input (active) and automated (passive). Active measurements: Use of stress assessment questionnaire and mood rating by the user. Passive Measurements: Automatic capture of mouse movements (clicks and wheel movement) and keyboard presses. *Monitoring:* E.g., When unusual keyboard usage is identified, the system pop-up with a message to alert the user and suggest engaging in stress management activities. [Just-in-time intervention] Daily mood changes and stress measurements are visualised using graphs.
Ability to visualise the upcoming activities, stress patterns, and stress management-related information	(Kocielnik & Sidorova, 2015; Maclean et al., n.d.)	Content provision on visualisation, notification, information and use of gamification.	Users can visualise stress patterns and mood changes over time. Intervention displays the upcoming event calendar. Send notifications/reminders based on previously planned activities. [Ecological momentary intervention] The To-do list is visible to the user with highlights of completed and pending activities. The expert support area provides the required stress and management information via credible sources.
Possibility of selecting and engaging in simple stress management activities	(Bendelin et al., 2011; Ebert et al., 2016; Eklund et al., 2018; Hoa et al., 2017; Howe et al., 2022; Villani et al., 2013)	Content provision and personalisation	An activity basket is provided to the user. Users may change the list and priority by dragging and dropping the items. All the activities provided are simple and can engage the user within a 1-5min time frame.
Possibility of collaborating with others to manage stress	(Cavanagh et al., 2018; Cunningham-Hill et al., 2020; Hoek et al., 2018; Leung et al., 2011)	Social connectedness, content provision	The system supports both peer-support and expert support platforms. Peer-support platform provides the ability to share, comment, and keep threaded communications with peers. Users can directly get in touch with selected peers through a chat system. Provide the ability to contact an expert through chat, audio, and video call.

Provide privacy, security, and confidentiality	(Cunningham-Hill et al., 2020)	Content provision and personalisation	Provide secure login through registered credentials. Users may decide what content to be published and share with others and what to keep secured. The data will not be accessible to outsiders.
Provide simple and user-friendly application	(Howe et al., 2022)	Aesthetics and simplicity	For some activities (e.g., breathing exercises), guided tours are provided to guide the user through various steps. A simple language style is used. Simple colours and fonts are used with consistent themes.
Easy Accessibility and interactive content	(Cunningham-Hill et al., 2020; de Witte et al., 2021; Hasson et al., 2010; Howe et al., 2022; Morrison et al., 2017)	Content Provision, Use of gamification, adaptive learning	The system will be available as a cross-platform application. Progress bars, dashboards and award/point systems motivate users to engage with the system. The system provides feedback on activities (e.g., when a user completes a breathing exercise system awards points). Automated chat is available to have a friendly conversation or seek support to manage stress. The system collects feedback from the user to learn the user preferences (e.g., at the end of a chatbot conversation, the user is asked to rate the discussion; and when the user completes a stress management activity, the system checks whether it is supportive in relieving stress or not). Tooltips are provided for each function so that users can understand the purpose of the activity/module. Reminders and notifications are provided to the user based on previously planned activities.

4. RESULTS

As a result of the first design iteration, 'Felix the DigiBud' (Felix) has emerged. This was developed as a clickable prototype that renders an interactive experience similar to the final application. The name selected for the application is 'Felix'. This was formed as an abbreviation of 'Feeling Relaxed". Felix carries the meaning of happy, prosperous, and lucky in Latin, Greek, and the Bible. The application was designed as a responsive web usable in any browser viewpoint to increase usability. Felix follows primary user interface (UI) design principles to render the best user experience to the user. The fonts, colours, and language are carefully selected to offer consistent user interfaces that make it more user-friendly and interactive. The icons, labels, images, and buttons were selected and placed in a way to increase the visibility and affordability of the UIs.

Felix presents a total of nine functional modules, including the settings module. The setting module is considered a standard unit available in Felix regardless of the user preferences. The settings module offers the facility to change the profile settings, customise preferences, backup, and restore contents. The other modules can be added to the application based on user preferences during the profile setup stage. Felix also comes with a desktop application that can run as a background to capture the keyboard presses and mouse clicks to detect stressful behaviour.

Felix offers different login options. Users may create an account and log in using credentials or as a guest. It also provides the facility to log in using the user's social media accounts, such as Google, Facebook, or Instagram. At the first-time login, the user is asked to create a Felix system version based on their situation and preferences. During the account creation process, Felix asks questions to capture the user demographics, perceived stress levels, and activity preferences. Felix creates a customised version of Felix that best suits the user based on input. Felix also captures the user behaviours and preferences and tries to learn to offer personalised experiences. As an example, after a conversation with the Felix chatbot, Felix asks the user to rate the conversation so it could give an enhanced discussion in the future. After listening to a piece of music or watching a video clip, Felix will ask for user feedback. This feedback is used to let the system learn about the user preferences to render personalised content in the future.

The first functional module, 'Plan my Day,' focuses on planning and scheduling. This allows for setting up a to-do list and adding it to the calendar. Felix will track and alert users to complete the tasks. The task completion is also visible on the home page. This module also offers a reminder facility where users can set up notifications based on their preferences (e.g., Drink water, have lunch, Take a rest, etc.). The 'Assess My Stress' module offers both active and passive capture of stress measures. It has three sub-components: (1) My Mood- a symbolic grid is given to capture the current mood with the timestamp (2) PSS- A standard questionnaire (PSS) is provided to capture the stress levels and moods periodically (3) Stress Tracker- shows the automatic data captured from Felix desktop app. 'Stress Monitor' is a module that provides the ability to visualise the stress data captured through various inputs. It could visualise mood and stress level changes on a day and over a period using line graphs.

The 'Activity Bucket' shows a customisable list of recommended stress management activities. All the activities are considered digital micro-interventions since these activities could be performed in less than five minutes. The original list is created based on the input received during the account setup stage, which could be rearranged based on user preferences. In some activities like breathing exercises, step-by-step guidance is provided. Upon completing the activities, the user will receive points/rewards displayed on the home page.

The ‘social feed’ module creates an interface between the user and peer circle, providing a platform similar to Facebook. Users can see the posts and updates of their peers. The user also can create their posts, add comments to the other posts, and share and add symbolic reactions. My Diary module provides a journaling feature to the user that s/he could use to write down their feelings, emotions, and stories as a record in a digital diary. This module also enables publishing on the social feed or hiding for private use. ‘vmFriend’ module offers the messaging facility between selected individuals or through a chatbot. The individual messaging facility provides text and emoticons to continue with threaded communications. The chatbot feature enables automated conversations and acts as a mentor to guide during stressful situations. The final functional module of the application is the 'Meet an Expert'. This module provides professional support in two ways. First, users could contact stress experts/counsellors via audio and video conferencing facility or make an appointment to meet the professional face-to-face at a physical location. Once the appointment is made, it is automatically added to the calendar, and the user receives notifications accordingly. Secondly, it provides access to trusted web sources to read and learn about stress and stress management.

Table 3. Application of Gamification Techniques in the Intervention

Gamification principle	Operationalised Description	Examples from the intervention
Meaningful purpose	The intervention renders activities aligned with the user's interests and goals.	Each of the functions/activities is provided with a description of the purpose of the action.
Meaningful/freed om of choice	Provide the user with an opportunity to select activities and options based on their preferences	Activity Basket: A range of stress management activities is presented so that the users can prioritise/select based on their preferences.

Supporting player archetypes	Intervention is personalised based on user characteristics. E.g. Socialisers- interested in seeking support from peers Explorers- interested in seeking and experiencing new things.	'Socialiser' user types are given the option of a peer-support platform to support their stress management activities. Explorer-type users could be encouraged to browse trusted web sources to find stress management activities that suit their preferences.
Feedback	The user gets feedback on the activities and how it affects their goal	Soon after completing the breathing exercise user receives points. Users who actively support a social feed (peer-support platform) receive badges acknowledging their contributions.
Visibility of progress and path to the destination	Users can visualise the progression	On the home page, no. of items completed out of the planned activity set is visible to the user through a progress bar.

5. DISCUSSION

The study revealed a prototype of an ICT-supported intervention designed based on empirical and literature evidence. A multidisciplinary team was involved in the design process, and digital micro-intervention and gamification concepts were heavily used. The intervention consisted of nine functional modules targeting various stress management interventions.

The proposed intervention aimed to fill several research gaps identified in the previous literature. Firstly, the literature highlights that majority of the previous studies neglected to explicitly discuss the design process involved in the intervention design stage (Tveito & Eriksen, 2009). Thus, the proposed study extensively elaborates the design process applied in the intervention design. These elaborations of techniques and procedures involved in the design could motivate future designers to yield better design ideas. Secondly, research evidence that most studies have not used any theoretical stress model to explain the stress management mechanisms used in the intervention delivery (Oman et al., 2006; Tveito & Eriksen, 2009). This critique is answered in the proposed intervention by making a considerable effort to ground the concept of the transactional theory of stress and coping. Thirdly, the previous literature claims that the mismatch between user needs and interventional design has lessened the engagement and motivation to use the interventions (Weerasekara & Smedberg, 2019). Thus, this study attempted to address this gap by utilising empirical evidence from different stakeholders in a series of studies. The co-designed process carried out by a multidisciplinary team could also be considered a positive aspect of the interventional design. Much research discusses the importance of involving an interdisciplinary team in digital health care (Ariani et al., 2017; Weerasekara & Smedberg, 2019). Such cross-disciplinary involvement helped combine engineering with social sciences and search for the possibilities of using technology to render novel mental health care practices (Ariani et al., 2017). Another gap in the literature is not focusing on specific occupational categories while designing the intervention (Weerasekara & Smedberg, 2019). The occupational demands change over the spectrum of work environments, work cultures, etc. (Mustafa et al., 2015). The stress experienced by a bank officer or schoolteacher working in a city limit is quite different and higher than those experienced by their counterparts in a rural area. Thus, it is essential to investigate the specific user needs of the employees to create more customised solutions. During this study, the design explicitly targeted software employees and provided the opportunity to personalise their Felix version based on their characteristics and preferences.

Though the digital health interventions overcome the challenge of delivering quality service across geographical boundaries, there is a considerable challenge in keeping the users engaged and adherent to the intervention. There are shreds of literature evidence that if the intervention requests ample time or rigorous effort from the user, it tends to discourage the user and eventually leads to discontinuing the intervention (Eklund et al., 2018). This is evident in the current social context where users prefer micro-interventions over traditional interventions. Receiving answers to a question posted on Facebook regarding yoga is chosen over expert support received at a conventional yoga class. Thus, the proposed intervention also tried to use the concept of digital micro-interventions to deliver the interventional activities. All the activities were designed as well-focused functions in-the-moment activities where users must make minimum effort and time to complete a specific task. For example, if the user needs to record his current mood, he needs to use a couple of ticks. Then, the intervention takes the necessary steps to capture the timestamp location details to save for future reference.

Moreover, the system could prompt to take a quick breathing exercise based on a pre-defined time frame or a trigger. With the widespread technological advances, intervention developers get the opportunity of embedding a diversified range of digital micro-interventions into their SMIs. This digital micro intervention could increase access to mental health care by lowering the effort and time required to achieve the targets (Baumel et al., 2020). The use of digital micro-interventions and gamification could be considered a milestone in this interventional design. However, there is a critique that micro-interventions have proven to provide short-term benefits but not long-term benefits in mood and distress (Elefant et al., 2017). Thus, future studies need to carefully assess the impact of digital micro-interventions. The design team made a comprehensive effort to embed gamification principles into the intervention using numerous elements like awards, points, progression bars, dashboard, etc. Such elements helped store the intervention's gamification dynamics like interactivity, motivation, and engagement. Such dynamics helped create a more dynamic environment within the intervention and intrinsically motivated the users to increase their interactivity and engagement (Floryan et al., 2019).

Despite the mixed research evidence, ICTs hold promise in addressing mental health care challenges (Breslau & Engel, 2016). ICT has already created an avenue to reach the target audience without geographical and time barriers. It also helped to create alternative paths of seeking support and care. Such methods helped the patients gain the required information and make rational decisions leading toward patient empowerment (Barak et al., 2008). The prototype and the design process presented in this paper provide several favourable insights that could be embedded into digital stress management interventions. Such insights would lay a foundation for researchers, designers and developers to investigate possible technological solutions to connect better the intervention and the user in time to come.

6. CONCLUSION

The current study proposed a preliminary design of an ICT-supported intervention for occupational stress management called Felix the DigiBud. The design process was targeted to bridge various research gaps identified in existing digital stress management interventions. The micro-interventional design concept used in the intervention opened avenues to deliver complex interventions efficiently and effectively. This will limit the effort and time estimations from the user's end. The gamification principles and elements applied in the intervention helped overcome less interactivity and engagement challenges. Moreover, the study's findings could contribute to the already existing knowledge base and to the practitioner audience in designing and developing ICT interventions for occupational stress management. Since the preliminary prototype is now ready to be evaluated by the stakeholders, Felix will be demonstrated to different stakeholder groups in the next iteration to assess aspects like usefulness, usability and visual attractiveness. The evaluation findings could feed into the next design cycle of the Felix.

REFERENCES

Ariani, A., Koesoema, A. P., & Soegijoko, S. (2017). *Innovative Healthcare Applications of ICT for Developing Countries*. https://doi.org/10.1007/978-3-319-55774-8

Barak, A., Hen, L., Boniel-Nissim, M., & Shapira, N. (2008). A comprehensive review and a meta-analysis of the effectiveness of Internet-based psychotherapeutic interventions. In *Journal of Technology in Human Services* (Vol. 26, Issues 2–4, pp. 109–160). Haworth Press.

Baumel, A., Fleming, T., & Schueller, S. M. (2020). Digital Micro Interventions for Behavioral and Mental Health Gains: Core Components and Conceptualization of Digital Micro Intervention Care. *Journal of Medical Internet Research, 22*(10), e20631. https://doi.org/10.2196/20631

Bendelin, N., Hesser, H., Dahl, J., Carlbring, P., Nelson, K. Z., & Andersson, G. (2011). Experiences of guided Internet-based cognitive-behavioural treatment for depression: A qualitative study. *BMC Psychiatry, 11*(1), 107.

Breslau, J., & Engel, C. C. (2016). Information and Communication Technologies in Behavioral Health: A Literature Review with Recommendations for the Air Force. *Rand Health Quarterly, 5*(4), 17. https://www.ncbi.nlm.nih.gov/pubmed/28083427

Cunningham-Hill, M., D.-R., Z., W.-M., C., S., C., N., & M., Schueller, S. M. (2020). *Digital Tools and Solutions for Mental Health: An Health Employer ' s Guide* (Issue May). Northeast Business Group on Health and One Mind PsyberGuide.

de Witte, N. A. J., Joris, S., van Assche, E., & van Daele, T. (2021). Technological and Digital Interventions for Mental Health and Wellbeing: An Overview of Systematic Reviews. *Frontiers in Digital Health, 3*.

Ebert, D. D., Heber, E., Berking, M., Riper, H., Cuijpers, P., Funk, B., & Lehr, D. (2016). Self-guided internet-based and mobile-based stress management for employees: results of a randomised controlled trial. *Occupational and Environmental Medicine, 73*(5), 315–323. https://doi.org/10.1136/oemed-2015-103269

Eklund, C., Elfström, M. L., Eriksson, Y., & Söderlund, A. (2018). Development of the web application My Stress Control—Integrating theories and existing evidence. *Cogent Psychology, 5*(1), 1–19.

Elefant, A. B., Contreras, O., Muñoz, R. F., Bunge, E. L., & Leykin, Y. (2017). Microinterventions produce immediate but not lasting benefits in mood and distress. *Internet Interventions, 10*, 17–22.

Floryan, M. R., Ritterband, L. M., & Chow, P. I. (2019). Principles of gamification for Internet interventions. *Translational Behavioral Medicine, 9*(6), 1131–1138. https://doi.org/10.1093/tbm/ibz041

Hänsel, K. (2016). *Large Scale Mood and Stress Self-Assessments on a Smartwatch*. 1180–1184.

Hasson, H., Brown, C., & Hasson, D. (2010). *Factors associated with high use of a workplace web-based stress management program in a randomised controlled intervention study*. *25*(4), 596–607.

Hoa, K., Ly, A., & Andersson, G. (2017). A fully automated conversational agent for promoting mental well-being : A pilot RCT using mixed methods. *Internet Interventions, 10*(August), 39–46.

Howe, E., Suh, J., Morshed, M. bin, McDuff, D., Rowan, K., Hernandez, J., Abdin, M. I., Ramos, G., Tran, T., & Czerwinski, M. (2022, April). Design of Digital Workplace Stress-Reduction Intervention Systems: Effects of Intervention Type and Timing. *CHI 2022*. https://www.microsoft.com/en-us/research/publication/design-of-digital-workplace-stress-reduction-intervention-systems-effects-of-intervention-type-and-timing/

Kocielnik, R., & Sidorova, N. (2015). Personalized Stress Management : Enabling Stress Monitoring with LifelogExplorer. *KI - Künstliche Intelligenz*, 115–122. https://doi.org/10.1007/s13218-015-0348-1

Maclean, D., Roseway, A., & Czerwinski, M. (n.d.). *MoodWings : A Wearable Biofeedback Device for Real- Time Stress Intervention*.

Morrison, L. G., Hargood, C., Pejovic, V., Geraghty, A. W. A., Lloyd, S., Goodman, N., Michaelides, D. T., Weston, A., Musolesi, M., Weal, M. J., & Yardley, L. (2017). The effect of timing and frequency of push notifications on usage of a smartphone-based stress management intervention: An exploratory trial. *PLoS ONE, 12*(1), 1–16.

Mustafa, M., Illzam, E. M., Muniandy, R. K., Hashmi, M. I., Sharifa, A. M., & Nang, M. K. (2015). Causes and Prevention of Occupational Stress. *IOSR Journal of Dental and Medical Sciences, 14*(November 2015), 98--104.

Oman, D., Hedberg, J., & Thoresen, C. E. (2006). Passage meditation reduces perceived stress in health professionals: A randomised, controlled trial. *Journal of Consulting and Clinical Psychology, 74*(4), 714–719. https://doi.org/10.1037/0022-006X.74.4.714

Rodrigues, J. G. P., Kaiseler, M., Aguiar, A., Cunha, J. P. S., & Barros, J. (2015). A mobile sensing approach to stress detection and memory activation for public bus drivers. *IEEE Transactions on Intelligent Transportation Systems, 16*(6), 3294–3303. https://doi.org/10.1109/TITS.2015.2445314

Sharmin, M., Raij, A., Epstien, D., Nahum-shani, I., Beck, J. G., Vhaduri, S., Preston, K., & Kumar, S. (2015). *Visualisation o f Time - Series Sensor Data to Inform the Design of Just - In - Time Adaptive Stress Intervention s*. 505–516.

Tveito, T. H., & Eriksen, H. R. (2009). Integrated health programme: a workplace randomised controlled trial. *Journal of Advanced Nursing, 65*(1), 110–119. https://doi.org/10.1111/j.1365-2648.2008.04846.x

Villani, D., Grassi, A., Cognetta, C., Toniolo, D., Cipresso, P., & Riva, G. (2013). Self-help stress management training through mobile phones: An experience with oncology nurses. *Psychological Services, 10*(3), 315–322. https://doi.org/10.1037/a0026459

Weerasekara, M., & Smedberg, A. (2019). Design practices and implications in information and communication technology supported occupational stress management interventions. *2019 International Conference on Advanced Computer Science and Information Systems, ICACSIS 2019*, 285–294.

COULD MEDICAL APPS KEEP THEIR PROMISES?

Raina Samuel, Iulian Neamtiu and Sydur Rahaman
New Jersey Institute of Technology, USA

ABSTRACT

Medical mobile apps are already in wide use, and their use, as well as user base are projected to grow even further. However, it is unknown whether medical apps achieve their claimed behavior effectively and accurately. To determine potential gaps between app claims and app behavior, as well as between app claims and user expectations, we conducted a study on over 2,000 Android apps. We first developed an information retrieval approach that maps an app's description to medical (ICD) terms, hence delineate the app's medical scope and stated goals; our analysis has revealed that weight management, heart rate measurement, blood sugar measurement and hearing aids constitute the most common conditions apps claim to address. Next, based on app functionality, we categorize apps into (a) apps that measure or manage a physiological parameter, (b) apps that claim to treat conditions, and (c) apps for self- assessment. Within these three categories, we establish fine- grained subcategories and for each subcategory we compare apps' claimed behavior with realizable behavior. We found that app widely overstate their behavior and functionality. We also found that apps employ disclaimers and misleading terms to lure users into installing/using the app yet avoid responsibility. Finally, based on our uncovered app behavior and claims, we outline actionable findings w.r.t app claims and actual vs. stated function, meant to make users safer and apps more forthright.

KEYWORDS

Medical Apps, Android, Digital Health, Mobile Computing

1. INTRODUCTION

Medical mobile apps are integral to daily life, offering diverse functions from connecting to medical devices to tracking physiological parameters to condition assessment or diagnosis.

Due to their convenience and ubiquity, users trust medical apps and generally assume that apps are validated and accurate. However, there is no direct evidence on whether a medical app is performing its claimed functions. For instance, in a study regarding blood pressure monitoring apps, users liked the perceived accuracy; however, the app under-reported users' actual systolic pressure and provided inaccurate results which gave users a false sense of security (Plante, 2018).

We conducted a study on more than 2,000 Android apps collected from Google Play to understand (1) the medical conditions targeted by medical apps, and (2) the claims app make, e.g., regarding diagnosis or cures; hence consequently reveal and categorize lapses between app claims and actual functionality.

To define app behavior and nature, we map app metadata terms onto ICD-11 (International Classification of Diseases) codes. Using ranked retrieval text analysis, we were able to accurately shed light on common conditions applications may claim to treat or manage (Section 2). For apps that perform measurement and tracking, we found that most ICD codes were related to physiological management, such as weight loss or heart rate measurement. For apps that address conditions, we found that the most common conditions include elevated blood glucose level (MA18.0) and speech therapy (QB95.5); we present the findings in Section 3.

Next, we focus on exposing questionable claims found in app descriptions.

We classified apps into three main categories of claimed behavior: physiological (Section 4), treatment (Section 5), and self-assessment (Section 6). Focusing on app descriptions allows us to observe better what may possibly convince users into installing certain apps. We establish keywords and frequencies to categorize suspicious behaviors accordingly. Within each category, we investigate app claims and compare these claims with what is realizable with an app running on a smartphone; we found a wide gap between claims and attainable functionality.

Overall, we make the following contributions:

- A classification of app behavior based on medical conditions established by international standards (ICD-11).

- A classification of possible misleading claims found in Medical apps.
- A discussion of app disclaimers and misleading description terms.

Prior work Several studies investigated the accuracy and overall usability of mobile health apps. Coppetti et al. studied the accuracy of smartphone heart rate measurement apps and revealed that there were substantial performance differences between heart rate apps and clinical monitoring -- as much as 20 beats per minute (Coppetti, 2017). While their study focused on only 4 apps, it is safe to assume that this issue persists with other apps of this nature.

The importance of responsible app marketplace safeguards regarding health apps is discussed by Wykes et al. (Wykes, 2019); their work expressed concerns with the "overselling of health apps" and suggested a set of four principles that app marketplaces could use to guide the user to more sensible choices. Their study was conducted on four apps, rather than a larger dataset. Wisniewski et al.'s study on top-rated health apps confirmed our findings that most medical apps "continue to have no scientific evidence to support their use" (Wisniewski,2019); their study is based on manual analysis of 120 apps.

Other studies show that there is little evidence to whether health apps work, finding that only a small fraction of apps is tested (Byambasuren, 2018), leading to suggestions of "prescribed health apps", meaning having health apps vetted by medical professionals as a prescription rather than being able to be freely installed.

The reliability and safety of health apps is discussed by Akbar et al. with most concerns stemming from the quality of content presented in apps, such as presenting incorrect and incomplete information (Akbar, 2020).

Regarding weight loss apps, Zaidan et al. addressed the usability features of these apps by applying an evaluation framework (Zaidan, 2016). Their framework revealed that app marketplace search engines had biases towards certain titles and keywords that did not reflect the full functionality of the app and that the most popular apps are not necessarily the most effective.

Brown et al.'s review of 76 pregnancy apps regarding nutrition determined that such apps should not be considered as an appropriate resource for pregnant women due to unsound nutritional advice and overall unreliability (Brown, 2019).

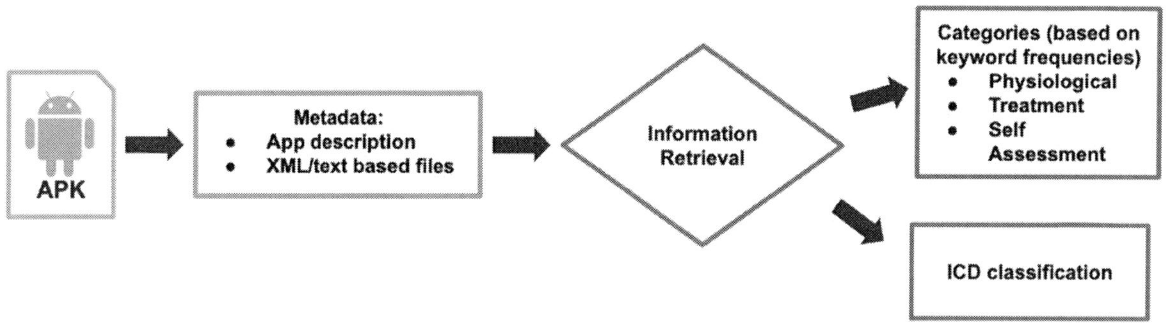

Figure 1. Overview of methodology

2. METHODOLOGY

We begin by describing our overall methodology illustrated in Figure 1. We acquired app descriptions and apps (APK files) from Google Play which correlated to the Medical and Health & Fitness categories. This resulted in a total of 2,339 apps that had installs over 1000 and were in the English language. From these apps, we used their descriptions and relevant text-based app metadata to determine ICD codes and observe claimed app functionality and misleading claims. We map the medical conditions apps may claim to treat (or monitor) onto an established ontology, ICD-11 codes. ICD -- the classification of diseases used by the World Health Organization -- provides an international standard for uniform naming of diseases and health conditions. Extracting ICD terms from app metadata not only enables us to identify possible conditions apps may claim to treat or monitor, but also (1) can reveal lapses in app descriptions regarding functionality and (2) helps us understand further the general medical app landscape. App descriptions and text metadata were processed, removing stop-words and irrelevant terms. We will next discuss how we managed to extract ICD codes and categorize app functionalities via information retrieval.

ICD code mapping challenges

We begin by describing the process and challenges faced when extracting ICD codes from apps. We used 106 ground truth apps as a basis to determine the score threshold for matched terms.

The first challenge was determining relevant app metadata. Using irrelevant terms and certain stop words can result in inaccurate ICD mappings or even no matches. We compared extracted keywords between the app description and XML files in order to understand common medical app functionalities and which source would be the most effective in mapping with ICD codes. We found that XML files provided less relevant results despite having more medically related keywords, especially for apps used as reference material, intended for patients, or to interact with patient portals. This is because XML files contain more fragmentation and individual words rather than cohesive sentences to provide any meaningful input.

The second challenge involved finding the best method to map and extract terms from app descriptions. We began with an initial mapping with a TF-IDF (term frequency – inverse document frequency) analysis, which showed that most apps correlated to the ICD code MA13.1 (*Finding of alcohol in blood*). However, when we attempted a ranked retrieval text analysis, we found much more accurate ICD terms mapped to keywords.

Table 1. Keyword Discrepancies in ICD Codes

App Name	TF-IDF	Ranked Retrieval	ICD Code
com.ebsco.dha	'management',**'difficulty', 'disorder', 'condition'**	'health','refer','clinic','care'	QB10 (Medical services not available in home)
com.pocketprep.nptepta	'disease', **'specified', 'defect', 'vertical'**	'brain','test','therapy', 'nervous system'	MB72 (Results of function studies of the nervous system)
com.ninezest.stroke	'therapy', **'devices', 'malignant', 'miscellaneous'**	'stroke','therapy', 'speech',**'enhance'**	QB95.5(Speech therapy)
com.srems.protocol	**'harm','malignant', 'classified', 'miscellaneous'**	'region','clinic','treatment', 'cardiac arrest'	MC82.1 (Bradycardic cardiac arrest)

Using a naive approach leads to inaccuracy in text extraction, as we show such discrepancies in Table 1. Here we compare the keywords extracted from TF-IDF analysis versus those from ranked retrieval; the text in bold indicates inaccurate or irrelevant keywords that do not map to the accurate ICD code displayed. We see that ranked retrieval provided the most accurate results. Thus, we used ranked retrieval to obtain each app's ICD code. We will discuss our results in Section 3.

Categorizing claimed app functionalities

Next, we will discuss how we categorized app behavior. Here we focused solely on app descriptions, as they are the initial reasons why users download applications. We used 33 apps as ground truth, which we had manually determined as potentially misleading due to specific terms in their descriptions. We focused on generic terms such as "diagnosis", "entertainment purposes", "instant", and "camera" and applied a TF-IDF analysis on the full dataset, resulting in a subset of 1250 apps matching these criteria.

Once the subset was established, we then manually reviewed common patterns based on general functionality to create a categorization. We developed another set of keywords to categorize the 1250 apps into three categories using TF-IDF analysis. Finally, to better refine our categories and the broad functions we found, we further characterize them into more specific subcategories. In doing so, we reveal possible lapses in claimed behavior and their legitimacy, especially in popular apps. We further discuss our findings in Section 4.

3. MEDICAL CONDITIONS

We will now present our findings. First, we will discuss the results of the ICD code analysis and the top codes found. Then we will describe the claimed app behaviors found in app descriptions.

3.1 Top ICD Conditions

ICD codes extracted from app descriptions help us ascertain whether descriptions accurately describe/explain app functionality. Table 2 displays the top ICD codes along with an explanation of how it is used and its categorization.

We found that most of the ICD codes relate to weight loss apps due to the frequency of the term: MG43.5 (*Excessive weight loss*) as we see in Table 2. We also see many ICD codes related to apps which connect to external medical devices, especially with pacemakers (QBB30.3: (*Adjustment or management of vascular access device*) and hearing aids (QB31.4: *Fitting or adjustment of hearing aids*). Among the top ICD codes, we find very few apps for professional use relate to any, if at all. This is because many app descriptions related to professionals or clinicians are either very vague or too complex to map correctly to a single specific ICD code. Nevertheless, we were able to accurately map ICD codes to medical conditions in apps that claimed to treat said diseases.

Table 2. Top 10 ICD Codes based on app descriptions

ICD Code	ICD Title	#Apps	Use
MG43.5	Excessive weight loss	511	Weight Control
MC82.1	Bradycardic cardiac arrest	255	Heart Rate Measurement
QB30.3	Adjustment or management of vascular access device	242	Pacemaker Management
QB31.4	Fitting or adjustment of hearing aid	232	Hearing Aid
MA18.0	Elevated blood glucose level	135	Diabetes Management
M54.5	Low back pain, unspecified	120	Pain Management
QA41	Pregnant State	104	Pregnancy Tracking
CA23	Asthma	84	Asthma Management
QB95.5	Speech Therapy	75	Speech Aphasia Treatment
H93.1	Tinnitus	70	Hearing Aid

3.2 Claimed App Behavior

We characterized app functionalities into three main categories: Physiological, Treatment, and Self-Assessment. Apps in these categories are examples of behavior that may potentially require regulations or further scrutiny and exemplify the need to categorize claimed app functionality. Apps should be clearer about their true functionalities in their descriptions while being explicit in their disclaimers; many times, disclaimers are hidden in the text or towards the end of the Google Play description; when the description is lengthy, users may end up ignoring or missing the caveat completely.

We will now describe each category and subcategory found in Table 3 starting with Physiological, Treatment, and Self-Assessment.

Table 3. Categories of app functionalities

Category	#Apps	%
Physiological	430	34
Heart Rate Measurement	115	9
Optometry	98	8
Blood Sugar Measurement	87	7
Hearing Test	42	3
Skin Cancer	41	3
Body Temperature Measurement	31	2
Weight Loss	16	1
Treatment	320	26
Natural Home Remedy	200	16
Hypnotherapy/Brain Wave Therapy	71	6
Pain Relief	49	4
Self-Assessment	500	40
Mental Health	309	25
Symptom Tracking	106	8
Pregnancy Quizzes	85	7

4. PHYSIOLOGICAL

Apps in this category claim to be able to measure certain physiological parameters such as heart rate or blood pressure, using the camera and other smartphone sensors. Concerningly, these apps claim to provide some form of diagnosis based on the measurement; furthermore, the apps claim that their measurements are accurate. We have found 430 such apps, categorized as follows.

4.1 Heart Rate

Heart rate-measuring apps use the smartphone camera's flash feature to measure a person's pulse. Measuring heart rates via a smartphone camera is not inherently inaccurate or deceptive, though a study has found differences between results obtained with apps versus results gather via clinical monitoring (Coppetti, 2017). However, users should not solely rely on such apps for diagnosis or treatment. For example, app **Cardiac diagnosis (arrhythmia),** with over 1,000,000 installations, states no disclaimers or recommendations to seek a medical professional or use an actual heart monitor along with the app. The accuracy is generally unknown, especially how the app manages to detect such conditions. Unless these apps work in conjunction with an external medical device, such as a blood pressure meter or heart monitor, the accuracy of such apps should not be relied on for diagnosis. Moreover, we believe that (1) such apps should include a disclaimer or recommendation to consult a medical professional, and (2) the term 'diagnosis' should be removed from apps' titles.

4.2 Optometry

Optometry apps claim to measure vision acuity by providing eye exams testing for astigmatism, near and far-sightedness or color blindness.

While these apps may provide a basic benchmark for vision, without a medical professional's diagnosis, the apps should not be used as a sole medical opinion. As a result, all apps with this functionality must include a recommendation to report their results to qualified ophthalmologists or optometrists before taking any sort of action.

4.3 Blood Sugar

Blood sugar apps claim to measure or track blood sugar. While many of these apps do have this behavior, as they work with a glucometer, many do not -- the apps simply serve as a journal.

Apps claiming to measure or track blood sugar without connecting to a glucometer or any sort of device can be misleading. Additionally, some apps whose name contains "Blood Sugar Test" have disclaimers stating the app cannot measure blood sugar but provides information on how to manage diabetes. Thus, these apps should modify their titles to better reflect app functionality, e.g., "Blood Sugar Tracking" or "Blood Sugar Log".

4.4 Hearing Test

Hearing test apps are different from hearing aid apps, which tend to connect to an external hearing aid device, serving as a remote control. These apps claim to provide (1) tests regarding tinnitus and (2) therapies for hearing issues; nevertheless, users need to see an ENT or audiologist for a reliable and accurate diagnosis.

4.5 Skin Cancer

Skin cancer apps use the device's camera to take pictures of skin and then use an AI algorithm to provide a preliminary diagnosis regarding skin cancer. Apps claiming to detect skin cancer solely through a device's camera and without a blood test are deceptive and misleading. An example would be the app **Medgic** which uses AI to check for dermatological conditions or diseases by using the device's camera. While AI algorithms have been able to detect conditions before, prognoses cannot be solely confirmed by a simple photo of one's skin -- other tests must be administered in order to make a conclusion. The app's description contains a disclaimer, albeit at the end, stating how the app is not a replacement for medical advice and that not all results are 100% guaranteed.

Another app, **Visus**, states that it is an experimental application that is publicly deployed and that its algorithm is "*30% more sensitive and precise than a conventional board-certified radiologist*".

4.6 Body Temperature

Body temperature apps claim to measure users' temperature, e.g., to detect a fever. However, this is ultimately misleading, as mobile devices do not have any means to measure temperature in their sensors. Instead, these apps serve as a mere journal to track user-inputted values for body temperature.

4.7 Weight Loss

Weight loss apps are numerous by nature, as seen in our ICD mapping. However, in this specific categorization we focused on apps that over-promise results within an arbitrary or unrealistic time frame or even "instant" results. We found that many apps do not urge the users to seek medical opinions prior to attempting weight loss. For example, the app **Lose Weight Fast at Home - Workouts for Women** with over 1,000,000 installs, claims that users following the app's regimen will lose weight in 30 days. However, there are no mentions in the app description of the influence other crucial factors such as diet, water intake, or genetic factors, have in weight loss.

5. TREATMENT

These apps claim to be able to cure diseases. We have found a total of 320 apps, falling into several subcategories.

5.1 Hypnotherapy/Brain Wave Therapy

These therapies are complementary forms of medicine (i.e., used to supplement traditional treatment methods). Apps in this category tend to not mention the importance of standard or clinically proven medical treatments to be used in conjunction with their suggested therapies. Hypnotherapy results are generally not clinically proven and may have adverse effects on users who are prone to epilepsy or other neurological conditions (Gruzelier, 2000). For example, the app **Atmosphere: Binaural Therapy Meditation** which has over 500,000 installations, states that it is able to "heal your DNA" with its guided breathing and meditation; nevertheless, the app description contains a disclaimer that the app is only for "entertainment purposes" and should not be a substitute for medical treatment.

5.2 Natural Remedy

These apps provide references to natural remedies, e.g., certain herbs or foods, to manage and treat specific diseases, such as skin diseases or even cancer. They also claim to help users "self-cure" certain conditions. Apps that provide home remedies are not malicious or intentionally misleading but should never be a replacement for actual treatment prescribed by a medical professional. While there are natural remedies to basic non-life-threatening illnesses or wounds, an app is not an alternative to prescribed treatment from a medical provider (Desmet, 2004). For example, the app **Doctor at Home**, which has over 100,000 installations, claims it can provide treatment for "110 diseases" and "cure diseases at home". Examples of three conditions -- cholera, angina, pneumonia -- and app-prescribed "cures" are shown in Figure 2. Additionally, the app states that the user can be a home doctor, defined as "you are yourself a doctor". Herbal treatments and reference material cannot replace professional diagnosis or treatment. While the app has useful tips for treating simple symptoms and issues, such as coughing and dandruff, it also has claims for treating more serious cases such as stomach ulcers and cholera.

5.3 Pain Relief

These apps rely on providing exercises and remedies to address various types of muscular pains or migraines. While such apps can offer a catalogue of exercises that can address certain types of pain, such apps should be used in conjunction with medical advice. Apps which work in tandem with qualified pain coaches can be a convenient way to help manage pain remotely. However, for many pain relief apps, pain is addressed through virtual exercises with claims that they are "proven to ease pain", such as in app **Lower Back Pain and Sciatica Relief Exercises**. Note that the issue is not whether exercises are effective or not; rather the issue is that app descriptions do not suggest seeking professional medical advice *prior* to app installation. Additionally, certain pain exercises, when performed incorrectly or without supervision, can lead to further damage and pain in many cases (Lubell,1989).

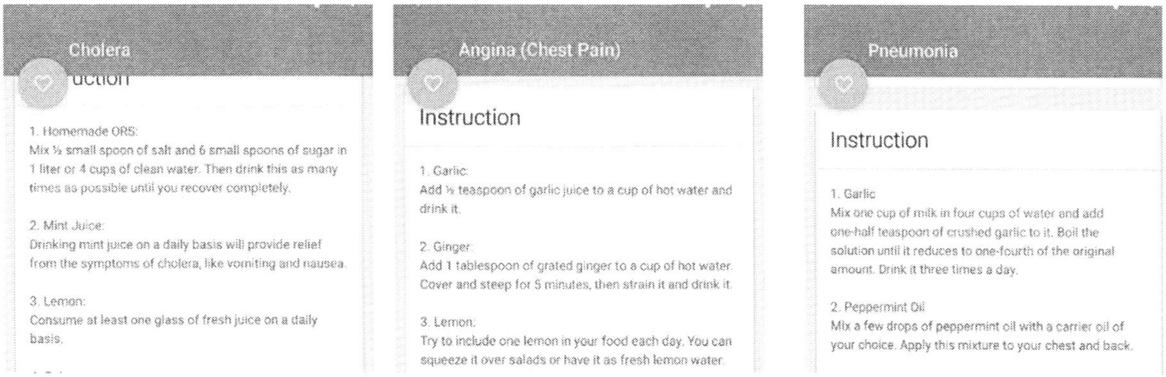

Figure 2. **Doctor at Home** claims to be able to 'cure' critical diseases naturally

6. SELF ASSESSMENT

We have found 500 apps which emphasize the use of assessments and self-help, categorized as follows.

6.1 Mental Health

Mental health apps rely on self-assessments without a professional entity providing feedback. Note that there is a lack of direct scientific evidence found in descriptions of apps that claim to help with mental health or behavioral patterns (Larsen, 2019). Many mental health apps do not provide confirmation or verification that the app is indeed vouched for by professionals. For example, the self-help app **MoodSpace** is focused on depression and mental well-being. While the description claims that the app is "a well-being app driven by research", there is no evidence of any research or authoritative proof accessible to users prior to installation. As with prior examples, the app's description contains a disclaimer and an emphasis that users should seek medical advice, but the disclaimer is found at the very end of the description, increasing the chance to be ignored by users.

6.2 Symptom Tracking

Symptom tracking apps are based on user input (rather than physiological measurements as prior discussed) to determine possible diagnoses. These apps are useful for a cursory understanding of certain symptoms but should not be used for a diagnosis. Many of these apps are extremely popular, such as **Ada-check your health**, with over 5,000,000 installations and classified as a Class I Medical Device, meaning it is considered as a device with low risk to the user in the European Union. While **Ada-check your health** is an example of a well-regulated medical app, there are many apps that claim to perform similar functions but are not as well scrutinized or moderated by government or marketplace entities, such as the **Disease Detector** which claims to detect diseases in a few seconds.

6.3 Pregnancy Quizzes

These apps ask a series of questions and claim to determine whether the user shows early signs of pregnancy. While a collection of certain symptoms can help determine the likelihood of pregnancy, it can only be validated through an actual physical pregnancy test. As a result, the framing and naming of these apps are misleading. An example was app **Real Pregnancy Test & Quiz** -- removed from Google Play during this research -- which suggested that it was "*an easy quiz for pregnancy. Just reply the quiz questions*".

7. CONCLUSION

Medical mobile apps are understandably convenient and appealing to users. However, app quality and app description quality remain sorely lacking. These lacunae are particularly concerning in this (medical) domain because app reliability can directly affect/impact user safety and well-being. Our approach and study found that the functionality landscape of medical apps is broad and varied; however, the functionalities claimed in app descriptions are not entirely reliable. Our findings show a need for better regulation and scrutiny of medical apps in-app marketplaces to better protect users and their health.

ACKNOWLEDGMENT

We thank the anonymous reviewers for their feedback. This material is based upon work supported by the National Science Foundation under Grant No. CCF-2106710.

REFERENCES

Akbar, S. et al. (2020) "Safety concerns with consumer-facing mobile health applications and their consequences: A scoping review", *J. Am. Med. Inform. Assoc.* Oxford University Press (OUP), 27(2), bll 330–340.

Brown, H., Bucher, T., Collins, C. and Rollo, M., 2019. A review of pregnancy apps freely available in the Google Play Store. *Health Promotion Journal of Australia*, 31(3), pp.340-342.

Byambasuren, O., Sanders, S., Beller, E. and Glasziou, P., 2018. Prescribable mHealth apps identified from an overview of systematic reviews. *npj Digital Medicine*, 1(1).

Coppetti, T., Brauchlin, A., Müggler, S., Attinger-Toller, A., Templin, C., Schönrath, F., Hellermann, J., Lüscher, T., Biaggi, P. and Wyss, C., 2017. Accuracy of smartphone apps for heart rate measurement. *European Journal of Preventive Cardiology*, 24(12), pp.1287-1293.

Gruzelier, J., 2000. Unwanted effects of hypnosis: a review of the evidence and its implications. *Contemporary Hypnosis*, 17(4), pp.163-193.

Larsen, M., Huckvale, K., Nicholas, J., Torous, J., Birrell, L., Li, E. and Reda, B., 2019. Using science to sell apps: Evaluation of mental health app store quality claims. *npj Digital Medicine*, 2(1).

Lubell, A., 1989. Potentially Dangerous Exercises: Are They Harmful to All?. *The Physician and Sportsmedicine*, 17(1), pp.187-192.

Plante, T., O'Kelly, A., Urrea, B., MacFarlane, Z., Blumenthal, R., Charleston, J., Miller, E., Appel, L. and Martin, S., 2018. User experience of instant blood pressure: exploring reasons for the popularity of an inaccurate mobile health app. *npj Digital Medicine*, 1(1).

Wisniewski, H., Liu, G., Henson, P., Vaidyam, A., Hajratalli, N., Onnela, J. and Torous, J., 2019. Understanding the quality, effectiveness and attributes of top-rated smartphone health apps. *Evidence Based Mental Health*, 22(1), pp.4-9.

Wykes, T. and Schueller, S., 2019. Why Reviewing Apps Is Not Enough: Transparency for Trust (T4T) Principles of Responsible Health App Marketplaces. *Journal of Medical Internet Research*, 21(5), p. e12390.

"tfidf: A Single-Page Tutorial - Information Retrieval and Text Mining" (no date). Available at: http://www.tfidf.com/.

Zaidan, S. and Roehrer, E., 2016. Popular Mobile Phone Apps for Diet and Weight Loss: A Content Analysis. *JMIR mHealth and uHealth*, 4(3), p. e80.

SINGLE MR IMAGE SUPER-RESOLUTION USING GENERATIVE ADVERSARIAL NETWORK

Shawkh Ibne Rashid, Elham Shakibapour and Mehran Ebrahimi
Faculty of Science, Ontario Tech University
Oshawa, Ontario, Canada

ABSTRACT

Spatial resolution of medical images can be improved using super-resolution methods. Real Enhanced Super Resolution Generative Adversarial Network (Real-ESRGAN) is one of the recent effective approaches utilized to produce higher resolution images, given input images of lower resolution. In this paper, we apply this method to enhance the spatial resolution of 2D MR images. In our proposed approach, we slightly modify the structure of the Real-ESRGAN to train 2D Magnetic Resonance images (MRI) taken from the Brain Tumor Segmentation Challenge (BraTS) 2018 dataset. The obtained results are validated qualitatively and quantitatively by computing SSIM (Structural Similarity Index Measure), NRMSE (Normalized Root Mean Square Error), MAE (Mean Absolute Error) and VIF (Visual Information Fidelity) values.

KEYWORDS

Imaging, Deep learning, Generative Adversarial Network, MR Image Enhancement, Single MR Image Super Resolution

1. INTRODUCTION

Higher quality Magnetic Resonance Images (MRI) are valuable for early detection and accurate diagnosis of various medical conditions. High spatial resolution of images is essential to provide detailed anatomical information and help radiologists with accurate quantitative analysis. Acquiring higher resolution MRI requires higher image acquisition times that can be costly and may not always be possible due to physical limitations. Super-resolution (SR) techniques are alternative ways to improve the spatial resolution of images by producing a High-resolution (HR) image given a Low-resolution (LR) one.

SR approaches to produce HR MRI are mostly categorized into reconstruction-based and learning-based methods (Van et al, 2012). Reconstruction-based techniques use interpolation filtering methods such as bilinear, bicubic or lanczos (Duchon, 1979). They are among the first methods to tackle single image super resolution (SISR) problems. However, interpolated images blur or degrade important edge and texture information of images.

SISR methods have been widely advanced by the breakthroughs in deep learning. Methods based on Generative Adversarial Networks (GANs) (Goodfellow et al, 2014) are promising approaches for image generation and have been also used for SR (Ledig et al, 2017). GANs-based models show the increasing performance for SISR. Different architectures and loss functions aimed at improving the quality of the generated images using GANs have been proposed (Metz et al, 2015, Arjovsky et al, 2017, Mao et al, 2017).

Recent advances in the Super Resolution Generative Adversarial Network (SRGAN) are aimed to recover fine texture details and edge information even at large upscaling factors (Ledig et al, 2017). This motivated us to apply a recent extension of the method called Real Enhanced Super-Resolution Generative Adversarial Network (Real-ESRGAN) that achieves high perceptual quality for 2D real-world images (Wang et al, 2021). Our specific focus in this work is to apply Real-ESRGAN (Wang et al, 2021) to resolution enhancement of 2D slices of 3D MR images. To the best of our knowledge, no studies have been conducted on the use of Real-ESRGAN (Wang et al, 2021) to validate SISR on MRI scans.

2. RELATED WORKS

The state-of-the-art methods with deep learning techniques have shown an increasing performance on producing SISR on 2D real-world images (Kim et al, 2016, Lai et al, 2017, Lim et al, 2017, Tai et al, 2017, Haris et al, 2018, Wang et al, 2018, Zhang et al, 2018, Dai et al, 2019, Wang X. et al, 2019). Furthermore, many of the current deep learning techniques for medical image enhancement typically rely on GANs. GANs-based models generate more realistic images (Tan et al, 2020). To produce SISR, a GAN pipeline usually consists of a single generator network that takes in the degraded/down-sampled LR image as input and directly outputs the reconstructed SR image. A photo-metric loss is calculated between the SR image and the ground truth and drives the network to recover realistic image details (Wang J. et al, 2019). Most of the proposed approaches apply SRGAN developed by Ledig et al (2017).

mDCSRN (Chen et al, 2018), Lesion-focussed GAN (Zhu et al, 2019), and ESRGAN (Bing et al, 2019) are GAN-based solutions tackling SR for medical images. In (Tan et al, 2020), a meta-upscale module proposed by Hu et al. (2019) is combined with SRGAN to create a network called Meta-SRGAN. GAN-based models including ESRGAN and CycleGAN are used in (Do et al, 2021) to generate HR MRI with rich textures. Their experimental results have been conducted on both 3T and 7T MRI in recovering different scales of resolution. The authors in (Sanchez et al, 2018) have applied the SRGAN-based model (Ledig et al, 2017) adopted to 3D convolutions to generate HR MRI scans. They have explored different methods for the upsampling phase to alleviate artifacts produced by sub-pixel convolution layers. Chen et al. (2018) have applied the Densely Connected SR Network (DCSRN) (Huang et al, 2016) for 3D brain MR image enhancement. Though, direct conversion into 3D may result in many parameters and thus faces challenges in memory allocation.

SRGAN (Ledig et al, 2017) applies a perceptual loss using high-level feature maps of the VGG network (Simonyan, and Zisserman, 2015, Sprechmann, and LeCun, 2016, Johnson et al, 2016) combined with a discriminator that encourages solutions perceptually difficult to distinguish from the HR reference images. However, the discriminator requires a more powerful capability to discriminate realness from complex training outputs, while the gradient feedback from the discriminator needs to be more accurate for local detail enhancement (Wang et al, 2021).

In Real-ESRGAN (Wang et al, 2021), the VGG-style discriminator in ESRGAN is developed via U-Net design along with spectral normalization (Ronneberger et al, 2015, Schonfeld et al, 2020) to increase discriminator capability and stabilize the training dynamics. As a result, Real-ESRGAN (Wang et al, 2021) achieves better visual performance making it more practical in real-world applications. This motivates us to apply and validate Real-ESRGAN (Wang et al, 2021) to produce SISR of 2D slices of 3D MR images. Real-ESRGAN (Wang et al, 2021) makes use of the RGB LR images. In our proposed approach, we modify the structure of the Real-ESRGAN to train the 2D MR images. The obtained results are assessed and compared quantitatively and qualitatively with the standard bilinear and bicubic interpolation methods.

3. METHODOLOGY

In this section, we explain the GAN model we have used to enhance the resolution of brain MRI images, along with the description of the dataset and the modifications we have made to the GAN model for our purpose of working with grayscale brain MR images, cost functions used and the parameter settings.

3.1 Dataset Preparation

We have used the BraTS 2018 dataset for the resolution enhancement experiment. The dataset consists of MRI scans of glioblastoma (GBM/HGG) and lower grade glioma (LGG) as native (T1), post-contrast T1 weighted (T1Gd), T2-weighted (T2) and T2 Fluid Attenuated Inversion Recovery (FLAIR) volumes. For the image resolution enhancement purpose, we have used both HGG and LGG T1 brain MR images. There is a total of 285 3D MR T1 weighted images, each consisting of 155 2D slices. The slices are grayscale images and are available in NIfTI format.

We have used 80% of the 3D images for training purpose and the rest of the 20% images for testing the GAN models. For dataset preparation, we have removed the blank MRI scans and have normalized the pixel values to lie in the range of 0 to 1. As this dataset is primarily created for segmentation, the dataset also contains the segmentation masks of the brain MR images. As our purpose is resolution enhancement, we have produced the low-resolution images by down-sampling the 2D MRI scans by a factor of four using bilinear interpolation method. There are some 2D slices that includes no brain tissue.

We have removed such blank MR images. The original size of the 2D images is 240x240. We use zero padding to change the size to 256x256 for training our models. For training, we get 285x0.8x155 = 35340 2D images and after excluding blank images, we have a total of 31322 images. There are a total 7823 images in the test set after the pre-processing steps.

Since we are using a resolution enhancement scale of 4 for low resolution image generation, our low-resolution images are of size 64x64.

3.2 Real-ESRGAN Architecture

Real-ESRGAN is a GAN with a Residual in Residual Dense Block (RRDB) based generator and UNET based discriminator. The generator consists of multiple RRDB blocks. This RRDB block is modified from residual block in SRGAN. Instead of using convolutional layers in Residual Blocks, they have used dense blocks (Huang et al, 2016). Each dense block consists of five convolutional layers accompanied with Leaky-RELU layers except for the last one. They have used residual scaling (Szegedy et al, 2016) in residual blocks where output from residual blocks is scaled down by multiplying them with a scale factor in range 0 to 1 (denoted as β in Figure 1). This helps when network becomes deeper and produces very large or small gradients. The RRDB blocks are followed by an upsampling block and two convolutional layers for the reconstruction. For the discriminator, they have used a U-Net based model with skip connections. To stabilize the training dynamics, they used spectral normalization regulation. They have also employed relativistic discriminator based on relativistic generator (Jolicoeur-Martineau, 2018).

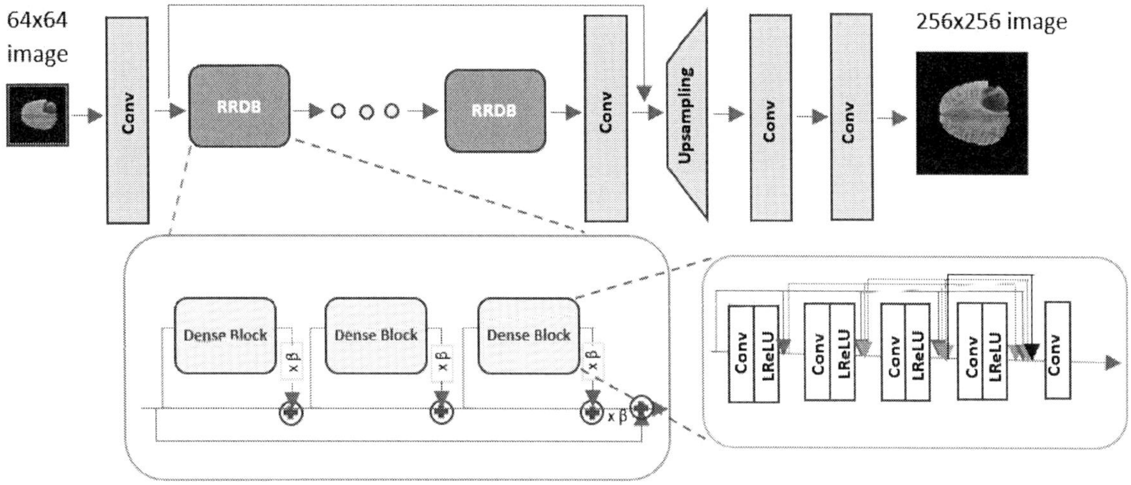

Figure 1. Generator Architecture of Real-ESRGAN

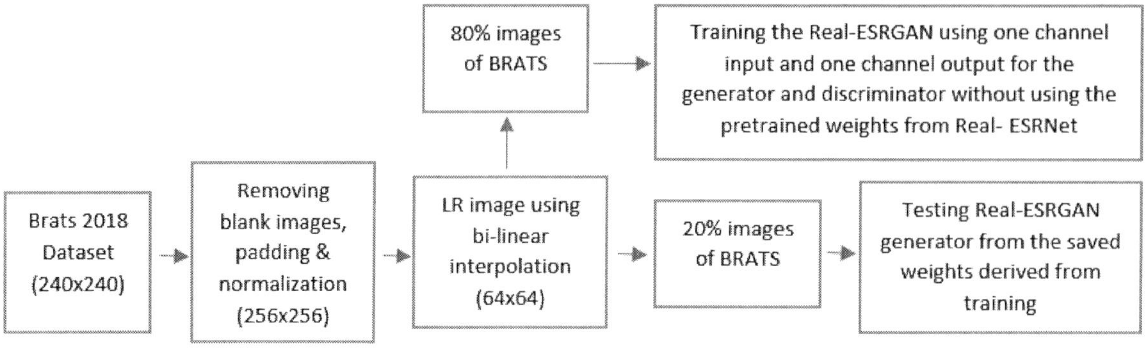

Figure 2. Summary of the training procedure

3.3 Loss Functions

We have used a combination of pixel loss (L1 loss), perceptual loss (Johnson et al, 2016), and GAN loss (Ledig et al, 2017) for training the generator and GAN loss for the discriminator. We have used a weight value of 1 for different losses meaning that we took the total loss values and combined them for training the generator. For discriminator, GAN loss was measured for both fake (generated from generator) and real images (ground truth) so that the discriminator can learn to distinguish between real and fake images. VGG19 weights are used for calculating perceptual loss.

3.4 Network Training

According to the Real-ESRGAN paper the training process is divided into two stages. The first part is training Real-ESRNet (Wang et al, 2021) with the L1 loss. When they trained Real-ESRGAN, they used the weights from the Real-ESRNet as the initialization point. In our experiment, we started our training by training the Real-ESRGAN without using the weights from the Real-ESRNet. The pre-trained weights from Real-ESRNet were obtained from training the model with real-world images. As we are working with brain MRI grayscale images, we started our training from scratch using only Real-ESRGAN.

Real-ESRGAN has three channel input and three channel output for the generator and three channel input for the discriminator. As brain MRI data is grayscale, we changed the input and output channel number to 1 for the generator and input channel number to 1 for the discriminator. We first train the model by copying the same MR image three times to match with the original input shape of the model and noticed variations among the three generated images from the generator (for three channel output). As a result, we changed the input shape and output shape of the model to facilitate one single channel training for the grayscale images.

3.5 Parameter Settings

We have used 23 Residual in Residual Dense Blocks (RRDB) in the generator. Each residual block has three dense blocks and five convolutional layers in each dense block. The initial input number of channels are set to 64 for these convolutional layers. Channels for each growth for the convolutional layers in the dense block is 32. The growth channel means the first convolutional layer will output 32 features, the second conv layer will output 32x2 = 64 features, the third one will produce 32x3 = 96 features and so on. In the upsampling block, there are two upsampling layers, each increasing the number of features by a factor of two using nearest interpolation method.

We have optimized the generator and discriminator weights using Adam optimizer with a learning rate of 0.0001 based on the loss function as described above. We have trained the Real-ESRGAN for 300000 iterations on the 80% of the BraTS dataset images. A summary of the training procedure is illustrated in Figure 2.

Table 2. The quantitative comparison

Method	SSIM	NRMSE	MAE	VIF
Bilinear Interpolation	0.92±0.03	0.04±0.01	0.011±0.004	0.55±0.07
Cubic Interpolation	0.93±0.03	0.04±0.01	0.010±0.004	0.64±0.07
Real-ESRGAN	**0.94±0.03**	0.04±0.01	**0.009±0.004**	**0.71±0.09**

Note: SSIM, NRMSE, MAE and VIF values of Bilinear Interpolation, Cubic Interpolation, and Real-ESRGAN generated high resolution images compared with ground truth images. The metric values are the mean values for 7823 images. We have also included the standard deviation for these metric values in the table as well. In terms of the metrices, higher value of SSIM and VIF and lower values of NRMSE and MAE denote better quality images.

4. EXPERIMENTAL RESULTS

We have compared the performance of Real-ESRGAN for the MR image resolution enhancement on BraTS 2018 dataset with bilinear and bicubic interpolation methods based on a variety of evaluation metrics including Structural Similarity Index (SSIM), Normalized Root Mean Square Error (NRMSE) (normalized based on the mean value of the ground truth image), Mean Absolute Error (MAE) and Visual Information Fidelity (VIF) (Sheikh, and Bovik, 2006). We have made qualitative and quantitative comparison among these different approaches.

We have run our experiments on a Linux based operating system with two 2199 MHZ processors, 13 GB RAM. We use a NVIDIA Tesla K80 GPU with 12 GB memory. We have used Python programming language and its libraries including OpenCV and PyTorch.

To compare Real-ESRGAN, and different interpolation methods, we have used 20% of the BraTS 2018 dataset. After removing the blank MR images, we obtain 7823 images for testing. We train the Real-ESRGAN for 300000 iterations with a batch size of one. The quantitative comparison can be seen in Table 1.

From Table 1, we can see that Real-ESRGAN produces MR images with higher resolution than other compared methods. Qualitatively speaking, we can see the same thing. Figure 3 shows some of the examples from the test set along with the produced higher resolution images from Bilinear Interpolation, Cubic Interpolation, and Real-ESRGAN.

5. CONCLUSION

The problem of generating and validating a single MR Image Super-resolution using Generative Adversarial Network was addressed in this paper. We utilized the Real-ESRGAN model on 2D MR Images available on the BraTS dataset to generate 2D HR MR Images. We demonstrated that the Real-ESRGAN model outperformed the bilinear and bicubic interpolation methods in restoring a resolution by a factor of 4. The generated images were perceptually and qualitatively superior compared to the interpolated ones.

From the obtained results, it can be observed that the interpolated images can be blurry with ghosts and shadows around the boundaries with suppressed sharp edge information. Future work will involve extending the model to generate the SR images with arbitrary zooming factors.

ACKNOWLEDGEMENT

This work was supported in part by the Natural Sciences and Engineering Research Council of Canada (NSERC).

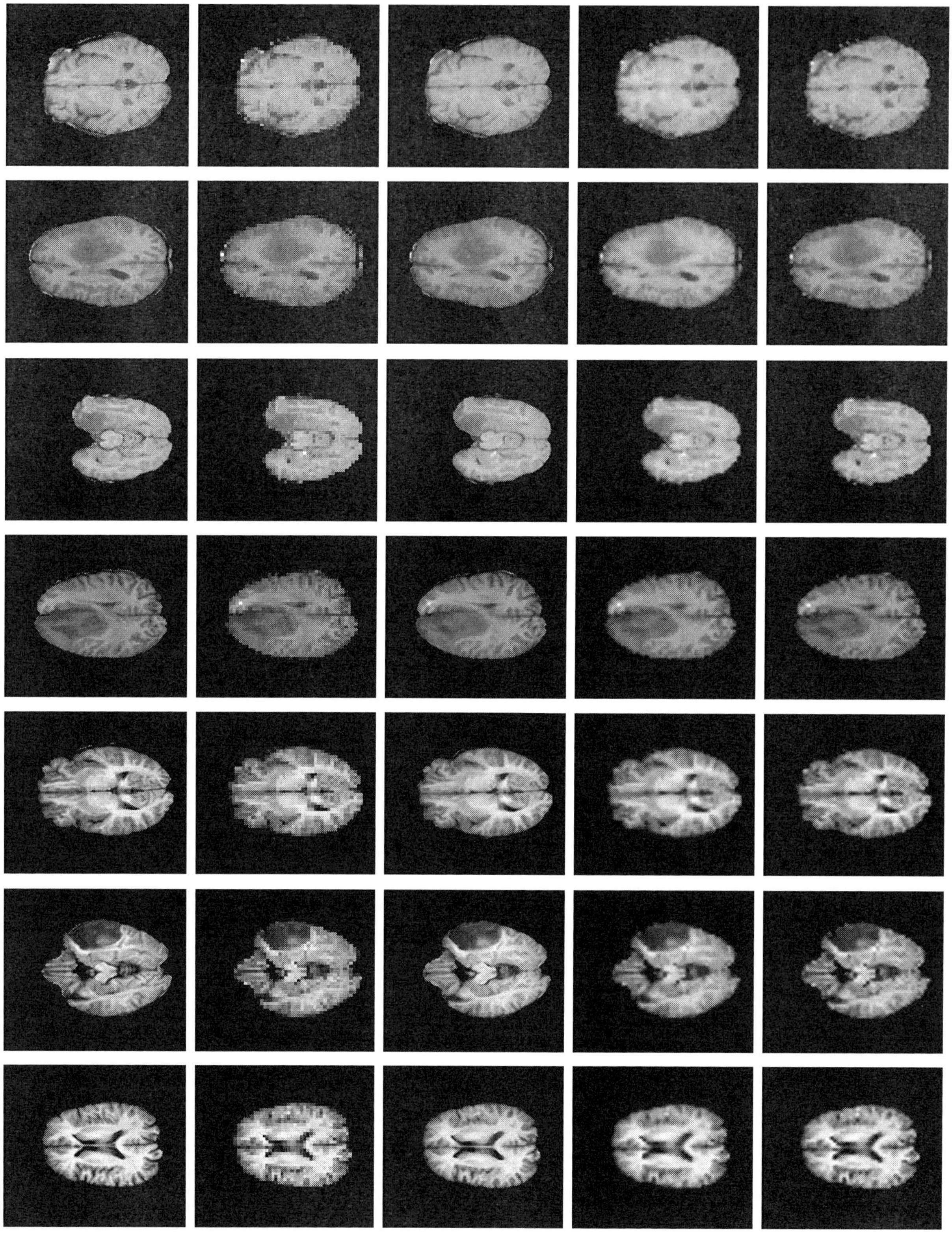

Figure 3. From left to right, Ground truth, Low-resolution image, Generated image based on the model, Bilinear interpolation, and Bicubic interpolation.

REFERENCES

Arjovsky M. et al, 2017. Wasserstein Gan. *arXiv*: 1701.07875.

Bing X. et al, 2019. Medical image super resolution using improved generative adversarial networks. *IEEE Access*, vol. 7, pp. 145030–145038.

Chen, Y. et al, 2018. Efficient and accurate MRI super-resolution using a generative adversarial network and 3D multi-level densely connected network. *CoRR abs/1803.01417*, *arXiv*: 1803.01417.

Chen Y. et al, 2018. Brain MRI super resolution using 3D deep densely connected neural networks. *IEEE 15th International Symposium on Biomedical Imaging*, arXiv: 1801.02728.

Dai T. et al, 2019. Second-order attention network for single image super-resolution. *CVPR*.

Do H. et al, 2021. 7T MRI super-resolution with Generative Adversarial Network. *IS&T Electronic Imaging 2021 Symposium*.

Duchon C. E., 1979. Lanczos filtering in one and two dimensions. In *Applied Meteorology*, vol. 18, pp. 1016–1022.

Goodfellow I. et al, 2014. Generative Adversarial Nets. *Advances in Neural Information Processing Systems (NIPS)*, pp. 2672–2680.

Haris M. et al, 2018. Deep backprojection networks for super-resolution. *CVPR*.

Hu X. et al, 2019. Meta-sr: a magnification-arbitrary network for super-resolution. *CoRR abs/1903.00875*, *arXiv*: 1903.00875.

Huang G. et al, 2016. Densely Connected Convolutional Networks. arXiv: 1608.06993.

Johnson J. et al, 2016. Perceptual losses for real-time style transfer and super- resolution. *European Conference on Computer Vision (ECCV)*, pp. 694–711.

Jolicoeur-Martineau A., 2018. The relativistic discriminator: a key element missing from standard gan. *arXiv*: 1807.00734.

Kim J. et al, 2016. Accurate image super-resolution using very deep convolutional networks. *CVPR*.

Kim J. et al, 2016. Deeply recursive convolutional network for image super-resolution. *CVPR*.

Lai W. et al, 2017. Deep Laplacian pyramid networks for fast and accurate super-resolution. *CVPR*.

Ledig C. et al, 2017. Photo-realistic single image super-resolution using a generative adversarial network. *IEEE Conference on Computer Vision and Pattern Recognition (CVPR)*, pp. 4681-4690, doi:10.1109/CVPR.2017.19.

Lim B. et al, 2017. Enhanced deep residual networks for single image super-resolution. *CVPR*.

Metz L. et al, 2015. Unsupervised representation learning with deep convolutional generative adversarial networks. *arXiv*: 1511.06434.

Mao X. et al, 2017. Least squares generative adversarial networks. *arXiv*: 1611.04076v2.

Ronneberger O. et al, 2015. Unet: Convolutional networks for biomedical image segmentation. *International Conference on Medical image computing and computer-assisted intervention*.

Sanchez I., and Vilaplana V., 2018. Brain MRI super-resolution using 3D generative adversarial networks. *arXiv*: 1812.11440.

Schonfeld E. et al, 2020. A U-NET based discriminator for generative adversarial networks. *CVPR*.

Sheikh H. R., and Bovik A. C., 2006. Image information and visual quality. In *IEEE Transactions on Image Processing*, vol. 15, no. 2, pp. 430-444, doi: 10.1109/TIP.2005.859378.

Simonyan K., and Zisserman A., 2015. Very deep convolutional networks for large-scale image recognition. *International Conference on Learning Representations (ICLR)*.

Sprechmann J., and LeCun Y., 2016. Super-resolution with deep convolutional sufficient statistics. *International Conference on Learning Representations (ICLR)*.

Szegedy C. et al, 2016. Inception-v4, inception-resnet and the impact of residual connections on learning. *arXiv*: 1602.07261.

Tai Y. et al, 2017. Image super-resolution via deep recursive residual network. *IEEE conference on computer vision and pattern recognition*, pp. 3147–3155.

Tan Ch. et al, 2020. Arbitrary scale super-resolution for brain MRI images. *arXiv*: 2004.02086.

Van Reeth E. et al, 2012. Super-resolution in magnetic resonance imaging: a review. *Concepts in Magnetic Resonance Part A*, vol. 40A(6), pp. 306–325.

Wang X. et al, 2018. Esrgan: enhanced super-resolution generative adversarial networks. *ECCV*.

Wang X. et al, 2019. Deep network interpolation for continuous imagery effect transition. *IEEE Conference on Computer Vision and Pattern Recognition*, pp. 1692–1701.

Wang J. et al, 2019. Enhanced generative adversarial network for 3D brain MRI super-resolution. *arXiv*: 1907.04835.

Wang X. et al, 2021. Real-ESRGAN: training real-world blind super-resolution with pure synthetic data. *IEEE/CVF International Conference on Computer Vision Workshops (ICCVW)*, pp. 1905-1914, *doi*: 10.1109/ICCVW54120.2021.00217.

Zhang Y. et al, 2018. Image super-resolution using very deep residual channel attention networks. *ECCV*.

Zhu J. et al, 2019. How can we make gan perform better in single medical image super-resolution? a lesion focused multiscale approach. *IEEE 16th International Symposium on Biomedical Imaging*.

DIGITAL SUPPORT ACTIVATES YOUNG ELDERLY TO HEALTH-ENHANCING PHYSICAL ACTIVITY

Christer Carlsson and Pirkko Walden
IAMSR and Åbo Akademi University
Gezeliusgatan 2, 20520 Åbo, Finland

ABSTRACT

There is consensus in health studies that regular physical activity contributes to better health in both the short and long term. In the ongoing *DigitalWells* research program, we focus on getting young elderly, the 60-75 years age group, to adopt physical activity as part of their daily routines. Growing evidence shows that running viable and effective combinations of regular physical activity forms are health-enhancing if they are exercised long enough and with sufficient intensity. We developed and implemented a digital support platform to guide young elderly towards health-enhancing physical activity and searched for drivers that could get both the platform and health-enhancing physical activity programs accepted and adopted for sustained use. In this paper we work out the design and functions of the platform and explore ways to combine technology- and psychology-based drivers to get physical activity adopted as daily routine and digital support technology accepted among young elderly.

KEYWORDS

Health-Enhancing Physical Activity, Digital Support Platform, Young Elderly

1. INTRODUCTION

There is growing concern about deteriorating health in the ageing population (European Commission, 2018) that now starts to be 18-23% of the population in most EU countries. The largest proportion of elderly people (65+) are found in Japan (28.79 %), Italy (23.37%) and Finland (22.49 %); there are 14 countries where the 65+ is over 20% of the total population, all but one is an EU country (https://aginginplace.org/fastest-aging-populations). Ageing citizens live longer but turn out to need more health and social care with increasing age that requires more resources and annually increasing costs. In Finland, where this study was carried out, there is a social commitment to take care of its ageing citizens, and the numbers are growing – in 2020 there were 1.3 million 65+ citizens (out of a population of 5.5 million), the annual health care costs for them were over 3.7 B€, and the predictions are that these costs will only grow for the next 20-25 years (Borodulin and Sääksjärvi, 2019).

The strategy and policy decisions on the use of resources to care for ageing citizens has broad political support in most EU countries and are anchored in an EU-level program (European Commission, 2018). Public health and social care resources are reserved for senior citizens (75+ years) who are ageing badly and need support - this is, of course, as it should be in a modern 2020s society. Nevertheless, arguments can be raised in support of measures for preventive programs for younger healthy elderly to keep them healthy and in better shape. This would in the long term require less net effort, less resources and less health and social care costs (Borodulin and Sääksjärvi, 2019) and will contribute to a better quality of life.

Preventive programs should start early enough – with the young elderly, the 60-75 yerars age group – to have sufficient effects in the long term, when young elderly become senior citizens. It appears that public policy ignores the young elderly (young elderly interviewed in the *DigitalWells* research program) – "we are too healthy, too active, with too good social networks to need any intervention or support from public resources"; and "we are too many", this – a bit sarcastically.

There is consensus that regular and systematic physical activity (PA) can serve as preventive health care. The Copenhagen Consensus statement notes that: "(i) being physically active is a key factor in maintaining health ... (ii) physically active older adults, compared with older inactive adults, show benefits in terms of physical and cognitive function ... (iii) physical inactivity in older adults is associated with a trajectory towards disease and increased risk of premature all-cause mortality ..." (Bangsbo et al, 2019). Regular PA at moderate intensity for at least 150 minutes per week will have positive health effects (Wallén et al., 2014); the Copenhagen Consensus finds that also less than 150 minutes could be sufficient for older adults, but as we will find out, there is quite some debate about what will be the "right PA", "the right amount of PA" or "the right intensity of PA" to actually get health effects – or "will the same PA work for all of us?"

We introduced PA, that builds and maintains stamina, muscle strength, agility, and balance; when several PAs are combined over time we have a *PA-program*; when the PAs in a program prove to have health-effects we get a *HEPA-program*, a health-enhancing physical activity program. The contribution of this paper is the development and implementation of a digital support platform to guide young elderly towards health-enhancing physical activity. As part of this work, we also searched for drivers that could get both the platform and health-enhancing physical activity programs accepted and adopted for sustained use.

Regardless of the evidence found in the literature, HEPA programs are not high in demand among young elderly. A synthesis (Wallén et al, 2014) of several studies of physical exercise among 2500 elderly Swedes (ages 65-84) found that 12%/14% (female/male) show no interest in PA; 69%/64% (female/male) mention regular PA at low or medium intensity. A Finnish study shows that in the age group of 30-54 years, only 30% spent several hours per week at regular physical activity; in the 55-74 age group, it decreases to 15%; in the 75+ age group only 7% are regularly physically active (Borodulin and Sääksjärvi, 2019). Moreover, another recent Finnish study found that only 34%/39% of (adult) women/men reach recommended HEPA levels (Keskimäki et al, 2019). Evidently, systematic action is needed to start getting changes.

Our context and our version of systematic action is a research and development program called *DigitalWells* (DW-program) that was running 2019-22 to activate young elderly in Finland towards trying out and adopting HEPA programs. As a research tool we developed a digital support platform and application (the most recent version is *DW-app 3.0*) that collects PA exercises, shows intensity and duration, logs the results to a cloud-supported database, builds graphical PA reports and shows (weekly, monthly) goal attainment for the user. The DW-program has so far attracted more than 1000 users, who have logged 294 140 PA events (by December 2021), from which we have collected PA and HEPA statistics and paired the results with cross-sectional and longitudinal data on experiences with the program (Kari et al, 2021b and Kari et al, 2021c).

The aim of the paper is to demonstrate the design and functions of the digital support platform and application, and to find out if technology- and psychology-based drivers combine to get physical activity adopted as a routine program with digital support as an accepted technology among young elderly.

The rest of the paper is organized in the following way: in section 2 we will work out physical activity, the DW-program and the design and implementation of the DW-app 3.0; in section 3 we will present some research results with the digital support platform; in section 4 we will connect the design and use of digital support platforms to classical DSS constructs, draw some conclusions and outline the next steps in the research program.

2. PHYSICAL ACTIVITY AND DIGITAL SUPPORT

Keskimäki et al, (2019) offer a couple of challenges: the proportion of young elderly who show physical activity that meets HEPA recommendations should be much higher, and PA-programs should be run to fulfil HEPA requirements (at least 150 minutes/week at moderate or 75 minutes/week at vigorous intensity).

Ainsworth et al, (2011) offer advice and guidelines for the design of HEPA programs. They quantify the energy cost of 821 specific activities in terms of metabolic equivalent of task (= MET), "the ratio of the rate at which a person expends energy, relative to the mass of that person, while performing some specific physical activity compared to the reference ... at 3.5 ml of oxygen per kilogram per minute (when sitting quietly)"; the MET for various PA has been validated with lab experiments.

The DW-program adopted the study by Ainsworth et al, (2011) as a baseline and worked out 48 PA that can be included in a HEPA program (Table1).

Table 1. A selection of physical activities, including CPA MET values

Physical activities (PA)	MET-Light	MET-Moderate	MET-Vigorous
Walking	2.8	3.5	4.3
Gym training	3.5	5.0	6.0
Home gymnastics	2.8	3.8	8.0
Swimming	3.5	6.0	9.8
Padel	4.7	7.3	10.0

PA recommendations for health effects (Bangsbo et al, 2019, Borodulin and Sääksjärvi, 2019) correspond to roughly 525-675 MET-minutes/week with PA exercises of different MET levels. The WHO (2020) offers a more demanding set of PA for health benefits: (i) older adults should do at least 150–300 minutes of moderate-intensity aerobic physical activity; or at least 75–150 minutes of vigorous intensity aerobic physical activity throughout the week; (ii) … should also do muscle strengthening activities at moderate or greater intensity on 2 or more days a week; (iii) … should do varied multicomponent physical activity that emphasizes functional balance and strength training at moderate or greater intensity, on 3 or more days a week. These activities would collect about 1800-2200 MET-minutes/week to get health benefits.

The HEPA recommendations apply to healthy adults and show PA that on average would have health effects. There are, by necessity, individual differences in the PA effects of PA programs and we must expect more deviations when we apply recommendations to young elderly: female/male, age groups, socio-economic background, physical demands from work, and HEPA capacity (decided by PA history and physical shape). In the DW-program we found the recommendations too vague to motivate sustained PA – "you cannot be sure that the time spent will actually give sufficient health effects" (Davis, 1989; Makkonen et al, 2021) – and we found out that this is a crucial point; the DW-app 3.0 (fig. 1 and 2) became a support instrument to give necessary answers.

The DW-app 3.0 for smart mobile phones (Android, iOS) registers the PA exercises carried out, calculates the MET-minutes for an activity and registers the results. The DW-app was built on Wellmo, a multi-purpose platform for mobile wellness services developed by Nokia and Mobile Wellness Solutions MVS Inc. (https://www.wellmo.com/platform). The platform provides the software infrastructure for the DW-app to define user profile and login setup functions including informed consent to allow PA data to be collected and analysed for research purposes (shown in green, figure 1). A user is known only through an 8-digit pseudo-code. PA data is processed in the DW-app part (shown in dark blue, figure 1) with its own cloud database (Heroku) and with secure links to a public database on wellness data to which the DW-app is synchronized. Wellmo is a well-tested, stable working platform with more than 10 years of active use.

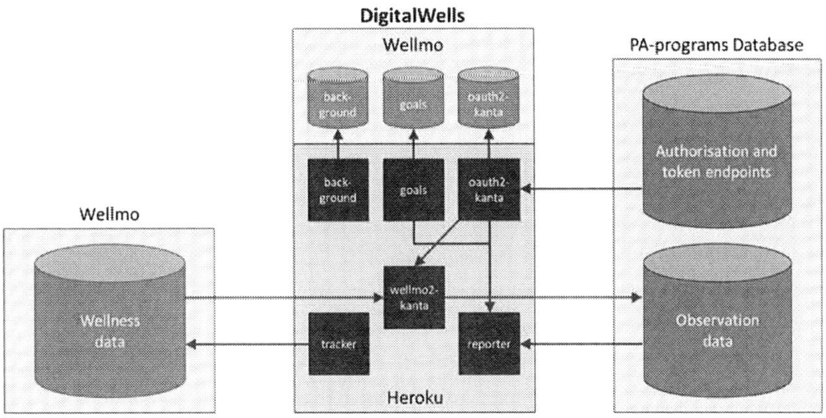

Figure 1. DW-app 3.0 application – basic platform structure

The logging of activities on the smart phone (if the user prefers to do it manually) is done in the left part of the screen (fig. 2): (i) the user selects the activity (gym training), (ii) the intensity (moderate), (iii) the date from the calendar, (iv) and the duration (1 hour), after which the app (v) calculates and shows the effect of the activity (330 MET-min, 398 kcal). If the PA exercise is logged with a smartwatch the manual operations are not used, the DW-app carries out the phases (i)-(v) automatically. The intensity of the PA exercise is shown through objective measurements with smartwatch sensors; this is better than subjective assessment - but the calibration of the PA is done for active, younger athletes and the intensity shown appears to be a bit low for young elderly.

The results of the PA exercise update the database (fig.1, light blue) on the user's individual 8-digit pseudocode. The data is used to produce individual reports on the user's smart phone (the fourth panel, figure 2): (i) the type of report is specified (weekly), then (ii) the reported week (10/2020), (iii) the user report is shown (MET-minutes/week; the red line is the weekly PA goal) and then (iv) further specified (MET-minutes/day). Further graphical reports are shown in the third panel that specify (for instance) MET-minutes per activity and Minutes per activity; the second panel shows the most recent (seven) PA exercises; weekly reports are supplemented with monthly and 6-monthly reports.

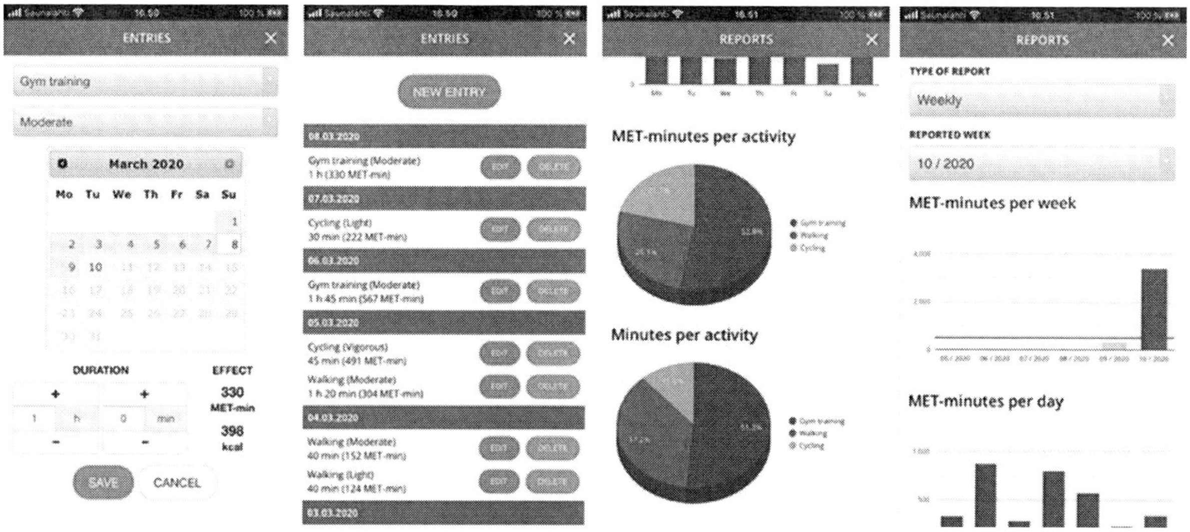

Figure 2. DW-app 3.0 logging and reporting of activities

The MET results which show the effects of individual PA exercises, may not be very precise if compared to lab experiments (Changizi and Kaveh, 2017) but will still motivate users. A user gets a basis to decide that "the time spent will actually give sufficient health effects" by registering intensity and duration for weeks and months, and then comparing the individual efforts with the recommendations we collected in the DW- program. We have collected selected PA experiences from DW-app users (Kari et al 2021b, c), here we will show just a few.

Among the participants in *DigitalWells* we found wide variations in PA exercises. A few 75+ aged marathon runners easily train at 2200-2500 MET minutes/week, but some people with limited PA experience find it challenging to reach 440 MET-minutes/week. Intensity is different for different types of PA exercise, which can be verified with the sensors of a sport watch (Changizi and Kaveh, 2017). Some DW-app users pointed out that home gymnastics requires much less effort than gym training, there is a difference in the quality of programs which should be better covered in the MET values. PA history turned out to be an important factor; physically active young elderly continues to be active on high levels (reaching 1500-2000 MET-minutes/week) into senior years. There is a golfer (75+) in the program who logs about 2500 MET-minutes per week with 3 rounds most weeks (18 holes, walking 10-12 km, 4.5-5.0 hours/round); he did not claim to be in good shape as golfing is aimed more at having a good time than at collecting MET-minutes/week.

Table 2 extracts some early, spontaneous comments collected over the participants' smart mobile phones using *LimeSurvey*. The comments were collected after the 12th month follow-up sessions with the participants; part of the survey aimed at finding and collecting technical errors and software misfunctions (that were corrected, and updated versions of the DW-app were distributed) another part collected open-ended reactions to the *DigitalWells*, which showed positive and negative comments (N = 241).

We found that PA programs are/can be part of weekly routines; for some participants there were no changes in their weekly habits, for others there were positive impacts with follow-up and reporting on systematic PA. For some participants the DW-app 3.0 was a nuisance (too simple and cumbersome); for some, PA programs should be group activities but for others only individual programs following their own schedule are viable. There are quite a few references to PA history and PA level – in terms of own proficiency and comments on others' (weaker) history (Makkonen et al, 2020 for details). The *DigitalWells* introduced the HEPA recommendations after this survey; later surveys showed that 96% of the participants reached a level of 675 MET-minutes/week after 3-4 months in the program (Carlsson 2022), which shows that HEPA exercise levels are quite possible; of course, there were some negative comments - a minority found weekly PA programs stressful. The key to the quotes in Table 2 is: [participant #] M or F/age/BMI.

Table 2. An extract of participant reactions to the *DigitalWells* after about 12 months

I regularly spend time with exercise; thus, the *DigitalWells* has not changed my exercise routines [#64] F/71/26.40
I spend as much time with exercise as feel good (quite much) [#65] M/74/25.86
This program does not motivate to exercise as such; it is too simple and cumbersome; I am in favour of technology that automatically registers exercise and other activities; MET points are not calibrated to active exercise nor to health exercise [#72] F/70/23.73
I left *DigitalWells* in mid-summer because my motivation was not sufficient to go on by myself [*covid-19 restrictions closed group activities*]; technology by itself is not enough as a motivation [#75] F/68/33.25
I follow up on my own exercise more than before and I have also been checking my results both from the tracker and from the [*app on the*] phone [#83] M/71/26.01
I increased exercise after I retired, because now I have more time for it; *DigitalWells* and the need to reduce weight have increased exercise; yet I have not decided on any goals [#120] M/63/34.48
[*DigitalWells*] has improved the follow up of different forms of exercise [#128] M/75/35.98
This [*DigitalWells*] is good for people that have not been active on exercise; for me personally there is not much effect [#134] F/66/23.59
The application has made me surer that my exercise activities are quite sufficient without any programs; I easily get 10 000 steps every day in my daily routine tasks [#141] F/67/23.38
Before the covid-19 restrictions I worked out in gym programs 3 times/week and spent 3 times/week in water aerobics or aqua-jogging; now everything is closed which reduced my exercises to yoga and qigong once a week [#164] F/71/27.83

3. SOME GENERALIZATIONS ON DIGITAL SUPPORT

The development part of the *DigitalWells* followed the design science research (DSR) paradigm, (Iivari, 2010) with successive iterations of improved constructs based on feedback from users (Makkonen et al, 2020b). Gradually, this process firmed up to a process of search for drivers that get potential users to adopt PA and/or HEPA exercises and turn them into (weekly) routines that extend to monthly or yearly sustained habits. Most studies focus on acceptance of digital technology (Venkatesh et al, 2016, Yuan, 2015) in the belief that intention to use a technology will drive adoption of PA/HEPA programs (Williams and French, 2011). This belief turns out to be a questionable assumption.

The field studies with the participants used several theory frameworks that offered different perspectives on activities (Unified Theory of Acceptance and Use of Technology (UTAUT), Self-efficacy and Self-Determination Theory (SDT). In a series of surveys, the *DigitalWells* found some support for UTAUT-based constructs (Venkatesh et al, 2016) to describe the acceptance of the DW-app: *performance expectancy*, *hedonic motivation*, and *habit* in one study with 115 participants (Kari et al, 2021b); in two studies with 91 participants where the statistically significant effects switched between *hedonic motivation* and *habit* in one longitudinal study (Makkonen et al, 2020a) and between *performance expectancy* and *effort expectancy* in a second longitudinal study (Makkonen et al, 2021). Besides the UTAUT2, the self-administered, short version of the International Physical Activity Questionnaire for the elderly (IPAQ-E) by Hurtig-Wennlöf et al, (2010) was used in a study with 294 participants (Kari et al, 2021a), on how PA choices are influenced by demographic backgrounds (*gender*, *age*, *education*, *marital status*) and the DW-app: walking and total PA increased between the baseline and a 12-month follow-up (Makkonen et al, 2020b); a further study with the

IPAQ-E showed that the changes were more substantial after 12 months than after four months (Makkonen et al, 2021). Self-efficacy (Bandura 1977) was tested as a probable theory framework with 165 participants that had been in the *DigitalWells* for 12 months and more (Kari et al, 2021b). The results showed that *performance accomplishment* was the main explanation for increased self-efficacy. The increase in self-efficacy for PA exercises is important for sustained PA and for sustained HEPA when it is reached. The results from field studies suggested – a bit surprising - to widen the search for theory frameworks beyond traditional Information Systems frameworks.

The Self-Determination Theory (SDT), (Teixeira et al, 2012) shows useful constructs for the *DigitalWells*: *intrinsic motivation* will get a participant to adopt a HEPA program because it gives inherent satisfactions (physical well-being, enjoyment, accomplishment, excitement, exercise of skills); *extrinsic motivations* initiate adoption for instrumental reasons (good health, social recognition, improved appearance, challenges to oneself). Extrinsic motives can be worked into HEPA programs through controlled forms of motivation, introjected regulation or through self-endorsed motivations. Teixeira et al, (2012) found that more autonomous forms of motivation support are more effective, identified regulation supports initial or short-term success, intrinsic motivation gives long-term success, multiple intrinsic motives strengthen success and competence satisfaction (cf. self-efficacy) and supports success.

The influence processes of the Elaboration Likelihood Model (ELM), (Petty and Cacioppo, 1986) can contribute to either intrinsic or extrinsic motivations to help build inherent satisfactions or instrumental outcomes that could motivate an adoption of HEPA programs. In the ELM framework *perceived usefulness* and *attitudes* to HEPA programs are intrinsic motivations for sustained use of HEPA programs. The perceived usefulness of HEPA programs builds on getting better and sustained health effects.

We find it doubtful that drivers that get *DigitalWells* participants to accept and use digital support platforms also make them adopt and use HEPA programs, and then continue to use the programs. Using synergistic combinations of theory frameworks, SDT+ELM with UTAUT for instance (Carlsson, 2022), improves the applicability of UTAUT for new and (so far) untried contexts. The drivers we are searching for should explain the sustained adoption and use of HEPA programs with useful and effective digital support platforms.

4. DIGITAL SUPPORT PLATFORMS AND CLASSICAL DECISION SUPPORT SYSTEMS

Series of field studies found explanations of why *DigitalWells* participants decided to accept the DW-app 3.0 as a support tool for PA. The explanations were sought from the UTAUT theory framework (Macedo, 2017) and followed the path of many similar studies (listed in Venkatesh et al, 2016). The UTAUT guides studies of the use of digital technology (Venkatesh et al, 2012) to find drivers for "intentions to use" digital technology.

The HEPA program invites ideas of coaching (Carlsson et al, 2021) to be included as part of the digital support platforms, i.e. users would be encouraged to increase their MET-minutes/week goals – which is in line with a health-enhancing program – by (i) switching to more demanding PA exercises, (ii) increasing the intensity of PA exercises, (iii) extending the duration of daily/weekly programs, or by (iv) finding some optimal combination of (i)-(iii). The first three parts are routine advice offered by a coach or personal trainer; many *DigitalWells* participants benefitted from the advice of volunteer coaches and trainers. The optimal combination of (i)-(iii) constitutes an optimal HEPA program, which requires input of analytics tools (Carlsson et al, 2021) and represents an attractive challenge for researchers. The DW-app 3.0 could be a useful platform for computational intelligence methods, machine learning, soft computing, and approximate reasoning. Outlining these possibilities in field studies showed that young elderly users were not amused at this prospect, which may be surprising.

Some explanations can be found from an earlier era - much before the digital technology – the decision support systems technology in the early 1980'es (Keen, 1981). DSS builders focused on users' priorities (not trying to push advanced technology because it would be "cool") and on service, fast delivery, ease of use, benefit, timely delivery, and user control; they emphasized "support to do a better job" as an informal DSS credo (Sprague, 1980, 1981). The DSS architecture built on three components: (i) an interface between the user and functional routines, (ii) a data manager, and (iii) functional routines; this same architecture prevails in our digital support platform (fig.1 and 2). The philosophy, attitudinal core of DSS is "support, not replace" – "coaching" was interpreted as having "replacing" features. Following Keen (1981) it is impossible to support young elderly HEPA programs if we do not know what young elderly do, how they think, what doing a "better job" means to them and what they need to build, use, and sustain the use of HEPA programs.

In the digital era there appears to be a "black box" ideology for "doing a better job". In case human cognitive ability is not enough, analytics will take over (to replace human cognition, if you like) and offer the best possible solution (optimal HEPA programs in our context). The algorithms are mostly beyond the knowledge and skills of platform users (Carlsson et al, 2021) who then do not see why offered solutions are the best possible – or even solutions at all. The DSS promoted managers' intuitive understanding and experience; the DW-app 3.0 took form in successive iterations with groups of users (using the DSR paradigm (Iivari, 2010); the iterative design is another DSS feature (Sprague, 1981). The lessons learned from the DSS era strongly suggest engaging the users in the design, testing, and use of digital support platforms; modern interface technology has gone through 4-5 generations since the DSS era and now supports interactive, intuitive work with the end-users in ways which were beyond belief in the 1980'es.

DigitalWells developed and implemented a digital support platform to guide young elderly towards health-enhancing physical activity; this attracted more than 1000 participants, many of whom stayed more than 24 months with the program and logged more than 294 000 PA events; this was rather successful. Several research projects used the material to search for drivers that could get both the platform and health-enhancing physical activity programs accepted and adopted for sustained use. In the next phase of *DigitalWells*, one part will test 10-12 selected HEPA alternatives in 12-month experimental programs (using test groups and control groups) to verify and validate that they are health-enhancing; the programs will then be offered as standard HEPA programs through digital support platforms; users will, of course, still be free to select any PA from the list of 48 alternatives and run the PA as they like (young elderly will not act against their will and experience).

REFERENCES

Ainsworth, B.E., et al, 2011. Compendium of Physical Activities: A Second Update of Codes and MET Values. *Med Sci Sports Exerc*. Vol. 43, No.8, pp. 1575-81. doi: 10.1249/MSS.0b013e31821ece12. PMID: 21681120.

Bandura, A. 1977. Self-Efficacy: Toward a Unified Theory of Behavioural Change, Psychological Review, Vol.84, No.2, pp.191-215.

Bangsbo, J. et al, 2019. Copenhagen Consensus Statement 2019: Physical Activity and Ageing. *Br J Sports Med* 2019;0: pp. 1–3. doi:10.1136/bjsports-2018-100451.

Borodulin, K. and Sääksjärvi, K.(eds.) 2019. FinHealth 2017 Study – Methods. Finnish Institute for Health and Welfare. Report 17/2019, pp. 1-132. Helsinki, Finland.

Carlsson, C. et al, 2021. Forming Sustainable Physical Activity Programs Among Young Elderly - A Combined ELM & UTAUT Approach. A. Pucihar et al (eds.) Proceedings of the 34th Bled eConference

Carlsson, C. and Walden P. (2021). Decision Support Systems: Historical Innovations and Modern Technology Challenges. Papathanasiou, J. et al (eds.), EURO Working Group on DSS. A Tour of the DSS Developments Over the Last 30 years. Springer-Verlag, Switzerland, pp. 1–14.

Carlsson, C. (2022). Self-efficacy improves UTAUT to describe adoption of Health-Enhancing Physical Activity Programs. A. Pucihar et al (eds.) Proceedings of the 35th Bled eConference

Changizi, M. and Kaveh, M.H., 2017. Effectiveness of the mHealth Technology in Improvement of Healthy Behaviours in an Elderly Population—A systematic Review. *mHealth*, Vol. 3, No. 51.

Davis, F. D., 1989. Perceived usefulness, perceived ease of use, and user acceptance of information technology. *MIS Quarterly*, Vol.13, No.3, pp.319–340. doi:10.2307/249008

European Commission, 2018. The 2018 Ageing Report: Economic & Budgetary Projections for the 28 EU Member States (2016-2070), Institutional Paper 079.

Hurtig-Wennlöf, A. et al, 2010. The International Physical Activity Questionnaire modified for the Elderly: Aspects of Validity and Feasibility. Public Health Nutr. Vol.13, No.11, pp.1847-54. doi: 10.1017/S1368980010000157. Epub 2010 Mar 3. PMID: 20196910

Iivari, J. 2010. Twelve Theses on Design Science Research in Information Systems, in: Hevner, A. and Chatterjee S. (eds.), *Design Research in Information Systems - Theory and Practice*, Springer, pp. 43-62

Jonasson, L. 2017. Aerobic Fitness and Healthy Brain Aging. Cognition, Brain Structure, and Dopamine, Doctoral Dissertation, Umea University.

Kari, T. et al, 2021a. Demographic Differences in the Effectiveness of a Physical Activity Application to Promote Physical Activity: Study Among Aged People. A. Pucihar et al (eds.) *Proceedings of the 34th Bled eConference*.

Kari, T. et al, 2021b. Using a Physical Activity Application to Promote Physical Activity Levels Among Aged People: A Follow-Up Study. *Proceedings of the HICSS-54*, [ScholarSpace]

Kari, T. et al, 2021c. Does Physical Activity Application Use Promote Self-Efficacy for Exercise? A Study Among Aged People, *Proceedings of the HICSS-55 Conference* [ScholarSpace]

Keen, P.G.W., 1981 Decision Support Systems – Lessons for the 80's. In: Young D., and Keen P.G.W. (eds.) *DSS-81 Transactions,* Atlanta, Georgia, pp.187-192

Keskimäki, I. et al, 2019. Finland: Health system review. Health Systems in Transition, Vol 21, No.2, pp. 1 – 166.

Kolu, P. et al, 2022. Economic Burden of Low Physical Activity and High Sedentary Behaviour in Finland. *Journal of Epidemiology and Community Health.* Published ahead of print.

Macedo, I. M. 2017. Predicting the acceptance and use of information and communication technology by older adults: An empirical examination of the revised UTAUT2. *Computers in Human Behavior*, Vol. 75, pp. 935–948. doi:10.1016/j.chb.2017.06.013

Makkonen, M. et al, 2020a. Applying UTAUT2 to Explain the Use of Physical Activity Logger Applications Among Young Elderly. A. Pucihar et al (eds.) *Proceedings of the 33rd Bled eConference*, 29.6.2020.

Makkonen, M. et al, 2020b. Changes in the Use Intention of Digital Wellness Technologies and Its Antecedents Over Time: The Use of Physical Activity Logger Applications Among Young Elderly in Finland. *Proceedings of the HICSS-54*, pp. 1262-1271, [ScholarSpace]

Makkonen, M. et al, 2021. A Follow-Up on the Changes in the Use Intention of Digital Wellness Technologies and Its Antecedents Over Time: The Use of Physical Activity Logger Applications Among Young Elderly in Finland. A. Pucihar et al (eds.) Proceedings of the 34th Bled eConference

Petty, R.E. and Cacioppo, J.T. 1986. *Communication and Persuasion: Central and Peripheral Routes to Attitude Change.* Springer-Verlag, New York.

Sprague, R.H. 1980. Framework for the development of decision support systems, MIS Quarterly Vol.4, No.4, pp. 1-26.

Sprague, R.H. 1981. Decision support systems: a tutorial. In: Young D, Keen PGW (eds) *DSS-81 Transactions*, Atlanta, Georgia, pp 193-203.

Talukder, M. S. et al, 2019. Acceptance and use predictors of fitness wearable technology and intention to recommend. *Industrial Management & Data Systems*, Vol.119, No. 1, pp. 170–188. doi:10.1108/IMDS-01-2018-0009

Teixeira, P.J. et al, 2012. Exercise, Physical Activity and Self-Determination Theory: A Systematic Review, *Int. Journal of Behavioral Nutrition and Physical Activity*, Vol. 9, No.78, pp. 1-30.

Wallén, M. B. et al, 2014. Motionsvanor och erfarenheter av motion hos äldre vuxna, Karolinska Institutet, Stockholm, March 2014

Wellmo, "Mobile health platform", https://www.wellmo.com/platform/, 2021.

Venkatesh, V. et al, 2012. Consumer Acceptance and Use of Information Technology: Extending the Unified Theory of Acceptance and Use of Technology. *MIS Quarterly*, Vol. 36, No. 1, pp.157-178.

Venkatesh, V. et al, 2016. Unified Theory of Acceptance and Use of Technology: A Synthesis and the Road Ahead. *JAIS*, Vol. 17, No. 5, pp. 328-376.

Williams, S.L. and French D.P. 2011. What are the Most Effective Intervention Techniques for Changing Physical Activity Self-efficacy and Physical Activity Behaviour— and are They the Same? *Health Education Research*, Vol. 26, No. 2, pp. 308-322.

World Health Organization, 2020. WHO Guidelines on Physical Activity and Sedentary Behaviour. https://www.who.int/publications/i/item/9789240015128, accessed 9.2.2021.

Yuan, S. et al, 2015. Keep Using My Health Apps: Discover Users' Perception of Health and Fitness Apps with the UTAUT2 Model, *Telemedicine and e-Health*, Vol. 21, No. 9, pp. 7335-741.

https://aginginplace.org/fastest-aging-populations/ Published on: Jan 21, 2022

https://www.wellmo.com/platform

CLINICAL DETERIORATION PREDICTION IN BRAZILIAN HOSPITALS BASED ON ARTIFICIAL NEURAL NETWORKS AND TREE DECISION MODELS

Hamed Yazdanpanah[1,2], Augusto C. M. Silva[1], Murilo Guedes[1], Hugo M. P. Morales[1], Leandro dos S. Coelho[3,4] and Fernando G. Moro[1]

[1]*Department of Research, Robô Laura, Curitiba, PR, Brazil*

[2]*Department of Computer Science, University of São Paulo, São Paulo, SP, Brazil*

[3]*Industrial and Systems Engineering Graduate Program, Pontifical Catholic University of Parana, Curitiba, PR, Brazil*

[4]*Electrical Engineering Graduate Program, Federal University of Parana, Curitiba, PR, Brazil*

ABSTRACT

Early recognition of clinical deterioration (CD) has vital importance in patients' survival from exacerbation or death. Electronic health records (EHRs) data have been widely employed in Early Warning Scores (EWS) to measure CD risk in hospitalized patients. Recently, EHRs data have been utilized in Machine Learning (ML) models to predict mortality and CD. The ML models have shown superior performance in CD prediction compared to EWS. Since EHRs data are structured and tabular, conventional ML models are generally applied to them, and less effort is put into evaluating the artificial neural network's performance on EHRs data. Thus, in this article, an extremely boosted neural network (XBNet) is used to predict CD, and its performance is compared to eXtreme Gradient Boosting (XGBoost) and random forest (RF) models. For this purpose, 103,105 samples from thirteen Brazilian hospitals are used to generate the models. Moreover, the principal component analysis (PCA) is employed to verify whether it can improve the adopted models' performance. The performance of ML models and Modified Early Warning Score (MEWS), an EWS candidate, are evaluated in CD prediction regarding the accuracy, precision, recall, F1-score, and geometric mean (G-mean) metrics in a 10-fold cross-validation approach. According to the experiments, the XGBoost model obtained the best results in predicting CD among Brazilian hospitals' data.

KEYWORDS

Clinical Deterioration, Vital Signs, Machine Learning, Artificial Neural Networks, Electronic Health Record

1. INTRODUCTION

Clinical deterioration (CD) is a physiological decompensation, and it happens when a patient undergoes deteriorating conditions or the onset of a severe physiological inconvenience. CD is a leading cause of mortality in hospitals and, in the case of late detection, it can cause organ failure and death (Fleischmann et al., 2016). Conventionally, early warning scores (EWS), such as the Quick Sequential Organ Failure Assessment (qSOFA) (Angus et al., 2016) and the Modified Early Warning Score (MEWS) (Subbe et al., 2006), use only vital signs to determine the CD risk of hospitalized patients.

However, Machine Learning (ML) models are able to utilize laboratory exams, Electronic Health Records (EHRs) data and demographic data in the model to attain more accurate predictions for CD by returning fewer false alarms and more precise detection (Al-Mualemi et al., 2021; Wyk et al., 2019; Wang et al., 2018; Deng et al., 2022). The easier access of EHRs data in hospitals and improved ML strategies encouraged researchers and clinical staff to predict CD in hospitalized patients using automated ML models.

Although there are significant advances in predicting CD by ML models, the majority of studies focused on Intensive Care Unit (ICU) data (Kong et al., 2020; Ibrahim et al., 2020; Moor et al., 2021; Selcuk et al., 2022). Therefore, in this work, ML models are designed to predict CD in patients hospitalized in departments different from ICU. To this end, 103,105 unique attendances out of ICU from thirteen Brazilian hospitals in different states are utilized in the ML models development. The collection period of these data is from March 2015 to July 2021.

In general, when dealing with tabular and structured data, popular ML models outperform artificial neural networks by providing better prediction, higher interpretability, and lower computational cost. In particular, tree-based models, such as Random Forest (RF), Light Gradient Boosting Machine (LightGBM), and eXtreme Gradient Boosting (XGBoost), absorbed more attention among ML models (Yuan et al., 2020; Zabihi et al., 2019; Lyra et al., 2019). Deep neural networks have provided outstanding achievement on unstructured data, such as images, video, audio, and text data (Bengio et al., 2017). However, recently, an extremely boosted neural network (XBNet) has been proposed by combining gradient boosted tree with a feed-forward neural network (Sarkar, 2021). This model uses the feature importance of a gradient boosted tree to update the weights of each layer of the neural network.

For some data sets, it has been shown that this architecture can outperform the traditional ML models (Sarkar, 2021). Therefore, in this work, we compare the performance of the XBNet in CD prediction with some tree-based models, such as XGBoost and RF. Also, to the best of our knowledge, it is the first time the XBNet is applied to the EHRs data for CD prediction. Moreover, the Principal Component Analysis (PCA) is utilized to reduce the data set dimension and to avoid abundant information. Thus, considering the 95% cut-off threshold in the PCA, the number of features decreases from 113 to 73, and the performance of models is compared before and after dimension reduction.

The CD prediction in this work is a binary classification task, where one class is for survival (class 0), and the other one represents death (class 1). Classical metrics for classification tasks, such as accuracy, precision, and recall, are employed to evaluate the performance of models. Furthermore, since our data set is highly class-imbalanced (death rate is almost 4%), some useful metrics for class-imbalanced data set, such as F1-score and geometric mean (G-mean), are reported. Also, all metrics are evaluated in a 10-fold Cross-Validation (CV) framework to prevent overfitting.

The organization of the remainder of this article is as follows. Some related works to the CD prediction are highlighted in Section II. Section III describes a brief description of the employed classifiers and methodology. The experimental results and discussions are presented in Section IV. Finally, conclusions are drawn in Section V.

2. RELATED WORKS

Recent advances in the ML field, together with the availability of sources of health and hospital care data, such as EHRs, have generated opportunities for automated medical decision-making to avoid complex problems in health monitoring and to identify clinical deterioration (Ye et al., 2020; Zheng et al., 2017). To this end, several ML models were employed to extract and evaluate different features of EHRs data (Negro-Calduch et al., 2021; Pang et al., 2021).

EHRs data from ICU are utilized in an RF-based model for the early prediction of sepsis (Nakhashi et al., 2019). In (Abromavičius et al., 2019), the patients in ICU are divided into short (less than 9 hours), medium (9 to 60 hours), and long (more than 60 hours) stay in ICU. Based on the patient's stay in ICU, they extracted different vital sign features. Then, gentle adaptive boosting ensemble learning and random under-sampling boosting algorithms are used for sepsis prediction in ICU. Also, in (Oğul et al., 2019), model-based techniques are compared with the instance-based ones by employing elastic time series measures to measure similarity between different instances of vital signs and predict septic shock in ICU.

In (Hu et al., 2019), data are collected from neonatal ICUs and are transformed into images. Then, a convolutional neural network (CNN) is implemented to predict late-onset neonatal sepsis. In (Chang et al., 2019), a recurrent imputation for time series is used to input missing values in vital signs and lab measurements. Then, temporal convolutional neural networks are employed on the imputed data to predict the onset of sepsis. Moreover, in (Wyk et al., 2017), a CNN model is compared to a multilayer perceptron model to detect sepsis, and the CNN model obtained a higher accuracy.

Also, vital signs, laboratory and demographic data are utilized in the early detection of sepsis six hours ahead of time by a bi-directional gated recurrent units model (Wickramaratne et al., 2020). In (Roussel et al., 2019), it is assumed that the prediction of the evolution of vital signs is needed to predict sepsis accurately. Thus, a recurrent artificial neural network is used to predict the vital signs six hours ahead, then the prediction of sepsis in ICU is implemented. Also, in (Demirer et al., 2019), partially observed Markov decision processes are used along with vital signs, laboratory and demographics data to design an artificial intelligence-based sepsis warning system.

3. BRIEF DESCRIPTION OF THE CLASSIFIERS

In this section, the XBNet, XGBoost, RF, and PCA are briefly reviewed. These models will be used for the CD prediction in the next section.

3.1 XBNet

Recently, XBNet was proposed by combining gradient boosted tree and a feed-forward neural network (Sarkar, 2021). In this algorithm, trees are trained in all neural network layers, and feature importance is obtained by the trees. Then, weight defined by the gradient descent is employed to modify the weights of neural network layers where trees are trained. This approach makes the neural network model robust for tabular data regarding all performance metrics. The adopted optimization strategy in the XBNet is the boosted gradient descent, and it is initialized using the feature importance of gradient boosted trees. Thus, the weights of each layer are updated by the model in the following steps: (i) weights are updated by gradient descent; and (ii) weights are updated by employing the feature importance of gradient boosted trees.

3.2 XGBoost

Boosting is an ensemble algorithm that converts a set of weak learners into a strong estimator by sequentially training ML models, in which, for each iteration, the model tries to correct the previous iteration. In gradient boosting, each new iteration is optimized in the residual error of the previous iteration using gradient descent. A popular system for this method is XGBoost, described by (Chen et al., 2016). It introduces a number of novel strategies to optimize learning speeds, enabling it to run ten times faster than gradient boosting machine while maintaining state-of-the-art results.

3.3 Random Forest

RF classifier is an ensemble learning model that consists of numerous decision trees and was established by (Breiman, 2001). In order to define the split of each node, RF considers only a random sample of the features and calculates the optimal cut-off point for each subset, which results in an ensemble of less correlated decision trees, potentially improving the accuracy of the model. The prediction of an instance is calculated by combining the predictions of all trees (through averaging if the target is numerical or through a majority voting if the target variable is categorical).

3.4 PCA

The principal component analysis seeks to express the most significant possible variability of the original features, replacing them with a new smaller set of independent features known as components. These components are linear combinations of original features with the eigenvectors of the variance-covariance matrix of the original features. The eigenvectors point to the direction of more significant variability, which means that a few components are able to retain most of the original variability. More details about principal components analysis can be found in (Jollifa et al., 2016).

4. RESULTS ANALYSIS

In this section, the employed data set for generating and evaluating the CD classifiers are described briefly. Then, the results and discussions are presented.

The data set contains 103,105 unique attendances (samples) of patients hospitalized out of ICU. Also, to focus the CD prediction on adults, all patients younger than 18 years old are discarded. The data are collected by EHRs from thirteen Brazilian hospitals from March 2015 to July 2021. The data set contains 77 variables (columns), where 6 of them are categorical variables, such as gender, registered disease, and clinical specialty. Thus, after implementing one-hot encoding, the number of features increases to 113. Regarding vital signs, such as heart rate, temperature, respiratory rate, glucose, oxygen saturation, systolic and diastolic blood pressure, some features describe the last five collections of them. Also, age, days from the last hospitalization, and length of stay are reported in the data set. Moreover, it should be mentioned that the use of this data set is approved by the ethics committee of the corresponding hospitals under protocol number 99706718.9.1001.0098.

All samples in the data set have completed the consultation, i.e., their output is medical discharge or death. Moreover, the designed models are predicting CD events for 12 hours ahead. To this end, the last 12 hours of vital signs before the patient's outcome are removed for each patient. Since the last five collections of vital signs are reported in the data set, and there is a correlation between different collections, some statistical measures of the last five vital signs, such as minimum, maximum, mean, median, and standard deviation (STD), are used as additional features. For patients hospitalized in different hospitals for treatment, information about the time between hospitalizations is utilized. Also, the filling forward imputation technique is adopted to impute missing values using the last collected vital signs. Finally, Table 1 describes the mean, STD, and the missing value percentage of numerical variables of the Brazilian EHRs data set for all samples and different classes.

Table 1. The missing value percentage, mean and STD of numerical variables of Brazilian EHRs data set

Variables	Total	Survival	Mortality	Missing (%)
Heart rate	79.18 ± 15.06	78.60 ± 14.38	95.01 ± 22.55	11.34
Respiratory rate	18.11 ± 3.92	18.07 ± 3.88	19.19 ± 4.91	15.16
Diastolic blood pressure	71.96 ± 11.36	72.26 ± 11.10	63.76 ± 14.88	11.56
Systolic blood pressure	120.40 ± 18.19	120.90 ± 17.78	106.64 ± 23.09	11.53
Oxygen saturation	95.87 ± 2.92	95.99 ± 2.68	92.57 ± 5.84	16.12
Capillary blood glucose	137.66 ± 60.65	137.19 ± 59.40	145.94 ± 78.84	76.37
Temperature	36.06 ± 0.69	36.06 ± 0.68	36.20 ± 0.82	16.26
Days from entrance	5.20 ± 9.86	4.91 ± 9.04	13.47 ± 21.67	0
Age	56.05 ± 18.07	55.55 ± 17.93	70.55 ± 15.92	0

Different metrics are used to measure the classifiers' skills in CD prediction. In classification tasks, accuracy, precision, and recall are common metrics to evaluate ML models. Thus, these metrics are reported in this work. However, these metrics are suitable for symmetric data set. The mortality rate is low in our data set (almost 4%), and the data set is highly class-imbalanced. Thus, some appropriate metrics for class-imbalanced data sets such as F1-score and G-mean are reported too.

All metrics are computed in the stratified 10-fold CV approach to avoid overfitting. This approach helps in measuring the ability of ML models with lower bias. The data set is randomly divided into 10 stratified folds of almost the same size to execute CV. Then, the model is trained on 9 folds and is validated on the remaining one fold. This process is repeated until all folds appear one time as a validation set. Finally, the 10-fold CV outcomes are reported via the selected metrics' mean and STD. Furthermore, the hyperparameters of ML models are tuned by the grid search strategy.

In this work, the XBNet model has two hidden layers, where the input and output dimensions of the first layer are 8 and 4, respectively. Also, the input and output dimensions of the second layer are 4 and 2, respectively. For both layers, the bias is set as False. The loss function is the cross-entropy loss, and the optimizer is the adaptive moment estimation (Adam) with the learning rate equal to 3×10^{-4}. Moreover, the batch size is 32, and the number of epochs is 100.

For the XGBoost model, the number of boosting rounds is 100, and the maximum tree depth for base learners is 6. The subsample ratio of the training instance and the subsample ratio of columns when generating each tree are 0.7 and 0.75, respectively. The ℓ_1 and ℓ_2 regularization parameters are adopted as 5. The minimum sum of instance weight required in a child is 9. The minimum loss reduction needed to make a further partition on a leaf node of the tree is 0.3. Finally, the learning rate is chosen as 0.12.

For the RF, the number of trees is 100. The maximum depth of the tree is 18. Also, the minimum number of samples needed to be at a leaf node is 4. Furthermore, MEWS is selected as an EWS candidate for comparison with the ML models since it is conventionally utilized in CD prediction. Also, note that the learning curves in Figures 1, 2, 4, and 5 are generated by taking the average of learning curves between 10-fold CV outcomes.

Figure 1(a) shows the logistic loss function learning curves for the training and validation sets of the XBNet model during 100 epochs. Moreover, the accuracy learning curves of the XBNet model for the training and validation sets during 100 epochs are shown in Figure 1(b). Also, the logistic loss function learning curves for the training and validation sets of the XGBoost model are depicted in Figure 2(a). It can be observed that these two curves are extremely close to each other, and they can attain logistic loss values less than 0.1. Figure 2(b) presents the accuracy learning curves of the XGBoost model for the training and validation sets. It is worthwhile to mention that besides the visual distance between the training and validation sets learning curves, the overfitting in the model did not happen since the distance between two curves is less than 0.01 (1% of accuracy). In other words, the difference of accuracy between the training and validation sets is less than 1%.

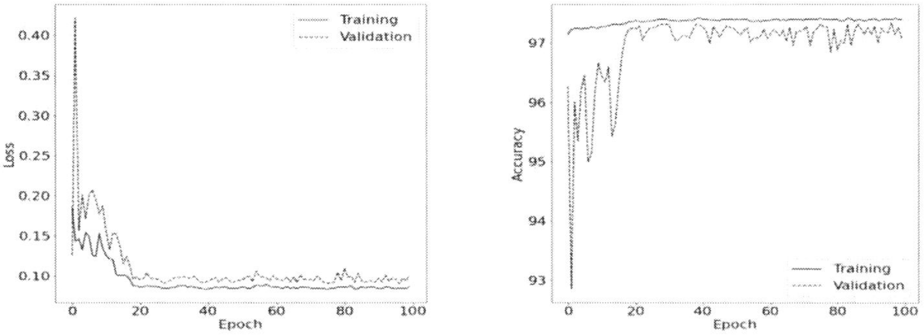

(a) (b)

Figure 1. The learning curves of the XBNet model for: (a) logistic loss function versus the number of epochs; (b) accuracy versus the number of epochs

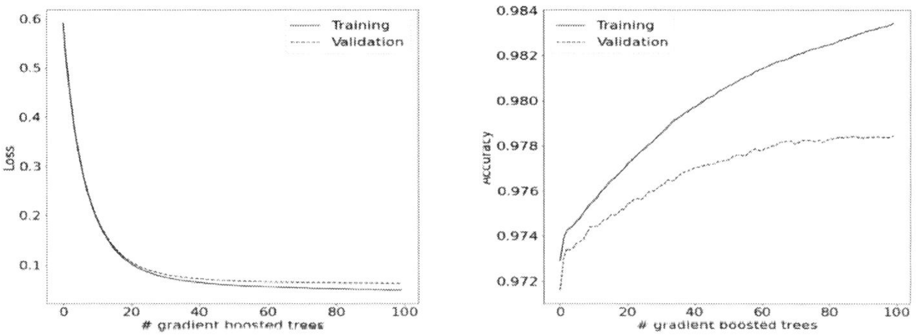

(a) (b)

Figure 2. The learning curves of the XGBoost model for: (a) logistic loss function versus the number of gradient boosted trees; (b) accuracy versus the number of gradient boosted trees

Furthermore, the PCA approach is applied to our data set, which contains 113 features, to reduce the number of features. Figure 3 shows how many principal components are required to explain the variance in the data set. As presented in this figure, 73 principal components out of 113 ones in the transformed space are required to explain 95% of variance in the data set. It shows that this data set contains informative variables, and most features provide unique information about the data set. Therefore, the first 73 principal components are used as the new data set for training and validating the XBNet and XGBoost models.

The logistic loss function learning curves of the XBNet model for the training and validation sets when using 73 principal components are depicted in Figure 4(a). Also, Figure 4(b) presents the accuracy learning curves of the XBNet model for the training and validation sets when 73 principal components are utilized. As can be observed in Figures 1 and 4, the XBNet model converges faster when it is applied to 73 principal components rather than 113 features. Also, the logistic loss function learning curves of the XGBoost model for the training and validation sets when employing 73 principal components are shown in Figure 5(a). In addition, when 73 principal components are used as inputs, the accuracy learning curves of the XGBoost model for the training and validation sets are presented in Figure 5(b). By comparing Figures 2 and 5, it is evident than the distance between the training and validation curves is higher in Figure 5 than that of in Figure 2. Therefore, overfitting is more likely when using the XGBoost on the principal components of the employed data set.

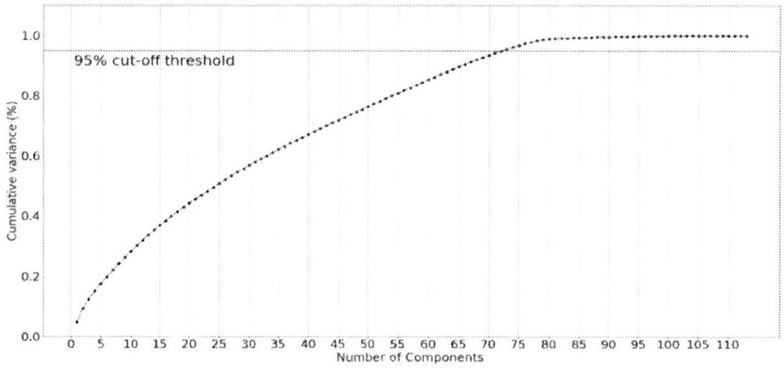

Figure 3. 73 of 113 principal components are required to explain 95% of variance in the data set

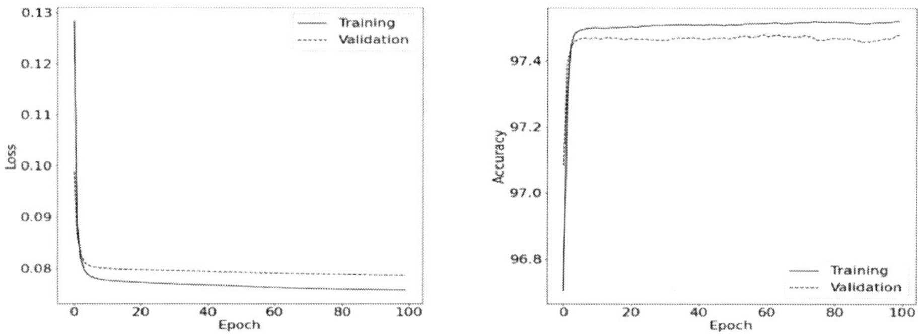

(a)

(b)

Figure 4. The learning curves of the XBNet model, when the number of features is reduced by the PCA, for: (a) logistic loss function versus the number of epochs; (b) accuracy versus the number of epochs

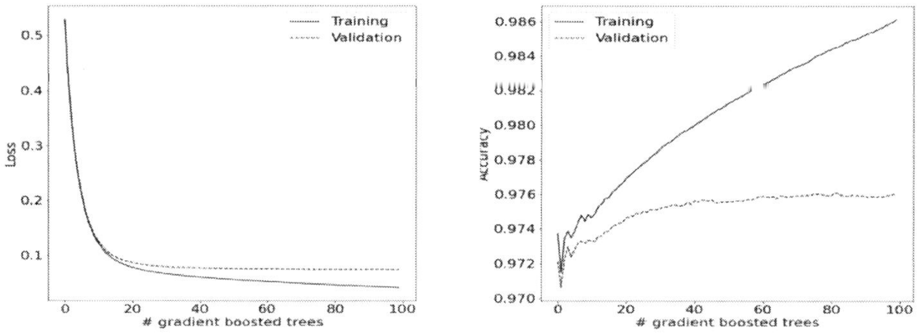

(a)

(b)

Figure 5. The learning curves of the XGBoost model, when the number of features is reduced by the PCA, for: (a) logistic loss function versus the number of gradient boosted trees; (b) accuracy versus the number of gradient boosted trees

Moreover, as another tree-based model candidate for predicting CD, the RF is applied to the data set before and after executing the PCA. For the XBNet, XGBoost, and RF models before and after applying PCA, and MEWS the mean and STD of accuracy, precision, recall, F1-score, and G-mean are reported in Table 2. As presented in this table, when the XGBoost model is applied to the original data set (without using principal components), it outperforms other tested models according to all the evaluated metrics, except the precision. In CD applications, having higher recall is more valuable than obtaining higher precision, and the XGBoost model can bring the highest recall among the tested models. Also, since our data set is highly class-imbalanced, the G-mean and F1-score values are of utmost importance to measure the superiority of models in CD prediction. Furthermore, it can be observed that the values of metrics for MEWS are significantly lower than those for the ML models, except the mean of G-mean for the RF+PCA.

Table 2. Accuracy, precision, recall, F1-score, and G-mean of algorithms for 10-fold CV

Algorithms	Accuracy		Precision		Recall		F1-score		G-mean	
	Mean	STD	Mean	STD	Mean	STD	Mean	STD	Mean	STD
XBNet	0.971	0.0057	0.635	0.1154	0.404	0.0381	0.485	0.0390	0.632	0.0289
XBNet + PCA	0.975	0.0009	0.755	0.0195	0.376	0.0380	0.501	0.0335	0.611	0.0308
XGBoost	**0.978**	0.0011	0.800	0.0229	**0.482**	0.0247	**0.601**	0.0237	**0.692**	0.0177
XGBoost + PCA	0.976	0.0011	0.774	0.0293	0.409	0.0289	0.535	0.0288	0.638	0.0227
RF	0.976	0.0011	**0.890**	0.0302	0.326	0.0289	0.476	0.0327	0.570	0.0250
RF + PCA	0.973	0.0006	0.874	0.0250	0.239	0.0160	0.375	0.0207	0.488	0.0166
MEWS	0.928	0.0023	0.187	0.0105	0.338	0.0192	0.241	0.0125	0.566	0.0159

The required execution time for the XBNet, XGBoost, and RF models are 23724, 18, and 29 seconds, respectively. As can be observed, the XGBoost model has the fastest execution time (18 seconds), and the RF requires a few more seconds to be executed. However, the execution time of the XBNet model is much higher than other models. Indeed, it needs more than six hours to be implemented.

5. CONCLUSION

In this paper, EHRs and demographic data of more than 100,000 samples of hospitalized patients in Brazilian hospitals have been utilized for CD prediction. All patients were hospitalized in departments different from ICU. The XBNet model, as a neural network candidate, has been employed for predicting CD using tabular data, and its performance has been compared to the XGBoost and RF models as two tree-based models. Also, the PCA approach has been adopted to reduce the number of features, and the XBNet, XGBoost, and RF have been trained on the transformed data set to verify whether the PCA technique can improve the performance of models. The XGBoost obtained the best results among all tested models by resulting in the highest accuracy, recall, F1-score, and G-mean. Furthermore, the XGBoost model had the minimum execution time compared to the XBNet and RF models, and the XBNet model needed much more time to be executed.

In this work, it has been shown that the XBNet model cannot hit the XGBoost model performance on Brazilian hospitals' data set. However, in future research directions, more effort will be put into the feature engineering task to verify if, with new and more creative features, the XBNet model can obtain a superior performance to conventional ML models.

REFERENCES

Abromavičius, V. and Serackis, A., 2019. Sepsis prediction model based on vital signs related features. *Computing in Cardiology (CinC)*, Singapore, pp. 1-4.

Al-Mualemi, B.Y. and Lu, L., 2021. A deep learning-based sepsis estimation scheme. *IEEE Access*, Vol.9, pp 5442-5452.

Angus, D.C. et al., 2016, A framework for the development and interpretation of different sepsis definitions and clinical criteria. *Critical Care Medicine,* Vol. 44, No. 3, pp. e113.

Bengio, Y., Goodfellow, I., and Courville, A., 2017, *Deep Learning*, MIT Press, Massachusetts, USA.

Breiman, L., 2001, Random forests, *Machine Learning*, Vol. 45, pp. 5-32.

Chang, Y. et al., 2019, A multi-task imputation and classification neural architecture for early prediction of sepsis from multivariate clinical time series. *Computing in Cardiology (CinC)*, Singapore, pp. 1-4.

Chen, T. and Guestrin, C., 2016, XGBoost: A scalable tree boosting system. *Proceedings of the 22nd ACM SIGKDD International Conference on Knowledge Discovery and Data Mining*, San Francisco, CA, USA, pp. 785-794.

Demirer, R.M. and Demirer, O., 2019, Early prediction of sepsis from clinical data using artificial intelligence. *Scientific Meeting on Electrical-Electronics & Biomedical Engineering and Computer Science*, Istanbul, Turkey, pp. 1-4.

Deng, H.-F., et al., 2022, Evaluating machine learning models for sepsis prediction: A systematic review of methodologies. *iScience*, Vol. 25, No. 1, pp. 103651.

Fleischmann, C., et al., 2016, Assessment of global incidence and mortality of hospital-treated sepsis. Current estimates and limitations. *American Journal of Respiratory and Critical Care Medicine*, Vol. 193, Vo. 3, pp. 259-272.

Hu, Y., Lee, V.C.S., and Tan, K., 2019, An application of convolutional neural networks for the early detection of late-onset neonatal sepsis. *International Joint Conference on Neural Networks (IJCNN)*, Budapest, Hungary, pp. 1-8.

Ibrahim, Z.M. et al., 2020, On classifying sepsis heterogeneity in the ICU: Insight using machine learning. *Journal of the American Medical Informatics Association*, Vol. 27, No. 3, pp. 437-443.

Jollifa, I. and Cadima, J., Principal component analysis: A review and recent developments. *Philosophical Transactions of the Royal Society A*, Vol. 374.

Kong, G., Lin, K., and Hu, Y., 2020, Using machine learning methods to predict in-hospital mortality of sepsis patients in the ICU. *BMC Medical Informatics and Decision Making*, Vol. 20, No. 1, pp. 1-10.

Lyra, S., Leonhardt, S., and Antink, C.H., 2019, Early prediction of sepsis using random forest classification for imbalanced clinical data. *Computing in Cardiology (CinC)*, Singapore, pp. 1-4.

Moor, M. et al., 2021, Early prediction of sepsis in the ICU using machine learning: A systematic review. *Frontiers in Medicine*, Vol. 8.

Nakhashi, M. et al., 2019, Early prediction of sepsis: Using state-of-the-art machine learning techniques on vital sign inputs. *Computing in Cardiology (CinC)*, Singapore, pp. 1-4.

Negro-Calduch, E. et al., 2021, Technological progress in electronic health record system optimization: Systematic review of systematic literature reviews. *International Journal of Medical Informatics*, Vol. 152, pp. 104507.

Oğul, H. et al., 2019, On computer-aided prognosis of septic shock from vital signs. *IEEE 32nd International Symposium on Computer-Based Medical Systems (CBMS)*, Cordoba, Spain, pp. 87-92.

Pang, X. et al., Prediction of early childhood obesity with machine learning and electronic health record data. *International Journal of Medical Informatics*, Vol. 150, pp. 104454.

Roussel, B., Behar, J., and Oster, J., 2019, A recurrent neural network for the prediction of vital sign evolution and sepsis in ICU. *Computing in Cardiology (CinC)*, Singapore, pp. 1-4.

Sarkar, T., 2021, XBNet: An extremely boosted neural network. *ArXiv preprint,* arXiv:2106.05239.

Selcuk, M., Koc, O., and Kestel, A.S., 2022, The prediction power of machine learning on estimating the sepsis mortality in the intensive care unit. *Informatics in Medicine Unlocked*, Vol. 28, pp. 100861.

Subbe, C.P. et al., 2006, Validation of physiological scoring systems in the accident and emergency department. *Emergency Medicine Journal*, Vol. 23, No. 11, pp. 841-845.

Wang, X. et al., 2018, A new effective machine learning framework for sepsis diagnosis. *IEEE Access*, Vol. 6, pp. 48300-48310.

Wickramaratne, S.D. and Mahmud, M.S., 2020, Bi-directional gated recurrent unit based ensemble model for the early detection of sepsis. *42nd Annual International Conference of the IEEE Engineering in Medicine & Biology Society (EMBC)*, Montreal, QC, Canada, pp. 70-73.

Wyk, F., Khojandi, A., and Kamaleswaran, R., 2019, Improving prediction performance using hierarchical analysis of real-time data: A sepsis case study. *IEEE Journal of Biomedical and Health Informatics*, Vol. 23, No. 3, pp. 978-986.

Wyk, F. et al., 2017, How much data should we collect? A case study in sepsis detection using deep learning. *IEEE Healthcare Innovations and Point of Care Technologies (HI-POCT)*, Bethesda, MD, USA, pp. 109-112.

Ye, C. et al., 2020, Identification of elders at higher risk for fall with statewide electronic health records and a machine learning algorithm. *International Journal of Medical Informatics*, Vol. 137, pp. 104105.

Yuan, K.-C., et al., 2020., The development an artificial intelligence algorithm for early sepsis diagnosis in the intensive care unit. *International Journal of Medical Informatics*, Vol. 141, pp. 104176.

Zabihi, M., Kiranyaz, S., and Gabbouj, M., 2019, Sepsis prediction in intensive care unit using ensemble of XGboost models. *Computing in Cardiology (CinC)*, Singapore, pp. 1-4.

Zheng, T., et al., 2017, A machine learning-based framework to identify type 2 diabetes through electronic health records. *International Journal of Medical Informatics*, Vol. 97, pp. 120-127.

FEATURE UTILIZATION BY MACHINE LEARNING MODELS FOR COLON CANCER CLASSIFICATION

Douglas F. Redd[1,2,*], Qing Zeng-Treitler[1,2*], Yijun Shao[1,2,*],
Laura J. Myers[3,4,5,*], Barry C. Barker[3], Stuart J. Nelson[2,**] and Thomas F. Imperiale[3,4,5,***]
[1]*Washington DC VA Medical Center, Washington, DC, USA*
[2]*George Washington University, Washington DC, USA*
[3]*Richard L. Roudebush VA Medical Center, Indianapolis, Indiana, USA*
[4]*Indiana University School of Medicine, Indianapolis, Indiana, USA*
[5]*Regenstrief Institute, Indianapolis, Indiana, USA*
**PhD*
***MD, FACP, FACMI*
****MD*

ABSTRACT

Many machine learning methods are now available for classification and prediction tasks in the healthcare domain. Some traditional statistical methods, such as logistic regression, are much more readily interpretable than the newer machine learning models. While many prior studies compared machine learning performances in specific tasks, there is no standardized way to assess feature importance/contribution in machine learning models, and few compared the features utilized. This study compares four machine learning and statistical models: logistic regression, support vector machine, random forest, and deep neural network, in their performance in classifying colorectal cancer patients as well as the features used for classification.

KEYWORDS

Machine Learning, Colon Cancer, Feature Utilization

1. INTRODUCTION

Machine learning techniques have been widely used in clinical data analysis, with impressive results in many cases (Weng et al. 2017; Jiang et al. 2011; Garvin et al. 2018). However, one shortcoming to many machine learning techniques is the inability to identify the features that have the greatest contribution to the models' abilities to reach their conclusions. Among the factors that affect the adoption of these models by clinicians and clinical researchers, this lack of interpretability reduces the willingness and confidence of clinicians and clinical researchers (Diprose et al. 2020).

By finding the relative contribution of individual features, inferences can be made to try and understand the underlying mechanisms affecting the outcome. These inferences can then be used as the basis for hypothesis generation for studies to better understand the mechanism of classification. The use of logistic regression in clinical analysis is popular partly due to the ability to assess each feature's importance on the outcome. Many machine learning methods have outperformed logistic regression in different tasks, but they often lack the ability to provide explanations of how they arrive at their conclusions. This is especially pronounced with deep learning models, commonly referred to as "black boxes," without the ability to explain their decision process.

As part of the explainable deep learning research efforts, the development of methods such as Impact Scores for deep neural networks (DNNs) has addressed this shortcoming by assigning scores to each input feature to indicate the magnitude of impact on the outcome of the model (Shao et al. 2019; Lee et al. 2019; Shrikumar, Greenside, and Kundaje 2017; Ribeiro, Singh, and Guestrin 2016; Lundberg and Lee 2017). The DNN impact score is calculated for each input feature by changing each feature value to a reference value and observing the change in the outcome. It is interpreted analogously to feature coefficients in logistic regression models, where values near zero are interpreted as having little impact, while large values reflect the high impact and can be in the positive or negative direction. Impact scores have been demonstrated to provide similar explanations to logistic regression (Redd et al. 2020). Other machine learning models also have indicators of the impact of individual features. Random Forests have a mechanism where noise is introduced to each feature in turn, and the percentage increase in misclassification rate is measured (Breiman 2001a). If the increase in misclassification is large, this indicates high importance. Another mechanism for measuring the impact of features in random forests is Gini importance (also known as average impurity decrease), which is a side effect of the random forest training process. In Gini importance, each node in the binary trees is examined to find the split that produces the maximum separation of the classes (Menze et al. 2009). The Gini importance of a feature indicates the frequency that it is selected for a split. These measures both use a positive magnitude as the importance; thus, no direction of impact is indicated. In support vector machines with linear kernels, a hyperplane is calculated that provides the best separation, with the largest margins, between the points of the different classes (Cortes and Vapnik 1995). The weight given to each feature is an indicator of how important the feature is to the separation. Weights for non-linear kernels can be interpreted similarly, though the non-linear transformation makes it difficult to compare the weights' absolute values. Logistic regression coefficients are the best-known and understood of important indicators, where coefficients indicate both the direction and magnitude of the impact of each feature (Tolles and Meurer 2016). In some of these methods, the importance values must be interpreted relative to the other values for the same model. Comparisons of individual values between models are not always meaningful. But by representing the values relative to each other, e.g., linear model or rank order, some comparisons can be made.

It is not uncommon for clinical research studies to utilize and compare multiple machine learning methods, few however, compared the features utilized by the different methods. One reason may be that, as discussed above, there is no standardized way to assess feature importance. Working with clinicians, we have learned that it is desirable to explain why/how one model outperformed another. In this study, we compared the classification performance along with feature utilization of 4 different modeling methods in the use case of colorectal cancer (CRC) identification.

2. BODY OF PAPER

2.1 Methods

Data Set: Data for this study was derived from the structured and unstructured data from the Veterans Administration's Corporate Data Warehouse (CDW), which is a collection of medical data from over 150 Veterans Health Administration (VHA) medical centers nationwide. The CRC dataset contains three groups of patients: 1) Cases - patients diagnosed with CRC during 2008-2015; 2) Colonoscopy controls - patients underwent colonoscopy for diagnostic purposes (e.g., rectal bleeding) but were not diagnosed with CRC; and 3) Clinic controls – patients did not have colonoscopy or diagnosed with CRC but had at least one the primary care clinic per year for each of the two years immediately preceding the date of diagnosis of a matched case (index date). Colonoscopy controls and clinic controls were matched to cases by a facility in a 2:1 ratio (4 total controls per case). In the machine learning analysis, we merged the two control groups (colonoscopy controls and clinic controls) into one control group.

Chart Review: The CRC status was determined using chart review. Four trained research assistants specializing in CRC conducted manual data extraction from the VA's electronic medical record (CPRS). Cases initially identified from CDW administrative data were verified through manual review of pathology reports. Presentation symptoms (e.g., rectal bleeding, blood in stool, unexpected weight loss, change in bowel habits, or abdominal pain) and risk factors (e.g., high-risk family history or previously diagnosed inflammatory bowel disease) were checked to ensure CRCs diagnosed via average risk screening were excluded along with individuals with prior colonoscopies with no available data. Lifestyle factors (smoking, alcohol use, and exercise), education and employment, comorbidities, and family history (degree-related and cancer diagnosis) were obtained from notes available up to and including the index date. Lifestyle factors were preferably abstracted from 6-18 months prior to the index date. When not available within the preferred date range, values were obtained at the most recent reference from the index date, looking progressively back to the earliest available note. Physical measures, vital signs, medications, and laboratory values were administratively obtained 6-18 months prior to the index date. Missing values in administrative data were completed manually as available. Each case was reviewed by at least two research assistants. Discordant reviews were resolved through consensus. The final dataset consisted of 4,339 patients, 722 of which were cases, and 3,617 were controls.

2.1.1 Machine Learning

Features from Structured Data: A total of 52 features representing medications, diagnoses, procedures, and relevant document types that were present in greater than 10% of the patients were included. The total numbers of documents, diagnoses, procedures, prescriptions, and visits for each patient during the time period between one year prior to the index date and one month after the index date were also included.

Features from Unstructured Data: In order to create features representing data contained in unstructured and semi-structured clinical notes, we used an unsupervised method of topic modeling, the Latent Dirichlet Allocation (LDA) algorithm (Blei, Ng, and Jordan 2003). We generated 1,000 topics from clinical notes (n=33,135) collected from between one year prior to and one month after the index dates of the case and control patients (n=420). In addition, we added 2,576 documents from 70 patients diagnosed with CRC and 10,634 documents from 1,188 patients who received colonoscopies. In order to reduce spurious topics, documents were converted to lowercase, and words were excluded if they occurred in a stopword list or if they occurred ten times or less. The stopword list used was a general-purpose list of 524 common English words, to which we added the 25 words most common in our set of clinical documents: "patient", "pt", "patient's", "veteran", "vet", "veteran's", "g/dl", "mmol/l", "mg/dl", "k/ul", "date", "time", "date/time", "icd", "sig", "tab", "tablet", "n/a", "cap", "capsule", "soln", "inj", "iv", "mcg", and "h/o".

We applied the trained topic model to all clinical notes each patient received from one year prior to the index date to one month after. The proportions of each topic's representation for each patient were the average from all the notes of the patient.

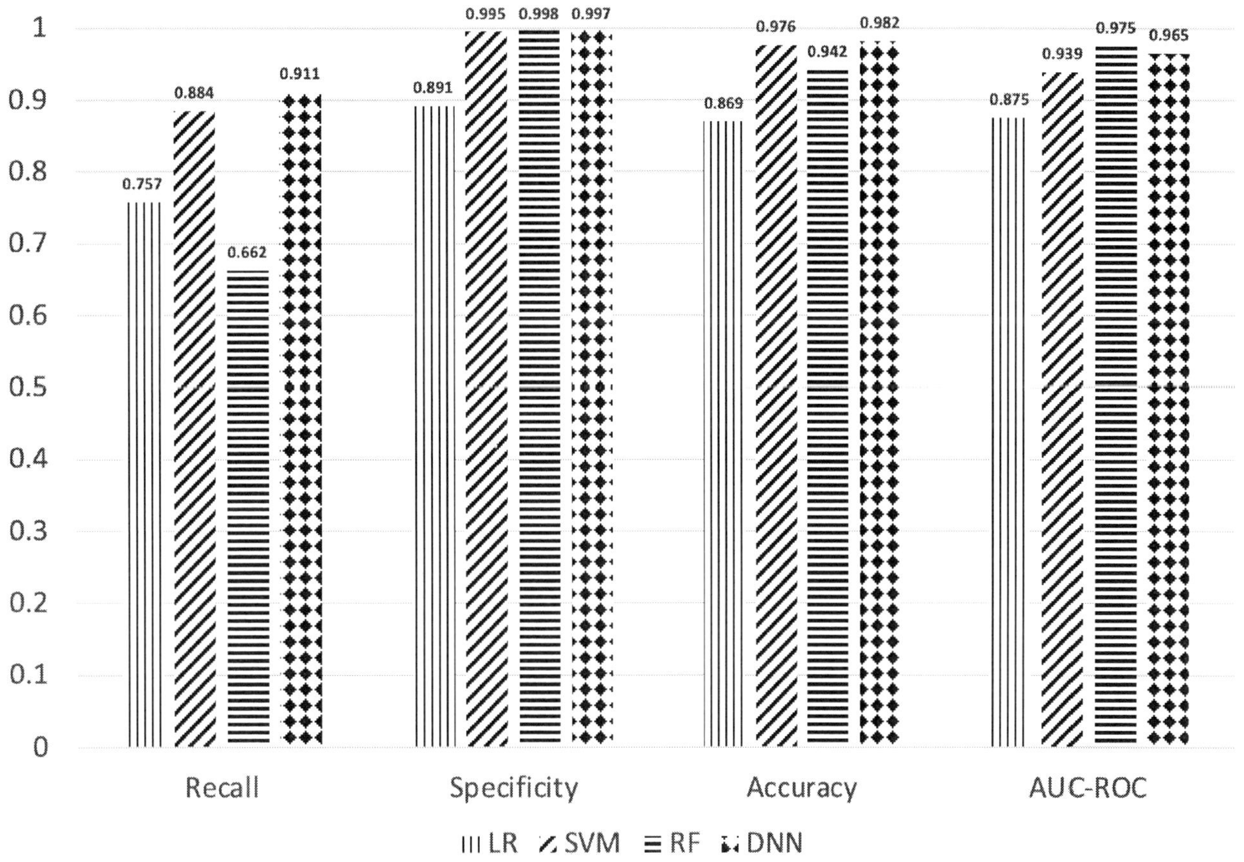

Figure 1. Performance of the 4 models

Machine Learning: We trained and tested Logistic Regression (LR), Support Vector Machine (SVM), Random Forest (RF), and Deep Neural Network (DNN) mechanisms to classify CRC cases vs controls. We used the LR, SVM, RF implementations and default parameters from Weka, an open-source machine learning software. The DNN model was implemented using a Python deep learning library called Theano (Bergstra 2010) together with a helper library called Lasagne (Dieleman 2015). Specifically, the DNN ha0d 5 hidden layers of sizes 200, 300, 200, 300, 200, and a single output using sigmoid activation, 300 epochs, batch size 100, and Nesterov momentum (Sutskever et al. 2013) with learning rate 0.001 and momentum 0.9. In all cases, training, validation, and testing groups were created using stratified sampling to ensure balanced case/control representation. The evaluation was performed with 10-fold cross-validation in all cases, each fold consisting of 3,905 training patients (650 cases, 3,255 controls) and 434 testing patients (72 cases, 362 controls).

Feature Contribution: The contribution of each feature was determined for each machine learning model. For logistic regression, we used the coefficients. For random forest, we used the Gini importance measure for each feature as determined by the random forest algorithm. In the case of SVM, we used the weights assigned to each feature. And for DNN, we used the Impact Score measurement, analogous to the coefficients in logistic regression (Shao et al. 2019; Redd, Goulet, and Zeng-Treitler 2020). The importance measure for random forests is always positive and indicates the magnitude of importance for each feature. To be consistent with this measure, we used the absolute values of the LR coefficients, SVM weights, and DNN impact scores. We calculated the Pearson correlation to find linear correlations and Spearman's correlation to indicate rank correlation.

2.2 Results

Figure 1. shows the 10-fold cross-validation results of the 4 models, all of which performed well. Overall, the DNN performed the best with a 0.965 AUC, 98.2% accuracy, 99.7% specificity and 91.1% recall.

When considering agreement of the contribution of the individual features for each model, linear (Pearson) and rank order (Spearman's) correlations are shown in Table 1. There is a very low correlation among the four models in all cases. Note that all correlations with RF were calculated based on the absolute values of the contribution measures (which can have negative values) to make them comparable to the RF importance measure (which do not have negative values).

The fifteen highest contributing features for each model are shown in 2, with very little agreement, as reflected in the small correlation measures. In all models, the most impactful feature was a topic.

Table 1. Pairwise correlations of the impacts of the individual features in the models. Linear correlation is shown by the Pearson correlation, and the rank order correlation is shown by Spearman's rank correlation

Pearson (Spearman)	DNN	RF	SVM
LR	-0.198 (-0.103)	0.021 (0.067)	-0.036 (-0.246)
SVM	0.113 (0.094)	-0.173 (-0.116)	
RF	0.036 (0.050)		

2.3 Discussion

This study trained 4 different models to classify CRC cases. All 4 models performed well in the 10-fold cross validation. Interestingly, the models are dependent on different feature sets to reach their conclusions. However, by and large, they reach the same conclusions with very good performance. To the best of our knowledge, prior research in the biomedical domain have not compare the features utilized by different machine learning models on the predication or classification task.

The differences in contributing features utilized by different models reflect the difference in machine learning algorithms, and on the other hand, may be due to the redundancy in features. For example, different models can use different but somewhat related features to reach the same conclusions. The highest contributing feature for SVM is very relevant on the surface, as the topic explicitly represents metastatic colon cancer. The highest contributing feature for DNN also appears relevant as it seems to represent colonoscopy procedures. For LR, the highest contribution feature is metastatic cancer, and the second is about biopsy resulting from colonoscopy. For RF, the highest contributing feature appears to be less directly connected to CRC, however, as it is about fall risk assessments. This feature may be used due to assessments being performed for hospitalized patients, with hospitalization being required for CRC treatment.

Although the features with the highest importance for RF appear the least related on the surface, RF obtained the highest AUC along with DNN. Both models reach over 96% in AUC, which raises suspicion of overfitting, with the use of spurious features grouping by chance in this particular data set. LR and SVM did not perform as well. It is also possible that, since steps were taken to avoid overfitting, RF and DNN identified previously unknown correlations.

The comparison of contributing features in this study presents a challenge in that the different models' contribution was calculated using different methods. These methods do not result in the same distributions in their results, making direct value comparison unreliable. Using Pearson correlation, these values can be compared, but only if they are linearly correlated. Another way to address this is to use rank-order correlation, as performed by Spearman's. This avoids the problem of result distribution shape, as long as there is reliable ordering within the results. In this study, however, neither of these methods gave meaningful correlations. SVM's weights may not be particularly meaningful after the non-linear transformation, while the odds ratio, importance score, and impact score have all been previously reported in clinical research studies (Redd et al. 2020; Li et al. 2020). In all cases, we are trying to identify the ground truth, and different models may have approached the ground truth in very different ways. The issue of having multiple well-performing models using different highly contributing features has at times been referred to as the "Rashomon" effect (Breiman 2001b), with the idea that different observers can reach the same conclusion using varied evidence.

Table 2. Fifteen highest contributing features for each of the four machine learning models. The contribution was measured by coefficient for logistic regression, weight for SVM, importance measure for random forest, and impact score for DNN

	LR	**SVM**	**RF**	**DNN**
1	Topic: liver lesions active status	Topic: colon cancer sigmoid colon	Topic: self-care level fall assessment	Topic: colon colonoscope tissue colonoscopy
2	Topic: occupational diagnosis ptsd impairment	Document Type: Pathology	Topic: military mental health traumatic	Topic: neg active day urine
3	Topic: active s/p folfox colon	Document Type: Gastrointestinal	Topic: pain assessment skin care	Topic: hud-vash apartment veteran's case
4	Topic: ref eval range urine	Prescription: Electrolytes/PEG-3350 (Colonoscopy prep)	Topic: procedure level monitor pain	Topic: pain colon morphine liver
5	Topic: tardive dyskinesia movements facial	Topic: rectal mass anal verge	Topic: released days supply expiration	Topic: history normal exam current
6	Topic: chemo liver rectal cancer	Diagnosis: Hemorrhage of rectum and anus	Topic: group discussion relapse addiction	Topic: ref eval range urine
7	Topic: heart failure shortness breath	Document type: Colon	Topic: hour total pain flu	Topic: call type identified phone
8	Topic: active pet basis fee	Diagnosis: Blood in stool	Topic: understanding evaluation verbalized patient/resident	Topic: caregiver support care assessment
9	Topic: score section nutrition weight	Topic: radiation therapy rectal cancer	Topic: active s/p folfox colon	Topic: program treatment housing services
10	Topic: diarrhea day cea chemo	Diagnosis: Other counseling NEC	Topic: barrier reported nurse discharge	Topic: refill qty days expr
11	Topic: pain treatment visit weight	Topic: tubular adenoma colonic mucosa	Topic: observation disturbances feel present	Topic: peripheral normal leg pain
12	Topic: cetuximab active liver colon	Diagnosis: Other specified counseling	Topic: axis mood disorder ptsd	Topic: pain inability worst assess
13	Topic: active inpatient supervising practitioner	Topic: colon cancer colonoscopy polyp	Topic: procedure sedation examination mental	Topic: days supply remaining refills
14	Topic: colon consent procedure sedation	Prescription: Magnesium Citrate (Colonoscopy prep)	Topic: pain level command discharge	Topic: occupational diagnosis ptsd impairment
15	Topic: program days past veteran's	Prescription: Bisacodyl (Colonoscopy prep)	Topic: colon consent procedure sedation	Topic: problems days reports past

Only limited hyperparameter tuning was performed on all of the ML methods used. We selected the initial parameters based on past experience, and since all ML reach a fairly high-performance level, there is no need for extensive tuning. Performance can likely be improved in all cases with additional tuning.

In future studies, dimensionality reduction methods and semantic analysis can be used to investigate whether different models used different features that actually represent the same underlying concept. This would be expected to produce a higher correlation. An important follow-on study will also examine the interactions between contributing features and if they indicate previously unidentified patterns. A future study might evaluate the recently proposed model class reliance (MCR) method, aiming to identify the range of explanations across multiple existing well-performing models (Fisher, Rudin, and Dominici 2019). In effect, MCR summarizes the other models and identifies features with high average contribution across all models. This study focused on the classification of CRC cases; however, an informative follow-on study could investigate predictive modeling. By restricting the time period of data to only include measurements from before the diagnosis, this may be accomplished. With a predictive model, it may be possible to identify features corresponding to risk factors. This could allow validation by features mapping to known risk factors and also identification of possibly unknown risk factors.

3. CONCLUSION

In summary, we trained 4 models to classify CRC cases. Methods to explain the models showed that the different machine learning models utilize alternate features, which may reflect the redundancy and correlation in features, modeling approach, and/or explanation methods.

REFERENCES

Bergstra, J., Breuleux, O., Bastien, F., Lamblin, P., Pascanu, R., Desjardins, G., Turian, J., Warde-Farley, D., & Bengio, Y. (2010). "Theano: A CPU and GPU Math Expression Compiler." Proceedings of the Python for Scientific Computing Conference (SciPy)

Blei, D. M, Ng,A. Y. & Jordan, M. I. (2003). "Latent dirichlet allocation." *Journal of machine Learning research* 3 (Jan):993-1022.

Breiman, L. (2001a). "Random Forests." *Machine Learning* 45 (1):5-32. doi: 10.1023/A:1010933404324.

Breiman, L. (2001b). "Statistical modeling: The two cultures (with comments and a rejoinder by the author)." *Statistical science* 16 (3):199-231.

Cortes, C., & Vapnik, V. (1995). "Support-vector networks." *Machine Learning* 20 (3):273-297. doi: 10.1007/BF00994018.

Dieleman, S., Schlüter, J., Raffel, C., Olso, E., Sønderby, S.K., Nouri, D., et. al. (2015). "Lasagne: First release." http://dx.doi.org/10.5281/zenodo.27878.

Diprose, W. K, Buist, N., Hua, N., Thurier, Q., Shand, G., & Robinson, R. (2020). "Physician understanding, explainability, and trust in a hypothetical machine learning risk calculator." *Journal of the American Medical Informatics Association* 27 (4):592-600. doi: 10.1093/jamia/ocz229.

Fisher, A., Rudin, C., & Dominici, F. (2019). "All models are wrong, but many are useful: Learning a variable's importance by studying an entire class of prediction models simultaneously." *Journal of Machine Learning Research* 20 (177): 1-81.

Garvin, Jennifer Hornung, Youngjun Kim, Glenn Temple Gobbel, Michael E Matheny, Andrew Redd, Bruce E Bray, Paul Heidenreich, Dan Bolton, Julia Heavirland, and Natalie Kelly. 2018. "Automating quality measures for heart failure using natural language processing: a descriptive study in the department of veterans affairs." *JMIR medical informatics* 6 (1):e5.

Jiang, M., Chen, Y., Liu, M., Rosenbloom, S. T., Mani, S., Denny, J. C., & Xu, H. (2011). "A study of machine-learning-based approaches to extract clinical entities and their assertions from discharge summaries." *Journal of the American Medical Informatics Association* 18 (5):601-606.

Lee, H., Yune, S., Mansouri, M., Kim, M., Tajmir, S. H., Guerrier, C. E., Ebert, S. A., Pomerantz, S. R., Romero, J. M., & Kamalian, S. (2019). "An explainable deep-learning algorithm for the detection of acute intracranial haemorrhage from small datasets." *Nature Biomedical Engineering* 3 (3):173.

Li, J., Y. Tian, Y. Zhu, T. Zhou, J. Li, K. Ding, & J. Li. 2020. "A multicenter random forest model for effective prognosis prediction in collaborative clinical research network." *Artif Intell Med* 103:101814. doi: 10.1016/j.artmed.2020.101814.

Lundberg, S. M, & Lee, S.-I. (2017). "A unified approach to interpreting model predictions." Advances in neural information processing systems.

Menze, B, H., Kelm, B. M., Masuch, R.,Himmelreich, U., Bachert, P., Petrich, W., & Hamprecht, F. A. (2009). "A comparison of random forest and its Gini importance with standard chemometric methods for the feature selection and classification of spectral data." *BMC Bioinformatics* 10 (1):213. doi: 10.1186/1471-2105-10-213.

Redd, D., Goulet, J. L., Shao, Y., Brandt, C. A., & Zeng-Treitler, Q. (2020). "Using Explainable Deep Learning and Logistic Regression to Evaluate Complementary and Integrative Health Treatments in Patients with Musculoskeletal Disorders." Proceedings of the 53rd Hawaii International Conference on System Sciences.

Redd, D., Goulet, J., & Zeng-Treitler, Q. (2020). "Using Explainable Deep Learning and Logistic Regression to Evaluate Complementary and Integrative Health Treatments in Patients with Musculoskeletal Disorders." Proceedings of the 53rd Hawaii International Conference on System Sciences.

Ribeiro, M. T., Singh, S., & Guestrin, C. (2016). "" Why should i trust you?" Explaining the predictions of any classifier." Proceedings of the 22nd ACM SIGKDD international conference on knowledge discovery and data mining.

Shao, Y., Cheng, Y., Shah, R. U., Weir, C. R., Bray, B. E., & Zeng-Treitler, Q. (2019). "Shedding Light on the Black Box: Explaining Deep Neural Network Prediction of Clinical Outcome." 21st International Conference on Health Informatics, Rome, Italy, January 17 - 18, 2019.

Shrikumar, A., Greenside, P., & Kundaje, A. (2017). "Learning important features through propagating activation differences." Proceedings of the 34th International Conference on Machine Learning-Volume 70.

Sutskever, I., Martens, J., Dahl, G., & Hinton, G. (2013). "On the importance of initialization and momentum in deep learning." *Proceedings of the 30th International Conference on Machine Learning, PMLR* 28 (3):1139-1147.

Tolles, J., & Meurer, W. J. (2016). "Logistic Regression: Relating Patient Characteristics to Outcomes." *JAMA* 316 (5):533-534. doi: 10.1001/jama.2016.7653.

Weng, S. F., Reps, J., Kai, J., Garibaldi, J. M., & Qureshi, N. (2017). "Can machine-learning improve cardiovascular risk prediction using routine clinical data?" *PloS one* 12 (4).

CHARACTERIZING THE CLINICAL LANGUAGE OF OPIOID USE DISORDER

Terri Elizabeth Workman[1], Joel Kupersmith[2], Cynthia A. Brandt[3], Christopher J. Spevak[2] and Qing Zeng-Treitler[1]

[1]*Biomedical Informatics Center, The George Washington University, Washington, DC, USA*
[2]*Georgetown University, Washington, DC, USA*
[3]*VA Connecticut, West Haven, CT, USA*

ABSTRACT

Opioid use disorder is a global crisis, afflicting U.S. Veterans at an increased rate. Unstructured data, in the form of clinical notes from electronic health records, can potentially elucidate opioid use disorder as documented by clinicians. The authors applied natural language processing, Word2Vec, and a specialized flood-fill algorithm to Veterans Health Administration clinical notes, and incorporated the output into graphs. This process enabled identification of themes characterizing opioid use disorder by time period and care site. This work can potentially inform clinicians treating opioid use disorder, and be generalized to research for other conditions.

KEYWORDS

Opioid Use Disorder, Natural Language Processing

1. INTRODUCTION

Opioid abuse and misuse is a global crisis (Martins, et al., 2015) resulting in addiction, suffering, and death. The Diagnostic and Statistical Manual of Mental Disorder V defines Opioid Use Disorder (OUD) as a problematic pattern of opioid use leading to clinically significant impairment or distress (American Psychiatric Association, 2013). In the 12-month period ending in April 2021, there were 75,673 deaths from opioid overdose in the United States, a 35% increase compared to the previous 12-month period (Centers for Disease Control and Prevention, 2021). OUD and opioid overdoses resulting in death resulted in accumulative costs of $471 billion and $550 billion, respectively (Luo F, et al., 2017).

Veterans of the United States Military are not immune to this problem. Overdose mortality rates among Veterans increased 23.7% between 2010 and 2015, and an additional 20.4% in 2016 (Peltzman et al., 2020). Risk factors among these Veterans include young age, male gender, mental health disorders, fewer years of education, and lower income levels (Edlund, et al., 2007; Rhee & Rosenheck, 2019). There is also an elevated risk of OUD for Veterans experiencing post-traumatic stress disorder (PTSD) (Bernardy & Montaño, 2019)

Clinical data addressing OUD are captured in electronic health records (EHRs) as structured data, such as diagnostic codes and lab values, and unstructured data, recorded as free text in clinical notes. Natural language processing (NLP) and machine learning techniques can provide insight into the contents of clinical notes. Word2Vec (Mikolov, et al., 2013) is an application that maps words in text to real number vectors, using a basic neural network and either a skip-gram or continuous bag-of-words architecture. Cosine similarity between vectors indicate which words are similar in terms of the context in which they appear. Word2Vec output can serve as input for flood-fill programming (Law, 2013). This technique can identify how these terms are connected in semantic space by building neighborhoods of contextually associated words. Figure 1 illustrates this idea.

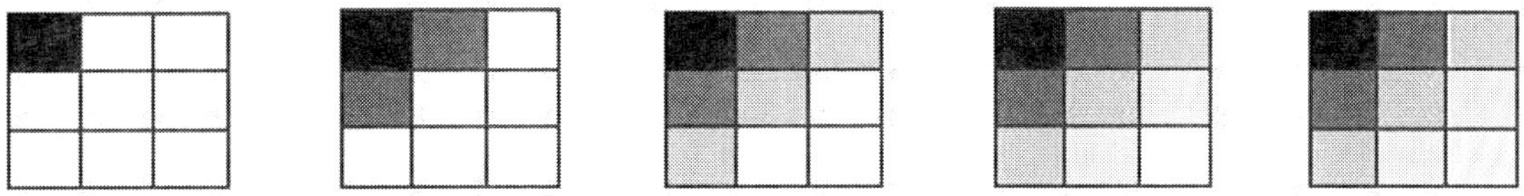

Figure 1. Defining a neighborhood with a flood-fill algorithm

Figure 1 illustrates how a flood-fill algorithm can define a neighborhood of connected entities, in this case cells within a grid. In the first image on the left, the upper left cell is designated in black as the first neighbor. In the next image to the right the algorithm identifies its neighboring cells in dark gray. Next, their neighbors are identified using a lighter shade of gray. This process continues to identify all neighbors, as illustrated in the last image. When the entities are terms in a corpus, the neighboring term pairs can be incorporated into graphs which can further illuminate how the identified terms characterize the clinical text. Graphs consist of vertices and edges. In this case, a vertex is a word, and an edge is the link connecting the vertex to the other word vertex in a word pair. Vertices in this graph would be connected to all other word vertices for which a word pair exists.

The Veterans Health Administration (VHA), within the United States Department of Veterans Affairs (VA), is the largest single healthcare system in the United States, providing care to over 9 million Veterans at 1,255 healthcare facilities (U.S. Department of Veteran Affairs, 2021). Patient data is stored in the Veteran Health Information Systems and Technology Architecture (VistA), one of the earliest and most widely used EHRs (Brown, et al., 2003). The EHR data, hosted in the VHA's national corporate data warehouse, are made available through the Veterans Affairs Informatics and Computing Infrastructure (VINCI) secure research platform.

The objective of this study was to answer the following questions:

- Are there notable themes in the documentation of OUD by time period, as represented by connected word pair vertices?
- Are there notable themes in the documentation of OUD by service region, as represented by connected word pair vertices?
- Will highly connected vertices (i.e. high degree vertices) and subsets in the graphs enhance OUD understanding?

We applied Word2Vec to clinical notes associated with OUD, and then used the Word2Vec output as input for a novel flood-fill algorithm that we developed. The output of the flood-fill algorithm was then used to produce graphs in order to better understand this disorder. We retrieved clinical notes using a relevant base string, according to time intervals before, at, and after the moment of a patient's first OUD diagnosis.

2. METHODS

2.1 Data Extraction

The authors identified 6841 VA patients who had received an opioid use disorder diagnosis (ICD-9-CM codes 304.70, 305.50, and all codes beginning with 304.0; all ICD-10-CM codes within the F11 range) in a VHA outpatient encounter between the years 2012 and 2021 at either a Washington DC VHA facility or a Baltimore, Maryland VHA facility. Select comorbidity data for the same time period and demographic attributes for these patients were also retrieved as descriptive data.

The authors retrieved notes containing the string 'opioid' for these patients according to time intervals: notes from the sixth month before the first OUD diagnosis, notes from the month of diagnosis, and notes from the sixth month after diagnosis. Limiting the notes to those containing this base string enabled retrieval of notes that more likely included content addressing OUD.

2.2 Unstructured Data Processing

Each corpus was first preprocessed by removing punctuation, removing words containing digits or other non-letter characters, and transforming all remaining words to lower case. The results were then processed applying a Word2Vec model consisting of 300 nodes, using the continuous bag-of-words architecture and 10 epochs, modeling words that occurred five times or more in the text.

The output of each Word2Vec model served as input to the flood-fill algorithm. Using the Word2Vec output for each corpus, in its first iteration the flood-fill algorithm identified the top *m* terms most similar to 'opioids' and added them to the list or "neighborhood" initially containing 'opioids', as its neighbors. In the next iteration, the algorithm identified the most similar *n* terms for each of these newly added terms. Of these *n* terms, those shared by at least 10% of terms added in the previous iteration were added as their neighbors. This process repeated until the algorithm converged, meaning that no more terms could be added. The algorithm used the Word2Vec model's vocabulary size to determine *m* and *n*; *m* is an integer rounded to equal 4% of the vocabulary size, and *n* is an integer equaling 2% of the vocabulary size. The discovered associations in each iteration were also recorded as word pairs. For example, if in a given iteration the word 'group' were a top *n* similar term to 'session' (and also 10% of other terms added in the previous iteration), the terms 'group' and 'session' would be recorded as a term pair.

High contextual similarity simply means that two words tend to appear in the same contexts in text. As a final step, the process determined which of the words in all the word pairs also co-occurred within a 10-word window in the given corpus. This final operation identified highly contextually similar, co-occurring term pairs for the graphs. Figure 2 illustrates the application pipeline as a whole.

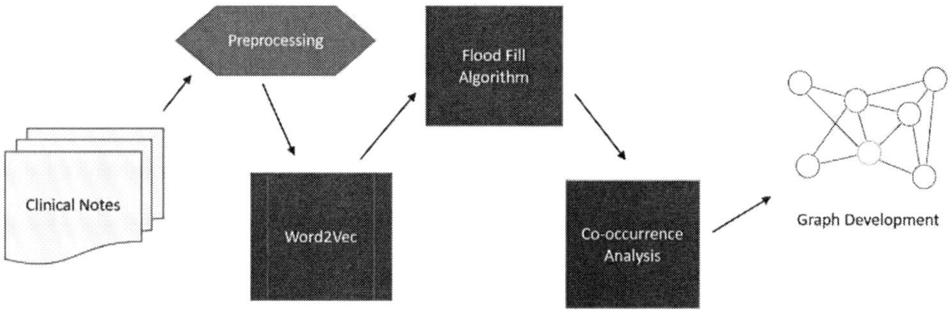

Figure 2. Application pipeline

3. RESULTS

3.1 Patient Descriptive Data

The demographic characteristics of the patients and select comorbidity values are in Table 1. Patients tended to be younger, more predominently male, and less likely to be married, when compared to Veterans in general (U.S. Department of Veterans Affairs, 2017a, U.S. Department of Veterans Affairs, 2017b).

Table 1. Demographics and select comorbidities on outpatient encounters

Variable	Percentage
Gender	
Female	7.3%
Male	92.7%
Age	
Median age at first OUD diagnosis	57
Race	
Asian	>1%
American Indian or Alaska Native	>1%
Black/African American	61.8%
Native Hawaiian/Pacific Islander	>1%
White	31.7%
Unknown/Declined to Answer	5.1%

Marital Status		
Married	25.6%	
Divorced	29.6%	
Never Married	27.7%	
Separated	11.9%	
Widowed	4.3%	
Unknown	>1%	
Ethnicity		
Hispanic or Latino	1.7%	
Not Hispanic or Latino	95.2%	
Unknown/Declined to Answer	3.1%	
Comorbidities		*Diagnostic Codes*
Other Substance Abuse Disorders	89.2%	ICD-9-CM codes (305, not including 305.5-); ICD-10 codes (F10, F12-F19);
Post-Traumatic Stress Disorder	46.3%	ICD-9-CM codes (309.81); ICD-10 codes (F43.1);
Mental Health Diagnoses including Depressive and Anxiety Disorders	75.5%	ICD-9-CM codes (296.2-3, 300.00-02, 300.09, 311-); ICD-10 codes (F32, F33, F41);

3.2 Clinical Note Attributes

Note counts are in Table 2. The majority of notes were generated during the month of diagnosis; the least during the sixth month before.

Table 2. Clinical note counts by site

	Sixth Month Before	Month of Diagnosis	Sixth Month After
Baltimore	1201	10673	3897
Washington DC	877	6217	1893

The top note types and their frequencies are in Table 3. Many note types for both care sites address mental health, emergency care, nursing care, and substance abuse. The initials SATP are short for substance abuse treatment program; SARP is for substance abuse rehabilitation program; PCC is for primary care clinic.

Table 3. Most common clinical note types and their frequencies by corpus

Baltimore, Sixth Month Before Diagnosis	Baltimore, Month of Diagnosis	Baltimore, Sixth Month After Diagnosis
Ed Nursing Triage Note, 145	Psychiatry Attending Note, 932	SATP Group Note, 299
Psychiatry Attending Note, 84	SATP Group Note, 415	Psychiatry Attending Note, 270
SATP Individual Note,77	Ed Nursing Triage Note, 409	SATP Individual Note,267
Primary Care Outpatient Note, 73	Mental Health Psychological Assessment, 362	Ed Nursing Triage Note,224
Nursing Admission Assessment, 44	Discharge Summary,359	Psychiatry Resident Note, 206
Washington DC, Six Months Before Diagnosis	**Washington DC, Month of Diagnosis**	**Washington DC, Sixth Month After Diagnosis**
PCC - Established - Problem Focused, 81	Nurse Practitioner Note, 400	SARP: MD Follow-Up Note,135
Emergency Department MD Urgent Note, 57	SARP: MD Initial Evaluation Note,381	PCC - Established - Problem Focused,134
Emergency Department RN Triage Note, 48	Psych: Inpatient Progress Note,344	Emergency Department MD Urgent Note, 102
SARP: MD Follow-Up Note, 35	Addendum, 243	SARP: Progress Note(E),81
Opioid Pain Medication Refill, 31	Emergency Department MD Urgent Note, 231	Nurse Practitioner Note, 76

3.3 Themes in Word Pairs

We examined word pairs according to similarity score, by time period and care site. Several common themes emerged (Table 4).

Table 4. Word pair themes, with examples

Sixth Month Before Diagnosis	Month of Diagnosis	Sixth Month After Diagnosis
Emotions: fear/shame, guilt/shame, hopes/rejoining, intense/shame, feelings/managing, *Physical Feelings:* feeling/sleeping, feel/tired, dizzy/feeling, having/lightheaded *Substances:* etoh/ivda, heroine (sic)/ivda, marijuana/mdma, beer/couple, smoke/using, illicit/substances, opioid/tapering *Recovery:* detox/seeking, reducing/slowly	*Substances:* bags/worth, beers/drinks, drinks/pints, dollars/vodka, bags/snorted, crack/habit, sniffed/using, opioid/opioids, benzo/opioids *Legal Issues:* arrested/possession, incarceration/prison, december/jail, jail/shot, fight/jail, duis/possession *Access:* bought/street, snorts/street, purchased/taking, *Violence:* cutting/hanging, fight/getting, killed/started, hanging/lethal, assault/fighting *Therapeutic:* coping/skills, relaxation/techniques, mindfulness/techniques, cope/deal *Emotions:* feelings/shame, feelings/guilt, memories/stressful	*Therapeutic:* group/session, going/trying, feelings/mindfulness, strategies/support *Emotions:* feelings/shame, guilt/shame, hypervigilance/irritability, feels/states *Substances:* liquor/whiskey, drugs/opioids, beer/vodka *Other People:* people/trust, people/talking, boundaries/relationships, people/places, hurt/people, people/tell *Recovery:* maintaining/sobriety, maintaining/managing

Overall, the results from each site expressed the same major themes, although there were differences in terms of the top themes by similarity scores. For the sixth month before diagnosis, the top ten results for Baltimore largely involved feelings (guilt/shame), whereas for Washington D.C. there was a mixture of therapeutic (psychotherapy/sessions), substances (drugs/illicit) and feelings (hopes/rejoining). Here, the initials 'etoh' indicate alcohol, 'ivda' indicate intravenous drug abuse, and 'mdma' indicate the illicit drug *ecstasy*. In the diagnosis month period, the top ten results were both dominated by substances (beer/vodka, bags/worth), but Washington D.C.'s top 10 also included references to legal issues and therapeutic measures. Here, 'duis' indicates episodes of driving under the influence of alcohol or another substance. For the sixth month after diagnosis, each site's results is a mixture of substances (liquor/whiskey), emotions and emotional states (guilt/sadness, hypervigilance/memories) direct and indirect references to relations with others (people/trust, avoidance/isolation; Baltimore), violence (hurt/kill; Washington D.C.) and time entities (february/january, saturday/sunday; Washington D.C.). Within the top results of each care site are also references to memories (intrusive/memories, hypervigilance/memories) which may reflect the significant percentage of patients who also have PTSD.

3.4 Highly-Connected Vertices and Subsets

To identify highly connected vertices and their subsets, we isolated all vertices whose connections to other vertices were within the top five percent of frequencies, for each graph. These findings are in Table 5.

Table 5. Graphs' vertices counts, most connect vertices excluding 'opioids', and path examples

Baltimore, Sixth Month Before Diagnosis	Baltimore, Month of Diagnosis	Baltimore, Sixth Month After Diagnosis
Total vertices: 257	Total vertices: 547	Total vertices: 372
Most connected vertices: reported, better, going, feeling, having	Most connected vertices: snorted, stopped, using, took, years	Most connected vertices: approx, thoughts, feelings, anger, managing
Highly-connected path examples: ivda-years-opioids-started reported-years-approximately work-going-feel	Highly-connected path examples: used-opioids-week stopped-started-snorted opioids-used-gets	Highly-connected path examples: having-persistent-nightmares anger-triggers-identifying racing-thoughts-talking

Washington DC, Sixth Month Before Diagnosis	Washington DC, Month of Diagnosis	Washington DC, Sixth Month After Diagnosis
Total vertices: 232	Total vertices: 425	Total vertices:304
Most-connected vertices: abuse, feels, using, substance, help	Most connected vertices: uses, going, taking, says, beer	Most connected vertices: help, support, recovery, states, says
Highly-connected path examples: substances-abuse-issues agreement-opioids-using opioids-dose-gradually	Highly-connected path examples: bags-used-taking says-began-opioids beer-drinking-opioids	Highly-connected path examples: group-support-recovery feels-getting-better help-know-people

Opioids is the most connected vertex in all the graphs, because it was the initial seed term.

3.5 Graphs

Figure 3 is an example of the graphs the application produced. It is the graph of the top five percent of vertices, by overall connection frequencies, for Washington D.C., for data from the sixth month after diagnosis.

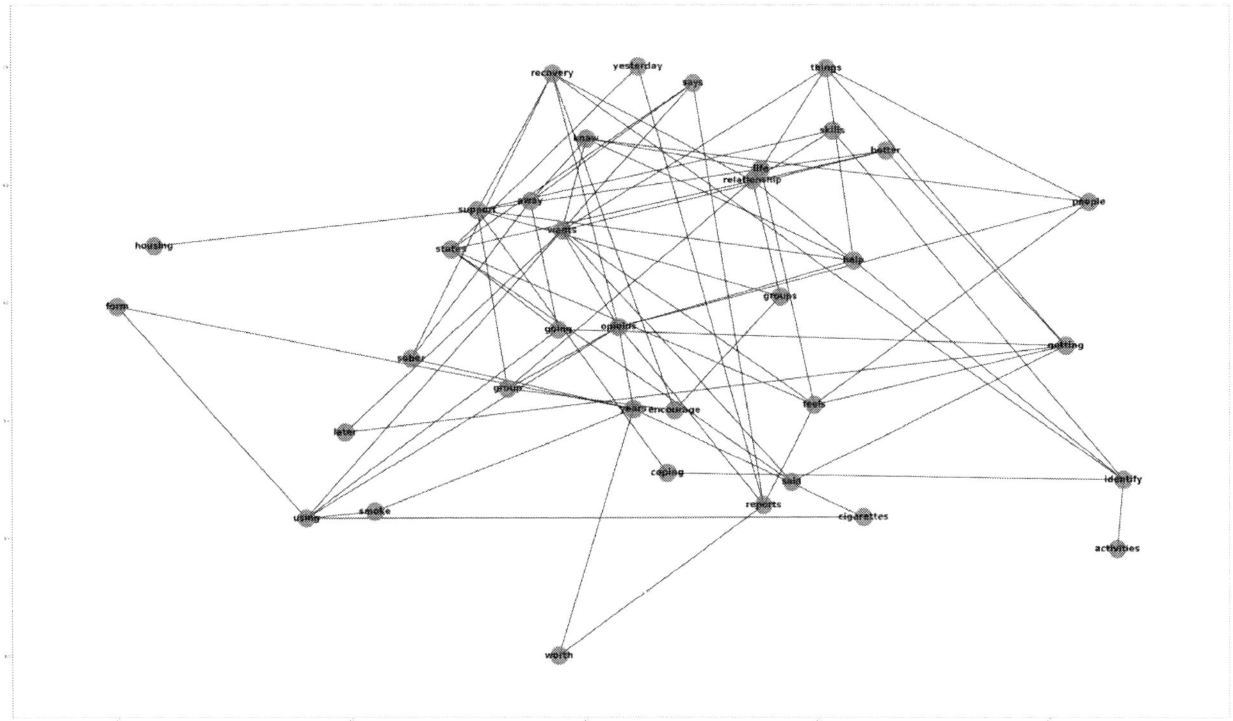

Figure 3. Washington D.C., sixth month after, top five percent of vertices

In Figure 3, one can observe the highly-connected paths noted in Table 5 (sometimes these paths also branch and/or extend further).

4. DISCUSSION

The collective findings address our original questions. Notable OUD themes emerged according to time period and region. The highly connected vertices and subsets in the graphs provided an enhanced understanding of OUD.

4.1 Clinical note and Patient Descriptive Data

The patient demographic data (Table 1) and frequent clinical note types (Table 3) provide an initial framework for understanding OUD. These patients had notable rates of the select comorbidities. A disproportion of patients were male (at least 9.4% of all U.S. Veterans are female (U.S. Department of Veterans Affairs, 2017a)). The median age of 57 years-of-age at first diagnosis was younger than the overall median age of approximately 64 years-of-age of all Veterans (U.S. Department of Veterans Affairs, 2017b). Patients in this group were less likely to be married i.e., 25.6% versus 64.7% (U.S. Department of Veterans Affairs, 2017b). Many of these values mirror prior research. Table 3 provides a temporal representation of clinical note types related to OUD. The prevalence of emergency department notes reflect the crisis nature of OUD. The sixth month before is the only time period including a prescription note type, suggesting increased awareness of OUD in the following time periods. The high frequency of notes recorded during the month of first diagnosis (Table 2) also suggests a correlating increase in care at this time.

4.2 Contextual Neighborhoods in OUD Documentation

The contextual neighborhoods characterize OUD documentation, and expand the initial framework. Incorporating their results into graphs enables the viewer to observe how the concepts in these neighborhoods interact. The emerging themes revealed by the connected word pairs in Table 4 demonstrate that issues of emotions, feelings, addictive substances, therapy, and recovery appear important as clinicians document this frequently for Veterans with OUD. The OUD experience overlaps with the use of other addictive substances. While themes addressing violence and legal issues emerge, so do others of positive recovery.

The outcomes of this work can potentially enhance understanding of OUD, as a temporal and regional phenomenon. The connected vertices "intrusive" and "memories", and those similar to them suggest that OUD is an issue for PTSD patients at both care sites. As indicated in Table 1, many of these patients experience PTSD, and other mental health-associated diagnoses. This is not surprising, since mental health issues are a known OUD risk factor for Veterans. The presence of other addictive substances suggests that OUD may be part of a more complex polysubstance addiction phenomena, especially since the majority of these patients also have received another substance abuse disorder diagnosis. The highly-connected path examples (Table 5) potentially characterize an extended experience for Veterans going through OUD and recovery, providing an enhanced understanding of OUD for the viewer.

To our knowledge, this methodology has not been previously applied. It is similar to topic modeling (Blei, et al. 2009), but allows the results to be focused on a seed concept, and the results can be incorporated into graphs for additional insight. Others have applied topic modeling to opioid-oriented content from social media (El-Bassel, et al., 2022; Chenworth, et al., 2021; Pandrekar, et al., 2018). Because of their source data, their results did not capture what clinicians record in clinical notes, so it is difficult to compare results. Chenworth et al. did find legal issues and treatment aspects within their results. El-Bassel found elements of recovery. Pandrekar et al. also found legal issues as a topic or theme. None of these studies included graph-centered analyses.

5. CONCLUSION

The complete application (Figure 2) provides a visualized characterization (Figure 3) of the major concepts emerging from OUD documentation, and how they interact, by incorporating the results into graphs. This can potentially inform clinicians who treat OUD patients. It can also be generalized to other topics in clinical care.

5.1 Future Work

Future efforts will include a more in-depth study of the graphs by analyzing network properties such as centrality and assortativity, and the vertex link distributions. The authors plan to apply the methodology using other seed concepts.

5.2 Limitations

This work demonstrates how clinicians document their findings and impressions from their treatment visits with patients with OUD. The voice of the patient is not incorporated, nor is the conversation of the patient with the provider. Recordings of clinical visits might be future considerations, as well as including the patient voice.

ACKNOWLEDGEMENT

The views expressed are those of the authors and do not necessarily reflect those of the United States Department of Veterans Affairs, the United States Government, or the academic affiliate institutions. This work was funded by Veterans Affairs Health Services Research and Development Services grant IIR 19-378 Assessing and Reducing Opioid Misuse Among Veterans in VA and Non-VA Systems: Coordination of Fragmented Care, and NIH National Center for Advancing Translational Sciences grant UL1TR001876.

REFERENCES

American Psychiatric Association. (2013). *Diagnostic and statistical manual of mental disorders (5th ed.).*

Bernardy, N., & Montaño, M, 2019. Opioid use among individuals with posttraumatic stress disorder. *PTSD Research Quarterly,* Vol. 30, pp. 1-8.

Blei, D.M., et al, 2003. Latent Dirichlet Allocation. *J Mach Learn Res,* Vol. 3, pp. 993-1022.

Brown, S. H., et al, 2003. VistA--U.S. Department of Veterans Affairs national-scale HIS. *Int J Med Inform,* Vol. 69, No. 2-3, pp 135-156. doi:10.1016/s1386-5056(02)00131-4

Centers for Disease Control and Prevention, 2021. Drug Overdose Deaths in the U.S. Top 100,000 Annually [Press release]. Retrieved from https://www.cdc.gov/nchs/pressroom/nchs_press_releases/2021/20211117.htm

Chenworth, M., et al, 2021. Methadone and Suboxone® Mentions on Twitter: Thematic and Sentiment Analysis. *Clin Toxicol (Phila)* Vol. 59 No. 11, pp. 982-991.

Edlund, N., et al, 2022. Harnessing the Power of Social Media to Understand the Impact of COVID-19 on People who use Drugs During Lockdown and Social Distancing. *J Addict Med,* Vol. 16 No. 2, pp. e123-e132

El-Bassel, M. J., et al, 2007. Risk factors for clinically recognized opioid abuse and dependence among veterans using opioids for chronic non-cancer pain. *Pain,* Vol. 129 No. 3, pp. 355-362. doi:10.1016/j.pain.2007.02.014

Law, G., 2013. Quantitative comparison of flood fill and modified flood fill algorithms. *International Journal of Computer Theory and Engineering,* Vol. 5 No. 3, pp. 503-508.

Luo F, et al. State-Level Economic Costs of Opioid Use Disorder and Fatal Opioid Overdose — United States, 2017. *Morbidity and Mortality Weekly Report,* Vol. 70, No. 15, pp 541-546. Retrieved from https://www.cdc.gov/mmwr/volumes/70/wr/mm7015a1.htm

Martins, S. S., et al., 2015. Worldwide Prevalence and Trends in Unintentional Drug Overdose: A Systematic Review of the Literature. *Am J Public Health,* Vol. 105, No. 11, pp. e29 – e49. doi: 10.2105/AJPH.2015.302843

Mikolov, T., et al., 2013. Efficient estimation of word representations in vector space. *arXiv preprint arXiv:*1301.3781. doi: 10.48550/arXiv.1301.3781

Pandrekar S., et al, 2018. Brief Report: Opioid-Involved Overdose Mortality in United States Veterans. *AMIA Annu Symp Proc,* Vol. 2018, pp. 867-876.

Peltzman, T., et al, 2018. Social Media Based Analysis of Opioid Epidemic Using Reddit. *Am J Addict,* Vol. 29, No. 4, pp. 340-344. doi:10.1111/ajad.13027

Rhee, T. G., & Rosenheck, R. A., 2019. Comparison of opioid use disorder among male veterans and non-veterans: Disorder rates, socio-demographics, co-morbidities, and quality of life. *Am J Addict,* Vol. 28 No. 2, pp. 92-100. doi:10.1111/ajad.12861

U.S. Department of Veteran Affairs, 2017a. *Women Veterans Report: The Past, Present and Future of Women Veterans.* Retrieved from https://www.va.gov/vetdata/docs/specialreports/women_veterans_2015_final.pdf

U.S. Department of Veteran Affairs, 2017b. *Profile of Veterans: 2017.* Retrieved from https://www.va.gov/vetdata/docs/SpecialReports/Profile_of_Veterans_2017.pdf

U.S. Department of Veteran Affairs, 2021. About VA. Retrieved from https://www.va.gov/about_va/

A PARSIMONIOUS MACHINE LEARNING APPROACH TO DETECT INAPPROPRIATE TREATMENTS IN SPINE SURGERY ON THE BASIS OF PATIENT-REPORTED OUTCOMES

Lorenzo Famiglini[a], Frida Milella[b], Pedro Berjano[b] and Federico Cabitza[a,b]
[a]*DISCo, Dipartimento di Informatica, Sistemistica e Comunicazione, University of Milano-Bicocca, Viale Sarca 336, 20126 Milan, Italy*
[b]*IRCCS Istituto Ortopedico Galeazzi, Via Riccardo Galeazzi 4, 20161 Milan, Italy*

ABSTRACT

Patient-reported outcome Measures (PROMs) are validated questionnaires or self-report instruments of the perception of patients about their own health status in response to a medical intervention. In the era of patient-centred care, consideration about the effectiveness of a medical treatment should be grounded not only on the physicians' assessment but also on the increase in PROM scores that is perceived as relevant by the patients, namely the minimum clinically important differences (MCID).
In this study, we collected data from the IRCCS Galeazzi Orthopaedic Institute (IOG) of Milan, Italy, to develop a preoperative machine learning model predicting the non-achievement of the MCID threshold in specific PROM scores 6 months after spine surgery, namely the Oswestry Disability Index (ODI) and the Physical Score of the Short Form 36 (SF-36). At IOG, nearly 39% of spinal surgeries do not achieve a minimum clinically important improvement and, of these, 22% are associated with negative outcomes. This is mainly due to the fact that IOG is a tertiary healthcare facility that receives the most critical cases from a vast territory (practically from all over the country) and it is known that spinal deformities and other related problems are difficult to solve, especially in an aging population. In this view, it is important to early identify those patients who will likely not benefit from treatment and, therefore, will not reach the MCID. This would help avoid overdiagnosis, reduce overuse and the related costs for unnecessary treatment, as well as optimise the allocation of resources and support more appropriate choices in shared decision-making.

KEYWORDS

Patient-Reported Outcome Measures, Machine Learning, Prediction, Appropriateness, Spine Surgery

1. INTRODUCTION

Patient-reported outcome Measures (PROMs) are validated questionnaires or self-report instruments (Field et al., 2019) of the perception of patients about their own health status (Dawson et al., 2010) in response to a medical intervention (Weszl et al., 2019). These measures are increasingly being used to supplement other data sources in assessing the outcomes of medical interventions (Baker et al., 2012). In the era of patient-centred care, consideration about the effectiveness of a medical treatment should be grounded not only on the physicians' assessment but also on the increase in PROM scores that is perceived as relevant by the patients, namely the minimum clinically important differences (MCID) (Copay et al., 2007).

IRCCS Galeazzi Orthopaedic Institute (IOG) of Milan, Italy, is a large teaching hospital in Milan (Italy) specialised in the diagnosis and treatment of musculoskeletal problems. Nearly 5000 surgeries are conducted at IOG each year, the majority of which are arthroplasty (hip and knee prosthetic surgery) and spine-related procedures. At IOG, nearly 39% of spinal surgeries do not achieve a minimum clinically important improvement and, of these, 22% are associated with negative outcomes. This is mainly due to the fact that IOG is a tertiary healthcare facility that receives the most critical cases from a vast territory (practically from all over the country) and it is known that spinal deformities and other related problems are difficult to solve, especially in an aging population (Armaghani et al., 2016). In this view, it is important to early identify those patients who will likely not benefit from treatment and, therefore, will not reach the MCID. This would help avoid overdiagnosis, reduce overuse (Philipp et al. 2011) and the related costs for unnecessary treatment (Langenberger et al., 2022), as well as optimise the allocation of resources and support more appropriate choices in shared decision-making (Fontana et al., 2019).

The importance of machine learning techniques for predictive analytics is increasing in medicine (Staartjes et al., 2019), in orthopaedics (Pedersen et al., 2020) and, in particular, in spinal surgery (Finkelstein et al., 2021). Some studies discuss the use of PROMs and machine learning (ML) approaches to predict whether an orthopaedic surgery will lead to meaningful improvement for patients (MCID). For example, in one study two machine learning models (i.e. artificial neural network (ANN) and logistic regression) were trained to predict the achievement of MCID in functional disability (Oswestry Disability Index -ODI scale) and in leg pain severity (Numeric Rating Scales -NRS scale) at 12 months post-surgery for patients who had undergone lumbar disc herniation (LDH) medicine (Staartjes et al., 2019). In another study, 5 machine learning models (i.e. artificial neural network, decision trees, random forest, boosted trees, support vector machine) and 2 conventional models (i.e. logistic regression and multivariate adaptive regression splines) were trained to predict whether the smallest relevant improvement would be reached in outcome measures (i.e. the EuroQol (EQ-5D), the Oswestry Disability Index (ODI), the Visual Analog Scale (VAS leg and back) and return to work) at 12 months post-surgery for patients who had undergone LDH (Pedersen et al., 2020). In another work, an elastic-net penalised logistic regression was trained to predict the achievement of MCID 12 months after surgery for patients who had undergone lumbar spine decompression surgery (Karhade, 2021).

To the best of our knowledge, few studies (i.e. Karhade, 2021; Pedersen et al., 2020; Staartjes et al., 2019) in spinal surgery focused on predicting whether or not the MCID will be reached in the chosen post-surgery outcome measures. No studies to date have been aimed at the task of identifying patients who will not benefit from the surgery. On the contrary, we believe that training a machine learning model that accurately predicts the probability that a patient will not improve after surgery is critical in a value-based healthcare context.

In this study, we report about the performance of a machine learning model that predicts the non-achievement of the MCID in two important PROM scores at 6-months postoperatively, namely the Oswestry Disability Index (ODI) (Fairbank & Pynsent, 2000) and the Physical Score of the Short Form 36 (SF-36) (Laucis et al., 2015). The SF-36 scale ranges from 0 (worst possible health condition) to 100 (best possible health condition). The ODI scale ranges from 0 (best possible health condition) to 100 (worst possible health condition). To this respect, we will use PROM data collected at the IOG by means of computer assisted telephone-interview or computer assisted web self -interview both before surgery (pre-operative) and at 6 months after spinal surgery.

2. METHODS

Both classification and regression tasks were addressed in our study. Specifically, we provided a realistic estimate of the efficacy of machine learning models at predicting specific scores 6 months after surgery, by exploring the balanced performance measures.

2.1 Data Analysis

The study encompassed patients admitted to IOG between November 2015 and March 2022. In total, data on 7112 patients who had undergone spinal surgery were included in the study. The data was extracted from the web-based PRO registry (SpineREG) that IOG established in November 2015.

As reported in Table 1, the mean age of patients is 53 years old (SD = 17.76). The majority of patients are female (60%). Patients without morbidity represent 61.5% of the total. Only 11.5% of patients had severe comorbidities. The remaining 26.8% had moderate comorbidity. The majority of cases were patients with lumbar arthrodesis (47%, n = 1301), lumbar hernia (16%, n = 446), idiopathic deformity (10%, n = 273), degenerative deformity (10%, n = 268), lumbar decompression (8%, n = 218) and cervical arthrodesis (7%, n = 191), other (2%, n = 92). Patients affected with kyphoplasty, vertebral tumour and cervical hernia were 59, 23, 10 respectively. Moreover, the sample mainly comprised no-smokers (n = 1503). The categorical variables were codified as follows: gender (0 female, 1 male), and smokers (1 yes, 2 no, 3 no valuable); morbidity was rendered in terms of an ordinal scale: 1 no pathology, 2 medium, 3 severe, 4 dying. After excluding missing values in regard to the variables of interest (SF-36 Physical Score and ODI Score at 6-month intervals), 2823 observations were available for training and validation. Moreover, variables were filtered based on the missing value content percentage. Specifically, the features that contained more than 80% of missing value were dropped, and at the same time, the instances with at least 25% of missing value were filtered off. Therefore 26 features were considered, that characterized a total of 2789 instances. As reported in Table 1, only few features presented a small number of missing values. The majority of missing values refers to the variable of smokers (15.6%). In addition to the features collected from the database, we evaluated up to other 50 features extracted from the physician's notes.

Table 1. Structured Score Features: missing values and descriptive statistics
T= Type of data (B = Boolean, C = Categorical, N = Numerical, O = Ordinal), NA: Not applicable
FABQ = Fear-Avoidance Beliefs Questionnaire

	Count	Mean	Std	Min	25%	50%	75%	Max	Missing Value (%)	T
Cervical arthrodesis	2789	NA	NA	NA	NA	NA	NA	NA	0	B
Lumbar arthrodesis	2789	NA	NA	NA	NA	NA	NA	NA	0	B
Kyphoplasty	2789	NA	NA	NA	NA	NA	NA	NA	0	B
Lumbar decompression	2789	NA	NA	NA	NA	NA	NA	NA	0	B
Degenerative deformity	2789	NA	NA	NA	NA	NA	NA	NA	0	B
Idiopathic deformity	2789	NA	NA	NA	NA	NA	NA	NA	0	B
Cervical hernia	2789	NA	NA	NA	NA	NA	NA	NA	0	B
Lumbar hernia	2789	NA	NA	NA	NA	NA	NA	NA	0	B
Vertebral tumour	2789	NA	NA	NA	NA	NA	NA	NA	0	B
Age	2789	53.18	17.76	7	43	55	67	88	0	N
Gender	2789	NA	NA	NA	NA	NA	NA	NA	0	B
Morbidity	2789	NA	NA	NA	NA	NA	NA	NA	0	O
Smoker	2353	NA	NA	NA	NA	NA	NA	NA	15.63	C
ODI_Total_PreOp	2789	43.58	19.99	0	30	44	57	100	0	N
Vas_Back_PreOp	2587	6.23	3.03	0	5	7	8	10	7.24	N
Vas_Leg_PreOp	2587	6.19	3.57	0	4	7	9	10	7.24	N
SF36_GeneralHealth_PreOp	2789	68.34	20.82	0	55	70	85	100	0	N
SF36_PhysicalFunctioning_PreOp	2789	46.22	27.11	0	25	45	65	100	0	N
SF36_RoleLimitPhysical_PreOp	2789	13.61	29.16	0	0	0	0	100	0	N
SF36_RoleLimitEmotional_PreOp	2789	43.64	44.06	0	0	33.33	100	100	0	N
SF36_SocialFunctioning_PreOp	2789	52.36	30.35	0	25	50	75	100	0	N
SF36_Pain_PreOp	2789	27.99	25.2	0	10	22.5	45	100	0	N
SF36_EnergyFatigue_PreOp	2789	47.99	26.38	0	30	50	70	100	0	N
SF36_EmotionalWellBeing_PreOp	2789	59.27	24.29	0	44	60	76	100	0	N
SF36_MentalScore_PreOp	2789	47.63	12.16	16.06	38.49	47.11	57.48	77.1	0	N
SF36_PhysicalScore_PreOp	2789	34.14	8.78	15.2	27.99	32.77	38.72	62.75	0	N
SF36_PhysicalScore_6months	2789	41.09	9.99	12.46	33.49	40.84	49.37	60.64	0	N
ODI_Total_6months	2789	23.72	20.07	0	8	18	36	97	0	N
FABQ_Work_PreOp	2763	19	20.68	0	0	11	39	66	0.93	N

For both binary and regression tasks, the data were partitioned into 90% training (and validation) and 10% (from the training and optimization procedures) test set to evaluate the performance of the developed models. Regarding the binary classification task, our class of interest was related to the patients who would not achieve MCID (Copay et al., 2007) ($\Delta = 5$ for SF-36 and $\Delta = 10$ for ODI Score). Indeed, our models were built to discover patients who would not have a significant improvement from surgery: for the SF-36 Physical Score improvement (0 improvements, 1 no improvement) and for the ODI Score (0 improvements, 1 no improvements), both six months after surgery, the target distributions are slightly affected from imbalanced data issues. Indeed, for the SF-36 Physical Score task, the positive class was 46%, while for the ODI Score, the positive class was 32% of the total distribution. Concerning the regression task, the target variable was the SF36 Physical Score and ODI scores obtained after six months from surgery.

2.2 Binary Classification Task

Several studies concerning patient improvement prediction focused only on finding whether a patient achieves the MCID or not: our approach probes the alternative approach to identify patients that do not benefit from the surgery. A more appropriate choice in decision making process would help to reduce overdiagnosis, minimize overuse (Philipp et al. 2011) and lower costs for avoidable treatment (Langenberger et al., 2022), especially if a Machine Learning model helps to discover more rare events than the frequent ones.

For the assessment of the binary classification task, the MCID was computed using one-half the standard deviation according to the distribution-based approach (Asher et al., 2018). Specifically, the thresholds identified for SF-36 Physical Score and ODI Score were 5 and 10, respectively. The binary targets were constructed on these thresholds. In regard to SF-36 Physical Score, if the difference between the post-operative score and the pre-operation score was lower than the above threshold, the label was set to 1, otherwise 0. On the other hand, for ODI Score, if the difference between the post-operation and pre-operation score was lower than minus the threshold (-10), the label was set to 0 (meaning an improvement in the disability index), otherwise to 1.

In regard to model development, as mentioned in the previous section, the training and hyperparameters optimization were performed on the entire training set. We exploited the Bayesian optimization procedure (Snoek et al., 2012) to find the sub-optimal set of XGBoost (XGB) (Hinterwimmer et al., 2022) hyperparameters. Specifically, a 6 folds cross-validation was performed by maximizing the precision-recall curve with a budget of 50. The search space is reported in Table S1 in appendix.

The training phase includes missing value imputation and resampling technique. In terms of imputation, the Bayesian Ridge estimator (Bishop, 2011) (a multivariate procedure for missing value imputation) was trained alongside the XGB model and applied to the validation folds. Furthermore, an oversampling technique (Mohammed et al., 2020) is applied to the minority class of the related training folds to reduce the imbalance bias. Indeed, the positive class was re-sampled 40% more than the negative for the ODI improvement task. In contrast, a small amount of the positive was re-sampled (10 %) for the Physical improvement prediction. The hyperparameters identified for SF36 Physical Score prediction are reported in Table S2 in appendix. The hyperparameters identified by the Bayesian optimization procedure for ODI Score prediction are reported in Table S3 in appendix.

2.3 Regression Task

Another aspect we wanted to investigate was the score prediction 6 months later surgery so that we would have a variable of interest not derived from an approach that might introduce systematic bias (Jiang & Nachum, 2020). To this aim, the Multivariate Adaptive Regression Splines (MARS) (Friedman, 1991) regression model was adopted. MARS is a non-parametric regression method that creates a set of models belonging to a certain range of values (called notches). This model also allows to detect nonlinear relationships by developing interactions of variables using different degrees of the polynomial. In our case, the maximum degree that a polynomial can have has been set to 3. The target distributions divided by the first quantile (vs others) of the patients' age for both SF-36 Physical score and ODI scores are shown in Figure S1. Furthermore, also for the regression task, we performed the Bayesian Optimization approach to discover the best hyperparameters referred to as the polynomial maximum degree (from 1 to 5) and the regularization term (from 0.5 to 5). As mentioned before, the budget for the optimization was set to 50, and the number of folds set was 6. The function to be maximized was the negative mean absolute error. The hyperparameters identified by the Bayesian optimization procedure for regression task are reported in Table S4 in appendix.

3. RESULTS

The results obtained on the test set (i.e., the 10% excluded from the training and validation steps) are reported in Table S5 (in appendix), in Table 2-3 and in Figures 1-2. The features importance for the classification task are reported in Figures S2-S3 in appendix.

3.1 Binary Classification Task

Table 2. High-Confidence (HC) Model performance on the test set grouped by feature sets. Binomial Confidence interval at 95% confidence level. SC=Structured Scores; PPV= Positive Predictive Value; AUROC (AUC) = Area Under the Receiver Operating Characteristic; AUPRC=Area Under The Precision-Recall Curve; ECE=Expected Calibration Error

Task	Features	HC-Balanced Accuracy	HC-Balanced Sensitivity	HC-Balanced F1 Score	HC-Balanced PPV	HC-AUROC	HC-AUPRC	HC 1-ECE
SF36 PS	SC	.78 [.70,.85]	.78 [.70,.85]	.78 [.70,.85]	.78 [.70,.85]	.78 [.70,.85]	.85 [.78,.91]	.91 [.85,.96]
ODI	SC	.85 [.77,.92]	.86 [.79,.92]	.86 [.79,.92]	.86 [.79,.92]	.85 [.77,.92]	.86 [.79,.92]	.98 [.96,1]

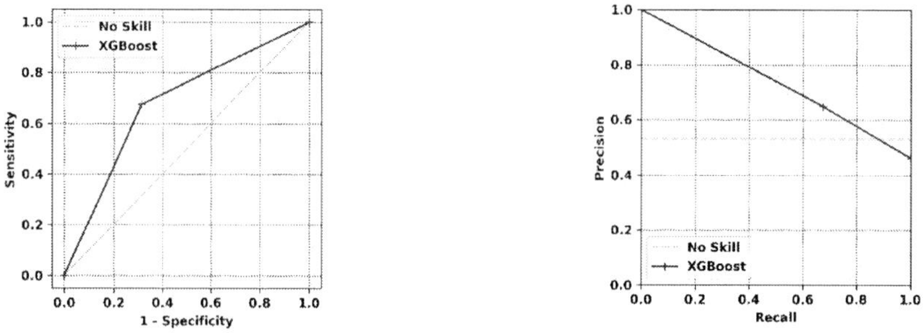

Figure 1. SF36 Physical Score ROC curve and precision recall curve (on test set)

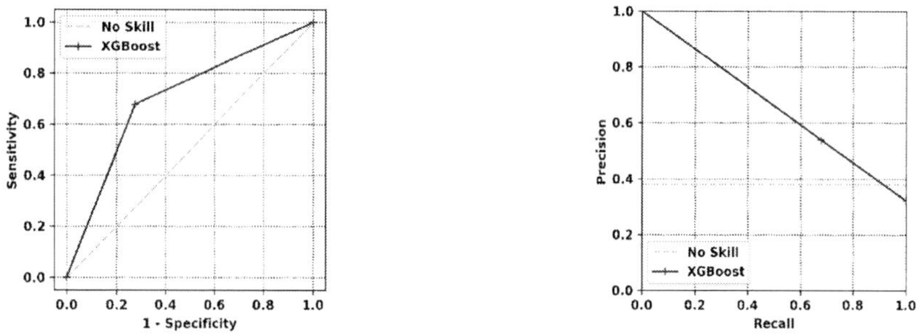

Figure 2. ODI Score ROC curve and precision recall curve (on test set)

3.2 Regression Task

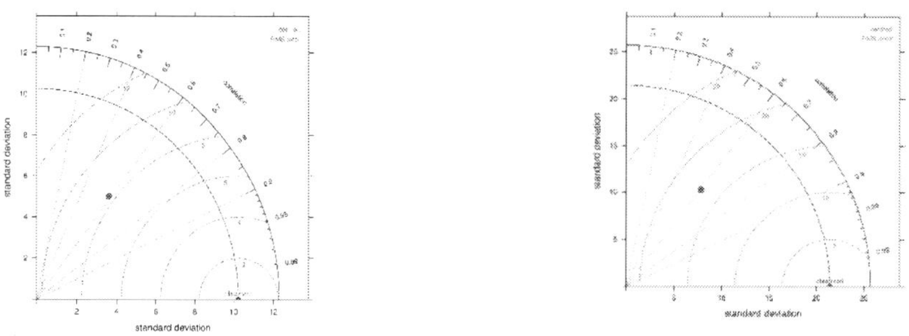

Figure 3. Taylor's Diagrams of the SF36 PS task (on the left) and the ODI task (on the right), respectively

Table 3. RMSE (Root-Mean-Square Error) and MAE (Mean Absolute Error): Regression task

TASK	FEATURES	RMSE	MAE	MAE/SD
SF36 PS	SC	8.29	6.79	.66
ODI SC	SC	17.03	13.51	.67

4. DISCUSSION

We reported the high-confidence results, that is, the results regarding the predictions that are associated with a confidence score equal to or higher than .75, as reported in Table 2. First, in regard to the binary classification task, we remark that we focused on the high confidence (HC) results. These results are particularly interesting, as they allow us to define three decision regions: confidence rejection of the non-improvement result (i.e., by achieving the MCID); high confidence acceptance of the non-improvement result (i.e., by non-achieving the MCID), and also an uncertain area where nothing can be said for sure and where doctors and patients have to decide considering other contextual factors. Indeed, the predictions associated with a high confidence score range between 40% to 70% of the test set. This means that the models have fair coverage on average (Cabitza et al, 2020). Regarding the binary classification task, our results show that HC-balanced accuracy achieves .78 [.70, .85] and .85 [.77, .92] for the structured scores in the SF36 PS and the ODI Score, respectively. The HC-balanced positive predictive values were .78 [.70 .85] and .86 [.79, .92] for the structured scores in the SF36 PS and the ODI Score, respectively. Furthermore, the HC-balanced F1 score and HC-Sensitivity achieve an upper bound of .85 and .92 for the SF36 PS and ODI Score, respectively, as shown in Table 2. Moreover, as reported in Table 2, the HC-AUROC of the model, namely the capability to discriminate between patients who will improve and patients who will not improve, is .78 [.70,.85] on the physical score and .85 [.77, .92] on the disability index, with strongly calibrated results (see the HC 1-ECE column in Table 2). Indeed, based on the Expected Calibration Error (ECE), the developed models produce calibrated confidence scores, where HC-1-ECE ranges between .91-.98. Calibrated models can estimate the probability that the worsening will actually occur reliably. This result is particularly interesting, because non calibrated model has the potential to be inaccurate and harmful in medical decision-making process (Van Calster et al, 2019). An important aspect to consider is the HC-AUPRC: estimating this measure is important to get performance estimates that are not affected by data imbalance and the rate of true negative cases (i.e., the number of patients who will get better). Indeed, as seen in several studies (Karhade, 2021; Pedersen et al., 2020; Staartjes et al., 2019), referring to only the AUROC could lead to partial interpretations of the model performance. Also, for the ODI task, the dummy classifier reaches an AUPRC of .38, while the XGBoost model with CS features has an average value of .86 [.79,.92]. Even more importantly.

These results can be considered acceptable as predicting a future outcome based on a few parameters collected before treatment is an intrinsically hard problem; the results given by our models can provide physicians and patients with useful pieces of information that must be interpreted in the light of other contextual (non-digitized) information in the shared decision-making process (Cabitza et al, 2020). Moreover, the performance is not too different from those reported by other studies, which nevertheless could be affected by data leakage, overfitting, and unbalanced problems. Indeed, regarding the binary classification task, the first issue encountered by analysing the other studies lies in identifying the right positive label and data imbalance problem. Specifically, the positive label is associated with the physical score or Oswestry Disability Index improvement (by reaching the MCID). From surgery, usually, we expect the improvement of the patient, not the worsening. Indeed, the class distribution is imbalanced concerning the positive label and not the negative one. This leads to produce measures that are too optimistic for the positive label. From Staartjes et al. (2019), the AUC scores range between .84 to .90 (based on the deep learning approach), while for the accuracy the values range from .75 to .87. The AUC is not directly influenced by the data imbalance issue (as the accuracy). However, if the model learns to predict well the positive majority class, this leads to inflated results (Saito & Rehmsmeier, 2015). Furthermore, the authors applied a deep learning model with a very low sample size that implies a high risk of overfitting (Ying, 2019). Pedersen et al. (2020) addressed the issue of data imbalance by applying the oversampling-based SMOTE (Synthetic Minority Over-sampling Technique) method. However, the authors report that SMOTE was applied to both training and validation, potentially leading to a data leakage and hence a risk of overfitting, which could induce overly optimistic results. Moreover, our model with the structured scores and the top 20 terms features shows better performance in regard to the minority class (no improvement) for the ODI MCID task. On the other hand, Karhade et al. (2021) validated the developed models by exploiting an external validation set. Even if the results are referred to COMI (Core Outcome measure Index) scores, the obtained performances seem to be more credible than above-mentioned studies due to the lack of the imbalanced data problem and independent dataset to support the results estimated during the training procedure: it is worthy of note that performance is relatively poor, with an AUC score of .63 and a sensitivity of .64, which makes our results even more interesting and promising for the preoperative shared decision making.

Data imbalance is a critical issue that can affect the reliability of ML models' performance estimates, as these latter can focus excessively on the subgroups with higher prevalence (i.e., those patients who improve their outcomes post-surgery), thus providing inflated estimates of accuracy. In this view, our work sheds light on the "less frequent event", namely the prediction of those patients who will not improve their health-related condition or, in other words, patients for whom spinal surgery will not be successful, or even detrimental. Predicting whether or not a specific medical treatment may be appropriate contributes to enhancing a more efficient resource allocation and fair treatment. We believe that this point plays a critical role in the era of value-based healthcare and it should be deepened (Holzinger, 2016). Regarding the regression task, our model shows that for the ODI Scores the MAE was 13.51 with the SC. The RMSE values were 17.03 with the SC. Moreover, we obtained that for the SF36 Physical Score the MAE was 6.79 with the SC. It is important to highlight that the ratios of the MAE over the standard deviation of the target distribution are ⅔ for both tasks: this implies that the models commit an error lower than the natural variation of the dependent variable on average. The RMSE values were 8.29 for the SF36 Physical Score with the SC. In addition, in Taylor's diagram (see Fig. 3), we noticed that the correlation between predictions and true values is strong for both tasks (~ 60%), while RMSE is higher for the ODI Scores task than for the SF36 PS task, as well as the variance (standard deviation) which in one case is 9.99, in the other about 20. This means that the variability for one task is much higher, as it is more difficult to make predictions about it.

To the best of our knowledge, no study exists that employed Machine Learning Regression models to predict continuous ODI Scores and SF36 Physical Scores within 6 months from spine surgery. In fact, Halicka et al. (2022) developed a regression model to predict the COMI score. If we compare the Pearson correlation between the observed values and the predicted ones, our model makes predictions more correlated with the observed target (ours Pearson ~.60 vs .38/.40) than Halicka et al. (2022). Nevertheless, the results of our model still present room for improvement and our study is affected by some limitations: as we have shown in Figure S1, the target variable, especially for SF36 Physical Scores, shows a different distribution among patients belonging to the first quartile of the age distribution (up to about 44 years) than among the remaining older patients. This, along with other confounders (such as comorbidities, and ICD classification), could affect the performance of the model itself. Future studies will therefore focus on the different possible subgroups within the general population and how they impact the model statistics. Further future work will consider different MCID thresholds, that is not only distribution based MCID (as in our case), but also anchor based MCID, and, more importantly, consider different MCIDs for different surgery types.

Compared with other studies in this field, we report results that are weighted in regard to the data unbalance issue, an oft-neglected issue that undermines the validity of many similar contributions.

5. CONCLUSIONS

In this study, we developed machine learning models to predict patients who will likely not benefit from spine surgery, according to the MCID threshold. It is not surprising that, due to the high impact that Machine Learning is having in the medical field, we are not the first researchers to address the problem of predicting PROMS outcomes after spine surgery (Karhade, 2021; Pedersen et al., 2020; Staartjes et al., 2019). However, there are still a few works in the literature in machine learning that aim to predict PROMs scores, specifically in the regression task. In addition, most of the works related to spine surgery present some gaps from the methodological point of view. Moreover, no studies to date have been aimed at identifying patients who will not benefit from the surgery. This study is aimed at filling this gap. More robust approaches that consider the data imbalance issue are needed to expand theoretical approaches into practical tools that could be used in the context of shared decision support in the treatment of spine problems.

REFERENCES

Armaghani SJ et al., 2016. Diabetes Is Related to Worse Patient-Reported Outcomes at Two Years Following Spine Surgery. *J Bone Joint Surg Am.* 2016 Jan 6;98(1):15-22.

Asher AL et al., 2018. Defining the minimum clinically important difference for grade I degenerative lumbar spondylolisthesis: insights from the Quality Outcomes Database. *Neurosurgical Focus.* 2018 Jan;44(1):E2.

Baker PN et al., 2012. The effect of surgical factors on early patient-reported outcome measures (PROMS) following total knee replacement. *J Bone Joint Surg Br.* 2012 Aug;94(8):1058-66.

Bishop CM, 2011. Pattern Recognition and Machine Learning, *Springer* Nature.

Cabitza F et al., 2020. All You Need Is Higher Accuracy? On The Quest For Minimum Acceptable Accuracy For Medical Artificial Intelligence. *Proceedings of eHealth, the 12th International Conference on e-Health.* 21 – 23 July 2020.

Copay AG et al., 2007. Understanding the minimum clinically important difference: a review of concepts and methods. *Spine J.* 2007 Sep-Oct;7(5):541-6.

Dawson J et al., 2010. The routine use of patient reported outcome measures in healthcare settings. *BMJ.* 2020 Jan 18;340:c186.

Fairbank JC, Pynsent PB, 2000. The Oswestry disability index. *Spine (Phila Pa 1976).* 2000 Nov 15;25(22): 2940-52;discussion 2952.

Field J, Holmes MM, Newell D, 2019. PROMs data: can it be used to make decisions for individual patients? A narrative review. *Patient Relat Outcome Meas.* 2019 Jul 29;10:233–241.

Finkelstein JA et al., 2021. Patient factors that matter in predicting spine surgery outcomes: a machine learning approach. *J Neurosurg Spine.* 2021 May 21:1-10.

Fontana MA et al., 2019. Can Machine Learning Algorithms Predict Which Patients Will Achieve Minimally Clinically Important Differences From Total Joint Arthroplasty?. *Clin Orthop Relat Res.* 2019 Jun;477(6):1267–1279.

Friedman JH, 1991. Multivariate adaptive regression splines. *Ann. Statist.* 19(1): 1-67.

Halicka M et al., 2022. Predicting patient-reported outcomes following lumbar spine surgery: development and external validation of multivariable prediction models. *Cold Spring Harbor Laboratory Press*, medRxiv.

Hinterwimmer F et al., 2022. Prediction of complications and surgery duration in primary TKA with high accuracy using machine learning with arthroplasty-specific data. Knee Surg Sports Traumatol Arthrosc. 2022 Apr 8.

Holzinger A, 2016. A. Interactive machine learning for health informatics: when do we need the human-in-the-loop?. *Brain Inf.* 3, 119–131.

Jiang H, Nachum O, 2020. Identifying and Correcting Label Bias in Machine Learning. *Proceedings of the Twenty Third International Conference on Artificial Intelligence and Statistics*, in Proceedings of Machine Learning Research, 108:702-712.

Karhade AV, 2021. Development of prediction models for clinically meaningful improvement in PROMIS scores after lumbar decompression. *Spine J.* 2021 Mar;21(3):397-404.

Langenberger B, Thoma A, Vogt V, 2022. Can minimal clinically important differences in patient reported outcome measures be predicted by machine learning in patients with total knee or hip arthroplasty? A systematic review. *BMC Med Inform Decis Mak* 22, 18 (2022).

Laucis, NC, Hays RD, Bhattacharyya T, 2015. Scoring the SF-36 in Orthopaedics: A Brief Guide. *J Bone Joint Surg Am.* 2015 Oct 7;97(19):1628-34.

Mohammed R, Rawashdeh J, Abdullah M, 2020. Machine Learning with Oversampling and Undersampling Techniques: Overview Study and Experimental Results. *11th International Conference on Information and Communication Systems (ICICS)*, 2020, pp. 243-248.

Pedersen CF et al., 2020. Applied machine learning for spine surgeons: predicting outcome for patients undergoing treatment for lumbar disc herniation using PRO data. *Global Spine J.* 2022 Jun;12(5):866-876.

Philipp LR et al., 2011. Achieving Value in Spine Surgery: 10 Major Cost Contributors. *Global Spine J.* 2021 Apr;11(1_suppl):14S-22S.

Saito T, Rehmsmeier M, 2015. The precision-recall plot is more informative than the ROC plot when evaluating binary classifiers on imbalanced datasets. *PloS one.* 10(3): e0118432.

Snoek J, Larochelle H, Adams RP, 2012. Practical bayesian optimization of machine learning algorithms. *Advances in neural information processing* systems, 25: 2960-2968.

Staartjes VE et al., 2019. Deep learning-based preoperative predictive analytics for patient-reported outcomes following lumbar discectomy: feasibility of center-specific modeling. *Spine J.* 2019 May;19(5):853-861.

Van Calster B et al, 2019. Calibration: the Achilles heel of predictive analytics. BMC Med 17, 230 (2019).

Weszl M., Rencz F, Brodszky V, 2019. Is the trend of increasing use of patient-reported outcome measures in medical device studies the sign of shift towards value-based purchasing in Europe?. *Eur J Health Econ.* 2019 Jun;20 (Suppl 1):133-140.

Ying X, 2019. An Overview of Overfitting and its Solutions. *J. Phys.: Conf. Ser.*. Vol. 1168:022022-.

A HUMAN-COMPUTER INTERACTION METHOD BASED ON U-NET CONVOLUTIONAL NEURAL NETWORK FOR TARGET MOLECULE OBSERVATION

Wenbin Yin, Xinfeng Zhang[1], Jinpeng Fang, Xudong Zhou and Bin Li

College of Information Engineering, Yangzhou University, Jiangsu Province Engineering Research Center of Knowledge Management and Intelligent Service, Yangzhou, Jiangsu, 225127 China

ABSTRACT

In order to accurately identify various physiological activities and movements in living bodies, we propose a U-Net-based method to identify all similar micro molecules such as cells and proteins in organisms. We first transform the molecular image to be observed into the feature space using the U-Net convolution network, and then use the target feature to match across the whole image to detect all similar targets of interest in the image. Extensive experimental results show that the proposed method can rapidly detect similar molecules of interest through a simple human-computer interaction and attain a more accurate detection performance than other approaches.

KEYWORDS

Deep Learning, U-Net, Target Detection, Human-Computer Interaction

1. INTRODUCTION

The observation of key molecules from a microscopic perspective plays an important role in the study of life sciences, e.g. in the observation of the molecules of interest, the number of molecules, the degree of aggregation, the degree of activity and other key information can be captured, which is the basis for the discovery of the intrinsic mechanism of life activities.

In recent years, the biological and medical image research based on deep learning has made remarkable progress. Ronneberger et al. (Ronneberger et al., 2015) present a FCN-based (Long et al., 2015) U-Net network to apply in medical image segmentation, which combines the features of high-level low-resolution and low-level high-resolution by upsampling. Mnih et al. (Mnih et al., 2014) introduce the attention mechanism into Recurrent Neural Network to improve the classification performance of digits images. Oktay et al.(Oktay et al., 2018) design a U-Net based on attention mechanism to increase the sensitivity of the model to foreground for improving the performance of pancreas segmentation.

However, almost all existing approaches for biological and medical image processing cannot meet the need for observation of various molecules, such as cells and proteins, in life science research. The main reasons are as follows: (1) These approaches are trained for target-specific detection or segmentation tasks. To make these approaches effective for other targets, the models need to be retrained with the new target data. However, for scientific tasks, the molecules of interest are often variable and sometimes undiscovered. (2) For images where various types of molecules are present, it is sometimes necessary to observe all targets belonging to certain classes, rather than individual specific target.

To meet the changing demands in scientific research, we propose a human-computer interaction method based on U-Net convolutional neural network for observing the target molecules. The proposed method enables understanding of the researchers' intentions through simple human-computer interaction, and thus fulfills the task of detecting the molecules belonging to the same class, such as cells and proteins. The proposed method is trained with cell nuclear images and tested on multiple datasets of molecular images. The experimental results show that the proposed method performs well in both trained and untrained datasets (human protein and bacteria datasets) and is a general way to observe microscopic molecules in the life science field.

---

[1] Corresponding author E-mail address: zhangxf@yzu.edu.cn (Xinfeng Zhang)

2. SIMILAR TARGET DETECTION METHOD BASED ON HUMAN-COMPUTER INTERACTION

The human-computer interaction method realized in this paper first selects the interested targets in the image by manual interaction, then extracts the image features by using U-Net neural network and attention-based U-Net neural network, matches the interested parts globally, and finds out the targets whose similarity meets the requirements, which are considered to be similar targets.

2.1 U-Net Convolutional Neural Network

U-Net network is a semantic segmentation network based on full convolution neural network. The network structure is shown in the Figure 1. It consists of an encoder part (downsampling stage) and a decoder part (upsampling stage). The encoder part follows the typical convolution network structure. It includes the repeated application of two 3x3 convolutions. Each convolution is followed by a ReLU activation function and a 2x2 maximum pooling operation for downsampling. In each downsampling, the number of feature channels is doubled and the size of the feature graph becomes half of the original. The decoder part includes feature upsampling, and then 2x2 convolution to halve the number of feature channels. In the last layer, 1x1 convolution is used to map the eigenvector of each 64 component to the required number of classes. In addition, U-Net uses skip connection to fuse the location information of the underlying information with the semantic information of deep features, so as to reduce the loss of image details.

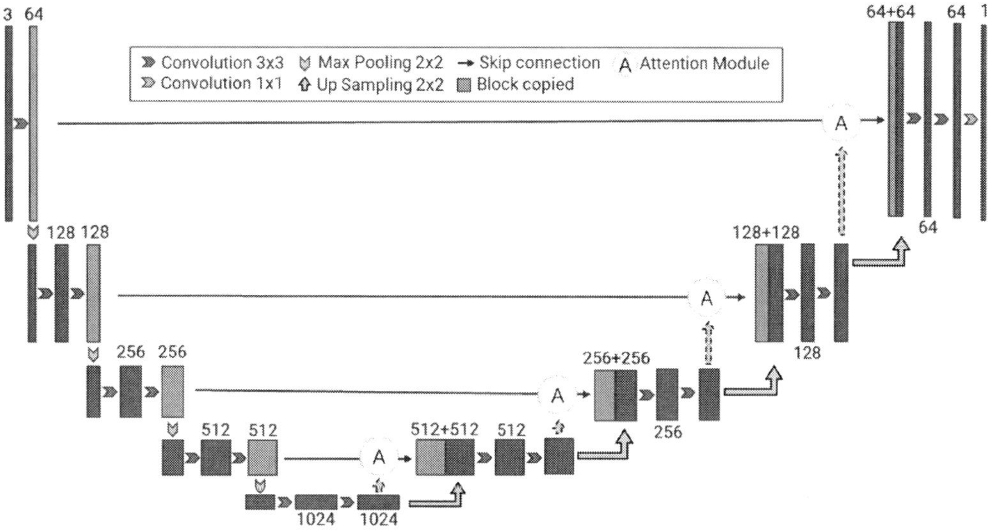

Figure 1. U-Net network structure

2.2 Attention Gate

In the classical U-Net, the features extracted from the coding layer are directly spliced to the corresponding decoding layer by means of skip-connection, so as to retain more detailed information. However, such a simple splicing not only brings more details, but also brings a lot of redundant information of low-level features (Fang et al., 2022). Therefore, this paper introduces the attention mechanism to process the low-level features, enhance the target region, inhibit the activation of irrelevant regions, and avoid the impact of redundant parts, so that the feature layer can get better results. The specific structure is shown in Figure 2.

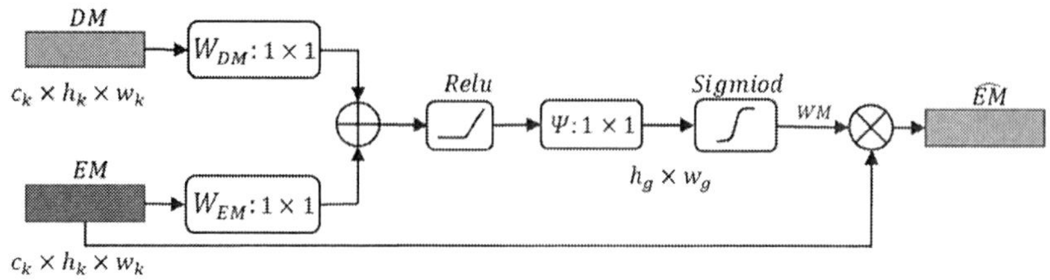

Figure 2. Attention Gate Structure

2.3 Loss Function

The training goal of this paper is image segmentation, which can be regarded as a binary classification task. The binary cross entropy can be simply used as the loss function of the network. The formula is as follows.

$$BCELoss = -\frac{1}{N}\sum_{i=1}^{N}\left(y_i log\widehat{y_i} + (1-y_i)log(1-\widehat{y_i})\right) \tag{1}$$

where, y_i is binary tag 0 or 1, and $\widehat{y_i}$ is the probability that the output belongs to tag y_i.

2.4 Detection of Similar Target Molecules

We detect multiple similar molecules in a picture by the following methods. In fact, different networks can complete the extraction of image features. Here, UNET is taken as an example.

Firstly, put the image into the trained model, use the U-Net model to conduct downsampling, upsampling, skip-connection and other operations, and take the depth feature of a decoding layer as the feature layer to be used. The target of interest (TOI) is selected in the sample box by human-computer interaction, and the coordinates of the position are mapped to the corresponding position of the feature layer to obtain the target features. Then, each region of the feature layer is scanned by sliding window, and the similarity with the target feature is calculated. If the similarity is greater than the set threshold θ, It is considered that the region exists in targets of the same kind as the target of interest; on the contrary, it is considered that there is no similar goal.

After these operations, we will get a number of areas that meet the requirements, but to ensure the accuracy of the results, the sliding window should not move too long each time, which will result in multiple stacked windows around a target. In this paper, we used Non-Maximum Suppression (NMS)(Rosenfeld and Thurston, 1971) to select the most similar and accurate target-containing boxes from the stacked borders of a target as the final exact boxes.

The specific steps of the method are as follows:

1) Set a resolution of $n \times n$ the pixel matrix of the picture is: I. Manually frame the object of interest, and the intercepted pixel block is IB_t. The coordinates on the image are recorded as (i, j, w_0, h_0), where i, j are the coordinates of the upper left corner of IB_t when the upper left corner of the original image is the coordinate origin; w_0, h_0 indicates the width and height of IB_t respectively.

2) Image P is processed by neural network to extract the feature layer F(the size of $C \times m \times m$), where C is the number of layers of the feature layer and m is the width and height of the feature layer.

3) According to the size of the TOI manually selected, the width and height of the sliding window on the feature layer F are w_F and h_F, which can be obtained according to the proportional relationship, $w_F = \frac{w_0}{n} \times m$, $h_F = \frac{h_0}{n} \times m$. Next, take the feature block FB on the feature layer corresponding to TOI (coordinates are (a, b, w_F, h_F), where $a = \frac{i}{n} \times m$, $b = \frac{j}{n} \times m$ is the coordinate of the upper left corner of the target feature block), and the size is $C \times W \times H$. It as a matching target feature.

4) Use the sliding window algorithm to intercept the feature block of size $(C \times W \times H)$ in the feature layer F and set it as FB_k, where k is the number of moves. The similarity of FB_k and FB is calculated using the cosine similarity method. If the similarity is greater than the threshold θ, it is retained in the set U, and the similarity of each window is recorded together. The window then slides to the next position, repeating the above matching until the sliding window traverses the entire feature layer.

5) Map the feature windows stored in the set U back to the original image proportionally, and get their position in the original image as a candidate box. The overlapping candidate boxes are screened using the NMS method, and after multiple iterations, each target corresponds to a single candidate box, and finally all targets of the same kind as to the TOI are obtained.

3. DATA ACQUISITION AND PROCESSING

There are three types of data sets used in this paper: Nucleus[1] , HPA[2], and Bacteria[3]. The details of the dataset are as follows:

Table 1. Information of datasets

Name	Format
Nuclei	The train set has 670 pictures, the test set has 65 pictures.
HPA	More than 120,000 pictures.
Bacteria	100 pictures.

As the only training set of the proposed method, the training set of nucleus data set contains 670 original pictures and segmented images of each nucleus in the picture, with several to dozens of nuclei in each. For the original image, it is divided into gray image and color image. Each picture of the training set corresponds to multiple masks, that is, there will be multiple nuclei in one picture.

Adjust the resolution of the original picture and mask to 265 × 256 pixels, which can improve the reasoning speed of the model while ensuring the effect of feature extraction. In practice, an image has multiple targets to be detected, so multiple masks of the same nucleus in the training set are combined to meet the requirements of task objectives. In addition, in the acquisition process of biological data sets, different substances are more prominent through dyeing, which makes different substances have the same dyeing color or the same substance has different dyeing colors. In order to avoid this situation, we treat the original image as a three-channel gray image. The training set and verification set are divided according to the ratio of 8:2.

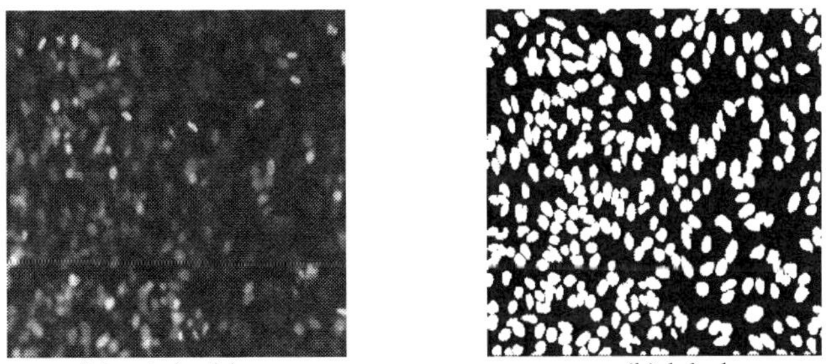

(a) original image (b) label

Figure 3. Nuclear samples in the training set

4. EXPERIMENT

In this paper, we randomly select 50 images from the test set of the three types of data sets as test images, and then manually selects a single target with obvious features on each image for frame selection as TOI. The purpose of deliberately selecting a target with obvious features is to consider that in the actual human-computer interaction process, users often want to obtain the best possible results, and selecting targets with obvious features is beneficial to improve the accuracy of recognition. Each network processes the test images and uses the method proposed in this paper to find other objects of the same type as the TOI.

[1] https://www.kaggle.com/c/data-science-bowl-2018/

[2] https://www.kaggle.com/c/human-protein-atlas-image-classification

[3] https://www.kaggle.com/longnguyen2306/bacteria-detection-with-darkfield-microscopy

4.1 Experimental Setup

1) Comparing network models: In order to verify the effectiveness of the U-Net network model, FCN32s(Long et al., 2015), CE-Net(Gu et al., 2019) is selected in this paper, and the attention mechanism is added to U-Net as the Attention U-Net, and the effects of the four are compared.

2) Training Settings: The network models used in this paper are based on the PyTorch implementation, the training initial dataset has a total of 670 pictures, and the training set and validation set are divided into 8:2 ratios. Select the Adam algorithm for end-to-end training, and the learning rate is set to 0.0001. The loss function of the training set uses the binary cross-entropy function. Batch-size is set to 8, the default number of training rounds is 50, and the MIoU of the validation set during the validation process is used as the indicator, and the model with the maximum value is retained as the final model.

3) Test Evaluation Metrics: The recognition frame obtained by the algorithm is compared with the real result to judge whether the recognition is correct. The results are mainly TP(true positive), FP(false positive) and FN(false negative)(Yang et al., 2021). The average index in the test adopts the most commonly used three quantitative evaluation indicators of $Precision$, $Recall$ and $F1$ value as the evaluation indicators of target detection accuracy.

4.2 Results

4.2.1 Precision Comparison Results

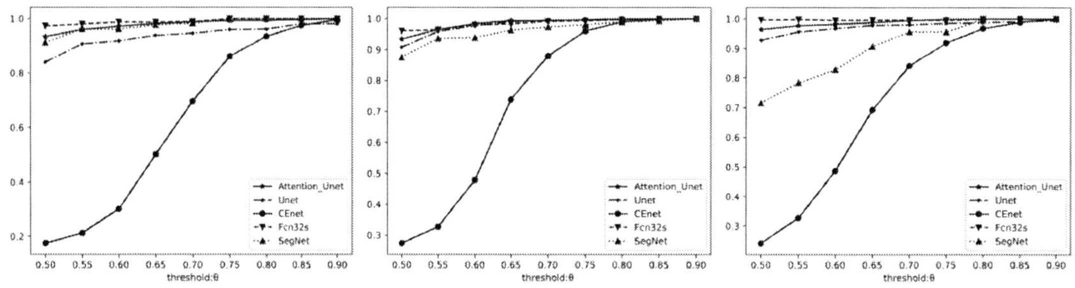

(a)The nuclear dataset (b)The HPA dataset (c)The bacteria dataset

Figure 4. The precision of each network detection on the dataset

It can be seen from Figure 4 that with the increase of the similarity threshold, the accuracy of each network on the three data sets is continuously improved and approaches 1. This is because after increasing the similarity threshold, only the targets with higher matching degree can be selected, which means that the algorithm is more reliable in picking out the targets in the target frame. The networks performed similarly in the three datasets (nucleus, HPA, bacteria), with FCN32s achieving the highest accuracy, followed by Attention U-Net. CE-Net performs the worst and only achieves better results at higher thresholds. U-Net and Attention U-Net are mostly above the curve, and the performance is almost the same as FCN32s at higher thresholds.

4.2.2 Recall Comparison Results

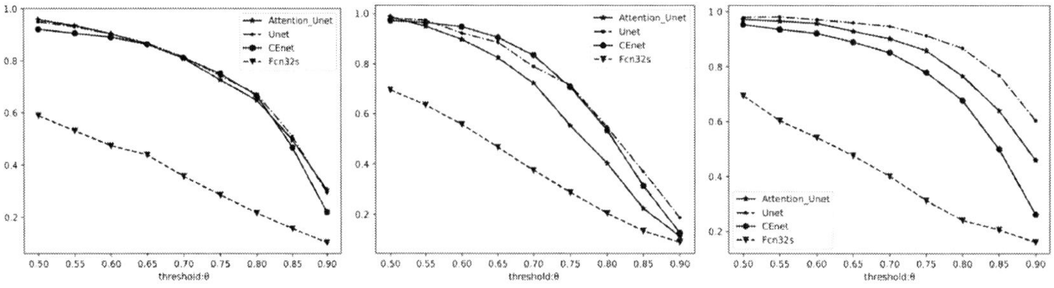

(a)The nuclear dataset (b)The HPA dataset (c)The bacteria dataset

Figure 5. The recall of each network detection on the dataset

It can be seen from Figure 5 that with the increase of the threshold, the recall rates of U-Net, Attention U-Net, and FCN32s on the four datasets all show a downward trend. U-Net performs the best in all three datasets, which shows that the network can capture the essential characteristics of the target. Attention U-Net performs the same as U-Net on the nucleus dataset, and is lower than U-Net on HPA and bacteria datasets. FCN32s has the worst performance in the three datasets, which means that the network does not learn the key features of the target well, and selects many wrong targets.

4.2.3 F1 Value Comparison Results

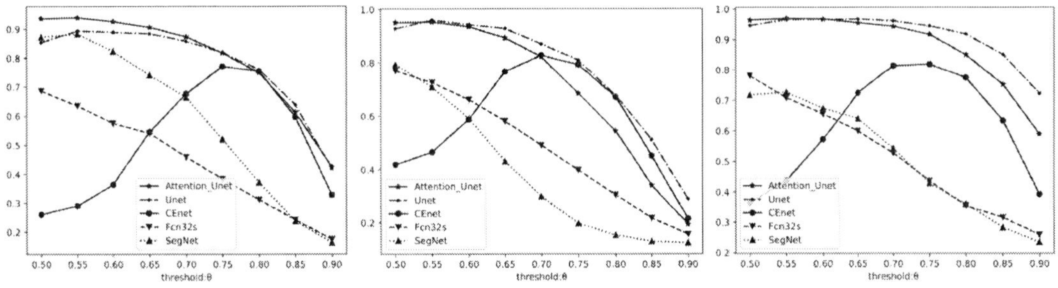

(a)The nuclear dataset (b)The HPA dataset (c)The bacteria dataset

Figure 6. The F1 value of each network detection on the dataset

It can be seen from Figure 6 that with the increase of the threshold, the F1 values of all networks except CE-Net show a downward trend. The F1 value is used to measure the degree of fit between precision and recall. The closer the two values are to the F1 value, the higher the value. In Figure 4, the precision of each network is increasing, while the recall except CE-Net is decreasing in Figure 5, which makes the value difference between the two increase and the value of F1 decrease.

For CE-Net, its proposed Dense Atrous Convolution (DAC) module can capture wider and deeper semantic features by injecting four cascaded branches with multi-scale atrous convolution, it can It performs very well in segmentation tasks. However, it is precisely because the extraction of target features is too fine that only targets with high similarity can be found. At low similarity, a large number of similar targets are lost, so the values of each item are relatively low.

U-Net performed the best, with the most balanced performance in terms of precision and recall, followed by Attention U-Net. The F1 value of FCN32s performs the worst among all datasets because it selects targets through a large number of selections, which have a large number of wrong targets, resulting in high precision, but low recall, resulting in the lowest F1 value.

4.2.4 Detailed Data of Experimental Results

The specific data of the experiment when the similarity threshold θ is set to 0.5 is shown in Table 2, the average performance of each neural network on all data sets is shown in Table 3. The data performance under other thresholds is roughly the same.

In Table 2, the best results are mostly achieved by Attention U-Net, which outperforms U-Net. Although the precision of FCN32s is highest, other metrics are lower than Attention U-Net and U-Net.

In Table 3, although Fcn32s has the highest precision of 97.7% on average, its recall is much lower than other networks, only 65.9%; the average F1 value is 0.746, which is only higher than CE-Net. The average precision of U-Net is 89.2%, the average recall is reaching 92.1%, and the average F1 value is 0.910, ranking second. The average precision of Attention U-Net is 94.4%, second only to FCN32s; the average recall is 97.2%, the average F1 value is 0.950, and the performance is the highest. The average precision of CE-Net is only 23.0%, which is much smaller than other networks, and the performance is the worst; the average recall rate is 94.9%, ranking third; the average F1 value is 0.347, which is also much smaller than other networks.

In summary, the performance of U-Net and Attention-U-Net networks is generally higher than that of other networks. The advantages of their common U-shaped structure can accurately extract the context information of the picture and retain the image by means of skip connection details; Attention U-Net is higher than U-Net in all indicators, indicating that adding attention module has improved this task, and it is more stable for the extraction of different types of image features. In addition, the model has not been specifically trained for protein and bacterial data sets, but it still has good results and its performance is higher than the rest of the networks, indicating that the U-shaped network structure is suitable for this task, and the model has high generalization ability. It can better achieve the same kind of target detection work in the untrained dataset.

Table 2. Experimental data results on each data sets

datasets	model	Precision / %	Recall / %	F1 value
Nucleus	Attention U-Net	93.4	**95.8**	**0.936**
	U-Net	84.1	95.0	0.854
	CE-Net	21.1	93.6	0.291
	Fcn32s	**97.3**	58.9	0.686
HPA	Attention U-Net	93.4	**98.7**	**0.950**
	U-Net	90.8	98.2	0.927
	CE-Net	26.8	97.8	0.410
	Fcn32s	**96.1**	69.5	0.770
Bacteria	Attention U-Net	96.4	97.1	**0.964**
	U-Net	92.8	**97.9**	0.946
	CE-Net	25.7	95.2	0.379
	Fcn32s	**99.6**	69.3	0.781

Table 3. The average value of specific experimental data on each data sets

model	Average Precision / %	Average Recall / %	Average F1value
Attention U-Net	94.4	**97.2**	**0.950**
U-Net	89.2	97.0	0.910
CE-Net	23.0	94.9	0.347
Fcn32s	**97.7**	65.9	0.746

5. CONCLUSION

In the detection of microscopic substances in living bodies, the use of deep learning methods to replace manual judgment has many advantages such as low cost, wide application range, high accuracy, and high efficiency. This paper uses deep learning technology to propose a detection method for microscopic substances based on human-computer interaction, and finds similar substances by matching the image feature layer. The conclusions are as follows:

1) Using the image depth feature layer for target detection, the proposed method does not require the label training set of specific substances, which greatly reduces the cost of deep learning work; it does not require training for specific targets in advance, and can be widely used in All kinds of substances or cells can still show good results for the observation of unknown substances;

2) In the experiment, the method proposed in this paper uses Attention U-Net as the feature extraction network (when the threshold is 0.5), the average precious on the dataset can reach 94.4%, the average recall can reach 97.2%, and the F1 value can reach 97.2%. 0.950, which can well meet the detection needs of life process changes in microscopic substances such as substances or cells in life science research.

In future work, we will study how to observe the motion or dynamic changes of microscopic molecules in real time throughout life activities to obtain more effective information.

ACKNOWLEDGEMENT

This work is supported by the National Natural Science Foundation of China (Grant No. 61801417 and 61802336), the Natural Science Foundation of the Jiangsu Higher Education Institutions of China (No. 18KJB520051) and Jiangsu Students' Innovation and Entrepreneurship Training Program (No. 202111117056Y).

REFERENCES

Fang, J., Zhang, X., Yang, B., Chen, S. & Li, B. An Attentionbased U-Net Network for Anomaly Detection in Crowded Scenes. presented at the 14th International Conference on Computer Research and Development (ICCRD 2022), 2022 Shenzhen, China.

Gu, Z., Cheng, J., Fu, H., Zhou, K., Hao, H., Zhao, Y., Zhang, T., Gao, S. & Liu, J. 2019. Ce-net: Context encoder network for 2d medical image segmentation. *IEEE transactions on medical imaging,* 38, 2281-2292.

Long, J., Shelhamer, E. & Darrell, T. Fully convolutional networks for semantic segmentation. Proceedings of the IEEE conference on computer vision and pattern recognition, 2015. 3431-3440.

Mnih, V., Heess, N. & Graves, A. 2014. Recurrent models of visual attention. *Advances in neural information processing systems,* 27.

Oktay, O., Schlemper, J., Folgoc, L. L., Lee, M., Heinrich, M., Misawa, K., Mori, K., Mcdonagh, S., Hammerla, N. Y. & Kainz, B. 2018. Attention u-net: Learning where to look for the pancreas. *arXiv preprint arXiv:1804.03999.*

Ronneberger, O., Fischer, P. & Brox, T. U-net: Convolutional networks for biomedical image segmentation. International Conference on Medical image computing and computer-assisted intervention, 2015. Springer, 234-241.

Rosenfeld, A. & Thurston, M. 1971. Edge and curve detection for visual scene analysis. *IEEE Transactions on computers,* 100, 562-569.

Yang, Y., Bingyan, S. & Yuqi, S. 2021. Research on anti-shadow tree detection method based on generative adversarial network. *Transactions of the Chinese Society of Agricultural Engineering,* 37, 9.

SHORT PAPERS

TESTING PUBLIC WARNING SYSTEM AT SCHOOL WITH USER INVOLVEMENT - CASE STUDY FROM A RURAL COMMUNITY

Anna Maria Urbaniak-Brekke[1], Øyvind Heimset Larsen and Ivar Petter Grøtte

Western Norway Research Institute, Røyrgata 4, 6856 Sogndal, Norway
[1]PhD

ABSTRACT

New, DAB public warning system is to be introduced in Norway, but firstly, it needs testing. The 'Selje school case' has a goal to evaluate this technology, with help of user involvement method applied to students in a small rural community of Western Norway. Results show the importance of such testing, both in terms of building emergency awareness and preparedness in the society, and for improvement of the system. This case is a part of a national research and innovation project supported by the Research Council of Norway.

KEYWORDS

User Involvement, Evaluation, Public Warning Systems, Norway

1. INTRODUCTION

The main task of public warning technologies is to inform people about danger, make evacuation easier, and guide them to a safer place (Bean and Botterell, 2019; Toyoda and Kanegae, 2014). An increasing number of new technologies is being developed, having the potential to improve existing warning systems. One of them is an innovative, DAB-based system developed by a Norwegian company called Paneda, together with its partner-companies of regional and national range. Digital Audio Broadcasting (DAB) is designed for delivery of high-quality digital audio programs and data services for mobile, portable and fixed reception from terrestrial transmitters in the Very High Frequency (VHF) frequency bands (WorldDAB, n.d.). And furthermore, using DAB makes it possible to 'take over' audio equipment, interrupt other broadcasting, and send a message or trigger specialised equipment (visual, sound, text, etc.) that will warn people about a threat (e.g., natural disasters or terror attack). However, this new technology needs structured user testing and evaluation before it will be introduced to a broader market. That is why a research project called Public Warning System (PWS) has been introduced, in order to test this new system's ability to inform different societal groups about dangerous events. Western Norway Research Institute has the responsibility of conducting tests and evaluation that will assure the usability of the new technology (Vestlandsforsking, 2020).

This paper's main goal is to present the results of applying a user involvement method to test and evaluate the new warning system, in the first case of the project, being a rural school premises in Selje, Western Norway. The questions to be answered are: 1. Is the proposed testing method suitable for assessing technology development involving a local user group of young pupils? 2. Which parts of this new technology need improvement?

2. BODY OF PAPER

2.1 PWS Project

"Public Warning System" (PWS) is a three-year national-range project supported by the Research Council of Norway, introducing and testing DAB-based warning technology for different societal groups and events (Urbaniak-Brekke et al., 2021). The need for new warning technology has been clearly demonstrated through many acts of terrorism, extreme weather, and environmental disasters. The European Union sets requirements for population alerts shared by Norway due to the EEA Agreement (European Union, 2018). According to them, "by June 2022, the European Electronic Communications Code (EECC) Article 110 requires all EU countries to operate a public warning system that can send geo-targeted emergency alerts to all mobile phone users located in the affected area during a natural or man-made disaster" ("Get ready for European regulations on public warning EECC Article 110," 2022). Existing systems, based mainly on alarm sirens and SMS warning, have several problems: the mobile network is often out of service or overloaded when it is needed the most, e.g. during extreme weather with power outages or large crowds. Another problem is how to reach people who do not listen to the radio, do not have their mobile phone with them or do not use it at all. In addition, there is a significant problem with distance/time-factor before the emergency services arrive, and how to evacuate large crowds immediately. The PWS project addresses these needs and develops and tests a new public warning system.

2.2 Selje Case

The test was carried out in June 2021 in Selje, a typical, rural environment in Norway. Its goal was the evaluation of the proposed DAB-warning system by school pupils. This exercise was conducted at the premises of Selje school, gathering pupils from the 1st to 10th grade in the Norwegian school system (age 6 to 15). The idea was to test the warning system at the school, using DAB signal to turn on specific equipment giving light, sound and picture/text effects in order to warn about an emergency (gas leakage) and support an evacuation action. The details about the idea and conduction of the test were described in an earlier paper (Urbaniak-Brekke et al., 2021).

2.3 Method

In this research project, we use an action research approach, where researchers are participants in the development and their conclusions are grounded in this action research (Greenwood and Levin, 2007). Case study research is a key element as "the study of the particularity and complexity of a single case, coming to understand its activity within important circumstances" (Stake, 1995, p. xi). Our first study is a single case where we look into a specific local community. According to Yin (1981, pp. 98–99) the strength of the case study is that it both covers a contemporary phenomenon and its context. Our findings are from a single case, but we claim that these findings also can be a "force of example" (Flyvbjerg, 2006) for other communities of the same nature (rural, school environments) facing similar challenges related to emergencies and population evacuation. These preliminary results will be further developed with new similar tests medio 2022.

2.3.1 User Involvement

User involvement methods are well-known in research and evaluation, and quite often applied (Beresford, 2002; Kushniruk and Nøhr, 2016; Thornicroft and Tansella, 2005). Both the advantages and disadvantages of this kind of evaluation methodology have been discussed broadly. However, counter views have so far tended either to be focused on perceived deficiencies in the methods and methodologies employed or not been clearly or publicly articulated (Beresford, 2002, p. 95). Nevertheless, user involvement has a significant potential impact on the system's success (Amoako-Gyampah and White, 1993). An essential element of the user involvement method is that the public (e.g. customers or patients) are not only participating in the research but that this research is done 'with' dem. This means that the users play a role in decision making during the research, and that their input influences the outcomes to a large extent (Auckland, 2010). This in turn contributes to the results being more suitable for the public and has a positive influence on their satisfaction from the offered services (Amoako-Gyampah and White, 1993).

2.3.2 Questionnaire

After completing the evacuation pilot at Selje school, students of the 9th grade distributed a questionnaire among their colleagues from 1st to 10th grade who participated in the test. The age of the pupils varies from 6 to 15 putting them on a different level of IT-skills and understanding of emergency situations. Selje school have students from different backgrounds, mostly Norwegians, but also children of work immigrants from Eastern Europe and refugees from the Middle East. The variety of the respondents' age, skills and background gave the test the opportunity to analyses trends that can occur in similar environments. The 9th grade students from Selje school were responsible for the local test implementation before, during and after the evacuation exercise. Firstly, supported by Paneda, they prepared and tested the functioning of the necessary equipment, then helped placing it in strategic points in and around the school premises, making the evacuation possible. Finally, they launched the evacuation exercise and, after colleagues had arrived at 'a safe place', distributed the survey among them. The questionnaire is built on 14 questions, and in 9 of them a simplified Likert scale with only 3 points (good, medium, bad) is used, in order to make it easy to understand for children. There is however a comment field left for each question, making it possible to leave a comment or suggestion for changes. Those questions concern the quality of placement and functioning of the equipment used for warning (e.g. screens, loudspeakers, lights). The other 5 questions focus on the general experience and ideas for improvement giving the respondents an open comment field. Community involvement plays a significant role in the real evacuation events (Ridzuan et al., 2017), and that is why engaging students and teachers in rehearsing an emergency scenario is very important.

2.4 Results

The case included two days of testing. On the first day, June 14, a technical test of equipment was carried out. It took place after the equipment was installed in various places in the school building, but before the user testing began. All elements of the equipment (i.e., 4 screens showing pictures and text, computer speakers, and a siren) were tested individually to check if they worked as planned. It must be emphasized that this was only one of the first tests of this technology with user involvement, and the number of warning elements (screens, typhoons, speakers) was not adapted to the size of the building. This caused e.g. that those who entered the main entrance of the school did not have the opportunity to see a screen, but only heard the alarm and information through typhoons. Also, there was no signal light tested, which normally is a part of the equipment, and will be implemented in the next test planned for this and next year.

The evacuation exercise was carried out in leisure time between classes. School employees were informed in advance, and it was considered that Selje school has students who do not speak Norwegian. Therefore, the communication through speakers and the text on the screens was prepared in three languages: Norwegian, English and Arabic. The information was also given in advance that a siren on the sports field would be used, so that the sound from it wouldn't frighten the youngest children unnecessarily.

Below (table 1) we present the evaluation results where the listed questions were asked to those who participated (were evacuated). We collected a total of 56 responses. From grades 1st, 2nd and 3rd we only received one scheme per class, since it was difficult for the youngest students to go through it alone. For the older children (4th, 8th and 10th grade) we collected one form per student. Grades 5, 6 and 7 were not at school that day. The number of responses also includes 4 forms filled out by students from 9th grade who joined an evacuated group and conducted a participatory observation, and 10 representatives of school staff who were invited to participate in the exercise. Numbers mean the number of answers to the various questions.

Table 1. User experience of the PWS technology at Selje school

The equipment	**Good**	**Medium**	**Bad**
Screens			
Could you read the text?	**36**	8	8
Did you see the picture?	**35**	9	4
Did you hear the audio message?	16	13	**23**
Did you see the arrow?	10	7	**28**
Did the screen information help you in the right direction?	14	**20**	15
Sound			
Did you hear the alarm and message?	16	14	**23**
Did the alarm and message make you go straight to the meeting place?	**24**	13	15
All components			
Did you see from one screen to the next?	5	**20**	14
Did you find the meeting place?	**40**	4	3
Question	**Comment / idea / presentation**		
Did you discuss the situation with others during the evacuation?	Most say 'yes'		
Did you go as a group or separately?	Most went as a group		
Do you have other ideas that can make this better or provide more information?	too low sound; another arrow on the screens; several screens; message twice in the same language		
How long did it take for you from the time the alarm went off until you were at the meeting place?	0,5 – 4 min		
Other input?	the sound is not clear with a hearing aid; the skull received the attention of the youngest; a lot of text; difficult to read for the youngest students and 'when you hurry past; wind 'took' the sound		

3. CONCLUSION

It is important to predict and avoid dangerous situations, so that there is as little need as possible for the evacuation. As Rød et al argues, the municipalities should use the historical natural damage data combined with an analysis of robustness in order to try to predict where and when the threats can occur (Rød et al., 2019). However, emergencies occur, especially when it comes to natural disasters in some areas (e.g. the risk of avalanches and tsunami waves in parts of Norway), and the warning will then have an essential role in evacuating and securing life of citizens. The national warning system in Norway today, based mainly on alarm sirens and mobile phone networks, should be supported by the innovative DAB equipment, making it more robust and resilient to potential threats setting human health and life in danger. That is in line with the EU Directive mentioned above (EECC) effective from December 2020, which requires that all domestic and in-vehicle DAB+ receivers in Europe must include DAB+ (WorldDAB, n.d.).

The questions we wish to discuss are, as mentioned in the introduction: 1. Is the proposed testing method suitable for assessing technology development involving a local user group of young pupils? 2. Which parts of this new technology need improvement?

The first question addresses the user involvement method applied in the evaluation process in the Selje case. It has many obvious advantages, as we know that the ultimate success of a technology means that it is understood and applied by the users (Amoako-Gyampah and White, 1993). Especially in the case of public warning systems, it is essential that the technology is adapted to the needs and perception of the potential evacuees, so that when a threat occurs, they react correctly. This is why structured testing of a technology (a pilot) should be performed in the situation that the solution is made for, but on a smaller scale (Eide and Ljunggren, 2018). In the case presented in this paper, the school in Selje represented a part of the infrastructure, which could also be any other venue both private and public. As Eide and Ljunggren further emphasize, it is important that the test involves potential or real customers, in this case residents of a community, that may in the future be exposed to a hazardous situation. The aim of such a pilot is receiving constructive feedback from users and others involved to try out the design, identify the need for redesign (improvement) or to consider stopping of further process (Eide and Ljunggren, 2018). It is however important to not be uncritical of the method and not base all the evaluation only on it (Beresford, 2002). The role of 'Selje case' was to take a first step in introducing an innovative technology to a broader public, asking them to evaluate it, and this task has been accomplished.

Despite the positive evaluation of the system as a whole, it turns out that for the youngest pupils, pictures are most important since not all of them can read. At the same time, graphics and symbols are important for all evacuees, because in a stressful situation people tend not to stand in front of a screen to read carefully, but preferably just move towards a safe place, only noticing simple icons or signs. The text must therefore be short, informative, supplied with commonly known graphics, and delivered in the languages the evacuated community is familiar with. What is more, the sound turns out to be more important than pictures and text, and it is therefore essential that the alarm can be heard well by everyone throughout the evacuated area (regardless of e.g. weather conditions, wind, etc.), and that the messages coming from speakers are short and in several languages. Interestingly, it turns out to be important that information is repeated twice in the same language. This confirms the findings from a previous test in Flåm, where audio messages also proved to be more important than text on the screens and text messages (Paneda, 2019).

The final conclusion is that planning, implementation, and evaluation of the 'Selje case' provided a lot of valuable data that forms the basis for the development of the technology, and future tests of warning systems. Schools seem to be a relevant arena for testing public alerts, as evacuation exercises can be disseminated to large parts of society that are often linked to schools, and residents build awareness around the possible need for evacuation early in life.

Furthermore, the success of this study has an actual impact on the progress of the introduction of the technology. In March 2022 a workshop was organised, gathering national and regional authorities, telecom providers, and other legislators and actors. During that meeting, the method was presented, and new field tests in the rural area of Western Norway were approved; in the largest shopping mall in the region, in a large factory and in an elderly care facility as well as another local school. Those coming tests will follow the presented method and will be adjusted according to the lessons learned.

ACKNOWLEDGEMENT

The authors have no conflicts of interest.

REFERENCES

Amoako-Gyampah, K., White, K.B., 1993. User involvement and user satisfaction. Inf. Manage. 25, 1–10. https://doi.org/10.1016/0378-7206(93)90021-K

Auckland, S. (Ed.), 2010. Involving users in the research process. A 'how to' guide for researchers.

Bean, H., Botterell, A., 2019. Mobile technology and the transformation of public alert and warning, First edition. ed, Praeger security international. Praeger Security International, Santa Barbara, California.

Beresford, P., 2002. User Involvement in Research and Evaluation: Liberation or Regulation? Soc. Policy Soc. 1, 95–105. https://doi.org/10.1017/S1474746402000222

Eide, D., Ljunggren, E., 2018. Testing som metode i innovasjonsprosesser. Et verktøyhefte med eksempler fra opplevelsesbasert reiseliv.

European Union, 2018. Directive (EU) 2018/1972 of 11 December 2018 establishing the European Electronic Communications Code.

Flyvbjerg, B., 2006. Five Misunderstandings About Case-Study Research. Qual. Inq. 12, 219–245. https://doi.org/10.1177/1077800405284363

Get ready for European regulations on public warning EECC Article 110 [WWW Document], 2022. . Ever Bridge. URL https://www.everbridge.com/products/public-warning/eecc/

Greenwood, D., Levin, M., 2007. Introduction to Action Research. SAGE Publications, Inc., 2455 Teller Road, Thousand Oaks California 91320 United States of America. https://doi.org/10.4135/9781412984614

Kushniruk, A.W., Nøhr, C., 2016. Participatory Design, User Involvement and Health IT Evaluation. Stud Health Technol Inform 222, 139–151. https://doi.org/0.3233/978-1-61499-635-4-139

Paneda, 2019. Paneda shows Emergency Warning system over DAB in Norway [WWW Document]. URL https://www.paneda.no/news/2019/12/05/emergency-warning-live-demo (accessed 12.5.19).

Ridzuan, A.A., Ungku Zahar, U.A., Mohd Noor, N.A., 2017. Association of Evacuation Dimensions Towards Risk Perception of the Malaysian Students Who Studied at Jakarta, Medan and Acheh in Indonesia. Malays. J. Geosci. 1, 7–12. https://doi.org/10.26480/mjg.01.2017.07.12

Rød, J.K., Opach, T., Scherzer, S., Setten, G., 2019. Er norske kommuner klare for en farligere fremtid? Plan 51, 6–11. https://doi.org/10.18261/ISSN1504-3045-2019-04-03

Stake, R.E., 1995. The art of case study research. Sage Publications, Thousand Oaks.

Thornicroft, G., Tansella, M., 2005. Growing recognition of the importance of service user involvement in mental health service planning and evaluation. Epidemiol. Psichiatr. Soc. 14, 1–3. https://doi.org/10.1017/S1121189X00001858

Toyoda, Y., Kanegae, H., 2014. A community evacuation planning model against urban earthquakes: A community evacuation planning model against urban earthquakes. Reg. Sci. Policy Pract. 6, 231–249. https://doi.org/10.1111/rsp3.12036

Urbaniak-Brekke, A.M., Grøtte, I.P., Larsen, Ø.H., 2021. A test case from Norwegian rural areas where structures user participation builds safer communitites. Multi Conference On Computer Science and Information Systems.

Vestlandsforsking, 2020. Public Warning System (PWS) [WWW Document]. URL https://www.vestforsk.no/en/project/public-warning-system-pws (accessed 3.18.22).

WorldDAB, n.d. DAB+ emergency broadcasting and warnings [WWW Document]. URL https://www.worlddab.org/dab/emergency-warning (accessed 3.18.22).

Yin, R.K., 1981. The Case Study as a Serious Research Strategy. Knowledge 3, 97–114. https://doi.org/10.1177/107554708100300106

AUTONOMY AND AUTOMATION: THE CASE OF CONNECTED AND AUTOMATED VEHICLES

Fabio Fossa
Department of Mechanical Engineering, Politecnico di Milano
Via Privata Giuseppe la Masa 1, 20156 Milan (MI), Italy

ABSTRACT

This short paper offers a preliminary inquiry into the impacts of driving automation on personal autonomy. Personal autonomy is a key ethical value in western culture, and one that buttresses fundamental components of the moral life such as the exercise of responsible behaviour and the full enjoyment of human dignity. Driving automation simultaneously enhances and constrains it in significant ways. Hence, its moral profile with reference to the value of personal autonomy is uncertain. Ethical analysis shows that such uncertainty is due not just to the complexity of the technology, but also to the multifaceted normative profile of personal autonomy, which offers reasons to support both conditional and full driving automation. The paper sheds light on this duplicity, underlines the challenges this poses to the ethics of driving automation, and advocates for further research aimed at providing practitioners with more fine-grained guidelines on such a delicate issue.

KEYWORDS

Connected and Automated Vehicles, Automation, Personal Autonomy, Engineering Ethics, Ethics of Technology

1. INTRODUCTION

The aim of this short paper is to shed light on how the ethical value of personal autonomy is impacted by driving automation. Its results, although preliminary, will hopefully contribute to raising awareness on the design and policy challenges that must be faced to effectively align future Connected and Automated Vehicles (CAVs) to such an important principle.

The quixotic relationship between personal autonomy and technological automation lies at the heart of several ethical quandaries across various AI-based applications (Laitinen & Sahlgren, 2021) such as, e.g., Autonomous Weapon Systems (Sharkey, 2019) and Recommender Systems (Varshney, 2020). Driving automation makes no exception (Chiodo, 2022). In this context, threats to and opportunities for personal autonomy are so numerous and deeply entangled with each other that much philosophical work is needed to clarify how personal autonomy is to be effectively pursued. The importance of such clarification should not be underestimated. Critically examining how widespread conceptions of personal autonomy apply to the case of CAVs is key to realise relevant ethical opportunities and risks. Effective design choices and policy decisions importantly depend on it.

The paper is structured as follows. Section 2 discusses how the ethical value of personal autonomy has been brought to bear on driving automation. Section 3 analyses the definition of personal autonomy proposed in the European report *Ethics of Connected and Automated Vehicles* (Horizon, 2020) and draws attention to its composite nature. Applied to the field of driving automation, personal autonomy is accordingly specified as self-determination of driving tasks and freedom to pursue a good life through mobility. As Section 3 shows, however, these two specifications turn out to support different driving automation models – respectively: conditional and full automation – thus leaving doubts on how to practically comply with the demands of personal autonomy. Based on these results, Section 4 claims that the principle of personal autonomy requires further ethical elucidation in order to inspire unambiguous design and policy choices.

2. PERSONAL AUTONOMY AND DRIVING AUTOMATION

The importance of aligning the design, deployment, and use of CAVs to ethical expectations concerning personal autonomy could hardly be belittled. Even though its philosophical status is controversial (Magnani, 2020), personal autonomy enjoys widespread sociopolitical recognition as a key ethical value. In essence, it characterises human beings as self-determining entities who, therefore, deserve respect and protection (Christman, 2020). As such, it buttresses fundamental components of the moral life such as the exercise of responsible behaviour and the full enjoyment of human dignity. Given its relevance, personal autonomy evidently qualifies as a value to be pursued through technological innovation as well.

The European approach to the ethics of driving automation confirms the latter statement. In 2020, an interdisciplinary group of fourteen experts appointed by the European Commission authored the report *Ethics of Connected and Automated Vehicles. Recommendations on Road Safety, Privacy, Fairness, Explainability and Responsibility* (Horizon, 2020). The document establishes an ethical framework for CAVs and offers concrete recommendations aimed at guiding stakeholders in the effort of aligning driving automation to relevant ethical values. In close connection with the European approach to trustworthy Artificial Intelligence (AIHLEG 2019), the report starts by identifying and describing the basic normative cornerstones of the framework. Acknowledging its relevance, the authors indicate personal autonomy as one of the eight overarching ethical principles for driving automation along with non-maleficence, beneficence, dignity, responsibility, justice, solidarity, and inclusive deliberation (Santoni De Sio, 2021).

According to the report, the principle of personal autonomy states that human beings are to be conceived as "free moral agents" (Horizon, 2020, p. 22) whose right to self-determination ought to be respected. In relation to autonomous driving, personal autonomy demands that CAVs are designed so to "protect and promote human beings' capacity to decide about their movements and, more generally, to set their own standards and ends for accommodating a variety of conceptions of a 'good life'" (Horizon, 2020, p. 22). As such, autonomy plays a crucial role in several recommendations, ranging from the protection of privacy rights and the promotion of user choice to reducing opacity and enhancing explainability.

From an ethical point of view, the insistence on protecting and promoting personal autonomy in the context of driving automation seems appropriate. Bypassing individual decision-making through technical means risks leading to situations where personal decisions are taken by actors (e.g., designers, engineers, manufacturers, policy-makers) who, however, have no right nor particular competence to do so. This state of affair is evidently incompatible with the individual right to self-determination on personal matters and should be carefully avoided when designing or deploying CAVs.

The protection of autonomy in driving automation is also critical to support other important moral values. Considering human beings as free moral agents by principle means, at the same time, considering them responsible agents as well, to the extent that they can exercise such freedom. This is a necessary presupposition to establishing who is responsible, and why, when harmful consequences follow from the use of CAVs (Nyholm, 2018).

The value of personal autonomy, then, is vital to the ethics of driving automation for many reasons. On the one hand, CAVs designed and deployed in ways that promote personal autonomy will meet demands grounded on the protection of human dignity, thus supporting social acceptance and trust. On the other hand, upholding personal autonomy is key to distributing responsibility in a clear and fair way while, at the same time, encouraging responsible behaviour. But how is personal autonomy to be pursued on a practical level?

3. ONE DEFINITION, TWO COMPONENTS

Whilst the ethical relevance of personal autonomy to driving automation is evident, it is difficult to specify how the principle is to be endorsed on a more tangible level. Driving automation, after all, consists in the delegation of driving tasks from human agents to digital systems. Arguably, constraints to personal autonomy are only to be expected. What needs to be further clarified, then, is how to automate driving functions without impacting too negatively on personal autonomy. This raises thorny practical questions. What aspects of human autonomy are relevant to driving automation? Which of them should be prioritised? What model of driving automation should be promoted through design and policy decisions? What guidelines should be offered to practitioners in this sense?

In order to answer these queries, a more fine-grained understanding is required of how personal autonomy is impacted by driving automation. In other words, it is necessary to identify which aspects of CAV users' experience qualify as expressions of their personal autonomy. These aspects, in turn, would serve as tangible constraints to driving automation: CAVs should be developed in ways that allow for their exercise. In sum, specifying how personal autonomy in driving automation actually looks like is a necessary step towards providing effective guidelines to stakeholders.

The definition of personal autonomy provided in the European report represents a good starting point to figure out more precisely what is at stake in this context. At a closer look, the definition exhibits two main components. They can be specified as (a) autonomy as self-determination of driving decisions; and (b) autonomy as freedom to pursue a good life through mobility.

(a) concerns the exercise of individual control over decisions that pertain to driving behaviour – i.e., to the ways in which the vehicle reaches its destination from its starting point. In the European report, this component is referred to when authors recommend to design, deploy, and use CAVs so to "protect and promote human beings' capacity to decide about their movements" (Horizon 2020, 22). In this sense, respecting CAV users' autonomy would mean to let them exercise some sort of control on the system operations that impact on their personal sphere.

(b), on the contrary, exhibits a wider scope. It refers to the freedom of pursuing happiness and to mobility as an important enabler of what makes life worth living. As stated in the European report, upholding personal autonomy also means to "protect and promote human beings' capacity to (...) set their own standards and ends for accommodating a variety of conceptions of a 'good life'" (Horizon, 2020, p. 22). In this sense, aligning driving automation to the principle of personal autonomy would mean to envision CAVs as means to support the individual pursuit of personal flourishing and well-being.

In what follows, the ethical challenges related to complying with these two forms of personal autonomy are outlined. It is shown that complying with (a) would pose significant obstacles to compliance with (b), and vice versa. The paradoxical outcomes of the analysis suggest that further ethical research is needed to understand how personal autonomy can be pursued consistently in the context of driving automation.

4. A CONFLICT OF AUTONOMIES

Let us start by considering how personal autonomy as in (a) could be promoted through driving automation. In this sense, human autonomy partakes in driving automation mostly as a threatened individual value that requires to be adequately safeguarded. Particular care is required since the exercise of personal autonomy is variously constrained by driving automation (Xu, 2021). The experience of driving is a complex one, composed by a myriad of decisions, some of which are personal decisions or might have a considerable impact on the moral sphere. The delegation of such decisions to automated systems poses the risk of bypassing human judgment in ethically problematic ways.

Threats to the exercise of personal autonomy might variously arise in the context of driving automation. At low levels of automation, a speed control system that could not be overridden by human intervention even in case of emergency might be considered as problematic with reference to personal autonomy (Schoonmaker, 2016). At the opposite extreme, suppose that full autonomous vehicles will be able to distribute harm during unavoidable collisions according to given ethical values. In this case, it might be problematic in terms of personal autonomy if said values were set not by passengers themselves, but rather by other stakeholders (Millar, 2016; Contissa *et al.*, 2017; Millar, 2017). Considering less futuristic scenarios involving high levels of automation, automated features concerning ethical driving behaviour – e.g., regarding the safety distance to be accorded to vulnerable road users or traffic etiquette at pedestrian crossings – might qualify as constrains to the exercise of personal autonomy. Finally, relying on CAVs would restrict the possibility of taking timely decisions concerning routes, which could variously impact the execution of self-determined intentions – e.g., staying away from given roads to protect one's privacy (Boeglin, 2015).

In light of the above, it seems reasonable to conclude that the rush towards full automation should not hinder limited forms of human control over driving tasks, at least when this would serve the legitimate expression of personal autonomy. As suggested, for instance, by the Meaningful Human Control approach (Santoni de Sio & van den Hoven, 2018), if some driving decisions are for users to make, then CAVs should allow for their personal autonomy to be expressed. Arguably, in a context of full automation this could only be accomplished indirectly, e.g., through the setting of user preferences. It is at least uncertain, however, whether this form of indirect control over system operations would satisfy the demands of the principle of personal autonomy. More likely, (a) seems to encourage the development of automated features that leave enough space for the exercise of user autonomy – as happens in conditional automation, where control over driving tasks is shared with the system rather than fully delegated to it.

The claim according to which personal autonomy would be better served by conditional automation is controversial, however, if the value is intended as in (b). Driving automation can indeed have beneficial impacts on human autonomy as the freedom to pursue a good life. At least two opportunities stand out: inclusive transportation and the reappropriation of travel time. Both importantly enable the possibility to fulfil personal needs and desires, thus increasing well-being.

On the one hand, CAVs could massively improve the autonomy of social categories that are currently excluded from manual driving because of physical and cognitive impairments. Independent access to transportation is critical for pursuing personal well-being and leading a satisfying social life. Since driving tasks would be automated, physical and cognitive impairments would no longer constitute an insurmountable barrier to the autonomous use of road vehicles (Lim & Taeihagh, 2018). On the other hand, driving automation could support the self-determined pursuit of a good life by allowing users to reclaim travel time. Freed from the burden of driving themselves, CAV users would be able to employ travel time as they prefer. In addition, autonomous decision-making on matters that importantly impact on individual well-being would also be supported. For instance, decisions about where to live would be less constrained by work locations and other circumstantial factors.

In both of the above cases, personal autonomy benefits entirely depend on full automation. As a matter of fact, individuals excluded from manual driving would be poor candidates for shared control as well (Goggin, 2019). Similarly, full delegation is necessary for CAV users to freely engage in other, more satisfying activities. In order to support autonomy as freedom to pursue one's own conception of a good life, then, human intervention and supervision should be increasingly automated away.

The contrast between a) and b) is evident. Supporting both partial and full automation, compliance with the ethical principle of personal autonomy steers in directions that are difficult to harmonise. Analogously, it is hard to realise how protecting the exercise of user self-determination over driving decisions can go hand in hand with protecting the right to a self-determined good life pursued through mobility. This ambiguity, that stems from the complexity of the notion of autonomy and competing expectations about driving automation, represents a barrier towards designing CAVs that protect and promote personal autonomy. Uncertainty on this matter leaves engineers with the puzzling task of figuring out in what sense personal autonomy can be a value to embed in driving automation, or how to do so.

5. CONCLUSION

When human autonomy meets driving automation, two of its essential components come into conflict. Fully delegating driving to CAVs would limit (or entirely bypass) personal autonomy as the self-determination of driving decisions, so that conditional automation appears to be the most promising option. However, autonomy as the pursuit of self-determined life preferences, interests, goals, and values is best supported by full automation.

The tension that obtains poses a most delicate issue to the ethics of driving automation. Aligning future CAVs to ethical expectations in terms of personal autonomy is an important task. Infringements in this sense are likely to generate distrust and public backlash. However, the paradoxical nature of the issue makes it complicated to move from abstract endorsements to more practical design and policy recommendations.

Future philosophical research must tackle this obscurity and provide less ambiguous accounts of personal autonomy in the context of driving automation. Meanwhile, ambiguity must be assumed as a given. Learning how to deal with it is then of utmost importance. The most urgent task on a design and policy level, then, likely consists in promoting reflection on possible threats to personal autonomy – however ill-defined the concept might be – and assessing solutions aimed at minimising potential harm. Such preliminary, applied ethics work might in turn offer precious help to further refine the notion of personal autonomy as it applies to driving automation (Fossa *et al.*, 2022).

To conclude, shedding light on what it means for driving automation to comply with the value of personal autonomy reveals a series of complicated issues that calls for further analysis. As a first step in this direction, the present paper has offered a preliminary contribution to the identification of such challenges and their origin. By doing this, it has set the stage for future research aimed at better defining the conceptual profile of personal autonomy as it concerns the ethics of driving automation.

REFERENCES

Boeglin, J.A., 2015. The Costs of Self-Driving Cars: Reconciling Freedom and Privacy with Tort Liability in Autonomous Vehicle Regulation. *Yale Journal of Law and Technology*, Vol. 17, No. 4, pp. 171-203.

Chiodo, S., 2022. Human autonomy, technological automation (and reverse). *AI & Society*, Vol. 37, pp. 39-48. https://doi.org/10.1007/s00146-021-01149-5

Christman, J., 2020. Autonomy in Moral and Political Philosophy. In: Zalta, E. (ed.), *The Stanford Encyclopedia of Philosophy (Fall 2020 Edition)*. https://plato.stanford.edu/archives/fall2020/entries/autonomy-moral/

Contissa, G., *et al.*, 2017. The Ethical Knob: ethically-customisable automated vehicles and the law. *Artificial Intelligence and Law*, Vol. 25, pp. 365-378. https://doi.org/10.1007/s10506-017-9211-z

Fossa, F., *et al.*, 2022. Operationalizing the Ethics of Connected and Automated Vehicles: An Engineering Perspective. *International Journal of Technoethics*, Vol. 13, No. 1, pp. 1-20. https://doi.org/10.4018/IJT.291553

Goggin, G., 2019. Disability, Connected Cars, and Communication. *International Journal of Communication*, Vol. 13, pp. 2748-2773. https://ijoc.org/index.php/ijoc/article/view/9021

HLEGAI – High-Level Expert Group on Artificial Intelligence, 2019. *Ethics Guidelines for Trustworthy AI*. https://digital-strategy.ec.europa.eu/en/library/ethics-guidelines-trustworthy-ai (last accessed May 3rd, 2022).

Horizon 2020 Commission Expert Group to advise on specific ethical issues raised by driverless mobility (E03659). 2020. *Ethics of Connected and Automated Vehicles: recommendations on road safety, privacy, fairness, explainability and responsibility*. https://op.europa.eu/en/publication-detail/-/publication/89624e2c-f98c-11ea-b44f-01aa75ed71a1/language-en (last accessed May 3rd, 2022).

Laitinen, A., and Sahlgren, O., 2021. AI Systems and Respect for Human Autonomy. *Frontiers in Artificial Intelligence*, Vol. 4, 705164, pp. 1-14. https://doi.org/10.3389/frai.2021.705164

Lim, H.S.M., and Taeihagh, A., 2018. Autonomous Vehicles for Smart and Sustainable Cities: An In-Depth Exploration of Privacy and Cybersecurity Implications. *Energies*, Vol. 11, No. 5, 1062. https://doi.org/10.3390/en11051062

Magnani, L., 2020. Autonomy and the Ownership of Our Own Destiny: Tracking the External World and Human Behavior, and the Paradox of Autonomy. *Philosophies*, Vol. 5, No. 53, pp. 1-12. https://doi.org/10.3390/philosophies5030012

Millar, J., 2016. An Ethics Evaluation Tool for Automating Ethical Decision-Making in Robots and Self-Driving Cars. *Applied Artificial Intelligence*, Vol. 30, No. 8, pp. 787-809. http://dx.doi.org/10.1080/08839514.2016.1229919

Millar, J., 2017. Ethics Settings for Autonomous Vehicles. In: Lin, P., Abney, K., Jenkins, R. (eds.), *Robot Ethics 2.0: From Autonomous Cars to Artificial Intelligence*. Oxford University Press, Oxford, pp. 20-34. https://doi.org/10.1093/oso/9780190652951.003.0002

Nyholm, S., 2018. The ethics of crashes with self-driving cars: a roadmap II. *Philosophy Compass*, Vol. 13, No. 7, e12506. https://doi.org/10.1111/phc3.12506

Santoni de Sio, F., 2021. The European Commission report on ethics of connected and automated vehicles and the future of ethics of transportation. *Ethics and Information Technology*, Vol. 23, pp. 713-726. https://doi.org/10.1007/s10676-021-09609-8

Santoni De Sio, F., and van den Hoven, J., 2018. Meaningful human control over autonomous systems: A philosophical account. *Frontiers of Robotics and AI*, Vol. 5, No. 15, pp. 1-14. https://doi.org/10.3389/frobt.2018.00015

Schoonmaker, J., 2016. Proactive privacy for a driverless age. *Information & Communications Technology Law*, pp. 1-33. https://doi.org/10.1080/13600834.2016.1184456

Sharkey, A., 2019. Autonomous weapons systems, killer robots and human dignity. *Ethics and Information Technology*, Vol. 21, pp. 75-87. https://doi.org/10.1007/s10676-018-9494-0

Varshney, L.R., 2020. Respect for Human Autonomy in Recommender Systems. *ArXiv, abs/2009.02603*. https://arxiv.org/abs/2009.02603v1

Xu, W., 2021. From Automation to Autonomy and Autonomous Vehicles. *Interactions*, January-February, pp. 49-53. http://dl.acm.org/ft_gateway.cfm?id=3434580&type=pdf&dwn=1

EXAMINING THE INFLUENCE OF ABILITY, TRUST, OPPORTUNITY AND MOTIVATION ON IOT SENSORS ADOPTION FOR PREVENTING FOOD WASTE

Yanqing Duan[1], Ram Ramanathan[2], Usha Ramanathan[3], Lakshmi Swamy[1] and Katarzyna Pelc[1]
[1]University of Bedfordshire, Luton LU1 3JU, UK
[2]University of Essex, Southend-on-Sea, Essex, SS1 1LW, UK
[3]Nottingham Trent University, Nottingham NG1 4FQ, UK

ABSTRACT

Preventing and reducing food waste in food supply chains will contribute to the United Nation's sustainable development goals on responsible consumption and production. Digital technologies, such as the Internet of Things (IoT) sensors, have seen several successful applications but have not yet been widely applied in food supply chains to help reduce food waste. However, food companies are still unsure about using IoT-based sensors and reluctant to adopt them for the purpose of food waste prevention. To address this problem, this study aims to examine the determinants of the intention to adopt IoT-based sensors for preventing food waste by the UK food companies. To achieve this aim, this research develops a comprehensive Ability-Trust-Opportunity-Motivation (ATOM) model that extends and contextualizes the original Motivation-Opportunity-Ability (MOA) model in the context of using IoT sensors in food supply chain companies for preventing food waste. The proposed ATOM model will be used to examine if and to what extent ability, trust opportunity, and motivation influence the managers' behavioral intention and actual behavior of using IoT sensors. The ATOM model will be tested using data collected from a questionnaire survey to be carried out with senior managers in the UK food sector. The final outcomes of this study will make valuable contributions to the theoretical development and practical understanding of the influence of ability, trust opportunity and motivation on IoT sensor adoption for preventing food waste.

KEYWORDS

Food Waste Prevention, Food Supply Chain, Internet of Things (IoT), IoT Sensors, MOA Model, Food Sustainability

1. INTRODUCTION

The effects of emerging Information and Communication Technologies (ICT), such as Big Data and Internet of Things (IoT), on business community and society as well as the interaction between ICT and human beings have attracted growing attention from ICT researchers and practitioners. IoT has been hailed as one of the most notable disruptive technologies of this century (Koohang et al., 2022; Nord et al., 2019) and we have recently witnessed the proliferation and impact of IoT-enabled devices. From 2012–2018, the use of IoT devices grew from 8.7 billion to 50.1 billion even though adoptions in homes and retail areas were still relatively sparse (Koohang et al., 2022). IoT describes the network of physical objects that contain embedded technology to communicate and sense or interact with their internal states or the external environment (https://www.gartner.com/en/information-technology/glossary/internet-of-things). The IoT-enabled devices and sensors have provided many new opportunities for organizations to revolutionize how they can connect, collect, analyze, and utilize data for transforming their business operations and processes towards achieving sustainable development goals. However, although IoT can create invaluable data in every industry, it faces many challenges (Nord et al., 2019). For example, the full benefits and effects of IoT sensors in food sector are still not fully explored. Food loss and waste are among the main contributors to global food security (Aamer et al., 2021) and 14% of world food production is lost before reaching retailers (FAO, 2020), but the uptake of IoT-enabled sensors for the purpose of preventing food wastes in food supply chains are still low and there is a lack of understanding on the determinants of food managers' intention and use of IoT sensors for food waste prevention. To address this knowledge gap, this research proposes a research question "what are the determinants of IoT sensors adoption for preventing food waste in food supply chains?"

Reducing food waste is of highest priority for EU as it can contribute to improving resource efficiency and food security at a global level. In its strategic policy documents, European Commission has confirmed ambitious targets for the EU to halve food waste by 2030 by focusing on all stages in the supply chain. The research reported in this paper builds on the work carried out by the REAMIT project consortium. REAMIT is a transnational European Territorial Cooperation project funded by Interreg North-West Europe (NWE) Programme 2014-2020. Though technologies exist to reduce food waste, their applications and impact are still limited. The REAMIT project aims to adapt and apply existing innovative big data and IoT technologies to food supply chains in NWE to reduce food waste and improve resource efficiency.

2. LITERATURE REVIEW

While IoT can have positive impact on reducing food waste, there are significant challenges. To avoid loss of food quality IoT sensors are being utilised to monitor and control environmental conditions in the post-harvest supply chain (Zhang et al., 2017); IoT has gained interest in intelligent packaging of food to track chemical changes, temperature etc. (Sohail et al., 2018) to provide traceability and tracking during FSC (Fuertes et al., 2016; Ghoshal, 2018); several applications of IoT for tracking and tracing are used during transportation of fresh food (Tsang et al., 2018), and some studies have looked at the IoT applications in fresh food and cold storage maintenance and control, to retain food quality (Onwude et al., 2020). Often FSC companies are mainly driven by inadequate information necessary to make decisions at the right time. They also often struggle with no or less information on inadequate packaging, lack of monitoring and temperature control during transportation, distribution and in storage facilities (Ben-Daya et al., 2019; Mogale et al., 2020). Lack of sufficient information to trace food quality is a significant challenge faced by FSC companies – this is either due to not having data or the right type of data with information to make appropriate decisions to reduce food waste (Cattaneo et al., 2021; Tsang et al., 2018). Among many challenges that the massive increase in IoT device use bring, Nord et al. (2019) identify privacy, security, and trust as three pervasive challenges. User awareness of IoT threats is evolving (Koohang et al., 2022; Nord et al., 2019), but there is no investigation if this concern is also prevalent in the IoT sensors adoption for food waste prevention. While some studies in the literature have demonstrated benefits of adopting IoT to reduce food waste in FSC, there is unfortunately, a lack of studies on understanding the key drivers for FSC company decision makers to adopt IoT device that provides real-time actionable information to prevent food waste.

The term food waste prevention is used in this research because feedback from our pilot case study FSC companies' suggest that the term "food waste reduction" may imply that the FSC companies have wasted food and need to reduce the waste. Instead, FSC companies prefer the word "prevention". EU also uses the term "Food loss and waste prevention" (https://ec.europa.eu/food/horizontal-topics/farm-fork-strategy/food-loss-and-waste-prevention_en).

3. RESEARCH MODEL

Over the last few decades, various theories and models have been developed and improved to explain and predict the acceptance and use of the new technologies (Cao et al., 2021). Among them is the Motivation-Opportunity-Ability (MOA) framework, which was first proposed by MacInnis and Jaworski (1989) in the context of information processing. The MOA model has been applied by scholars to explain various types of behaviour, such as: e-commerce adoption (Teh & Ahmed, 2011), green food consumption (Dong et al., 2022), energy-saving behaviours (Li et al., 2019), consumer behavior in reducing food waste every day (van Geffen et al., 2020). Based on the literature review of the relevant factors and inspired by the MOT framework, this study proposes a research model that extends MOA with an important additional dimension of trust. The operationalization of the MOA model is perceived to be difficult by MacInnis et al. (1991) due to the different conceptualizations of the constructs. The findings from the literature (e.g. Arfi et al., 2021; Jayashankar et al., 2018; Nord et al., 2019) and empirical evidence from pilot studies in our ongoing projects suggest that trust and data protection can act as a significant barrier to the IoT adoption due to the intrusive nature of the IoT devices. For example, one of the pilot companies withdrew its participation to the project due to the drivers' reluctancy to installing and facilitating the use the IoT sensors in their truck. They are seriously concerned about the close monitoring of their movement using the IoT device in their trucks. Therefore, this research extends the MOA framework by including trust to reflect the nature of IoT adoption in food supply chain companies. Figure 1 shows the proposed research model ATOM.

Ability – In the context of IoT adoption, ability refers to the skills and knowledge in using IoT sensors for preventing food waste. It includes action skills and Knowledge of issues related to the food quality and safety standard and regulations.

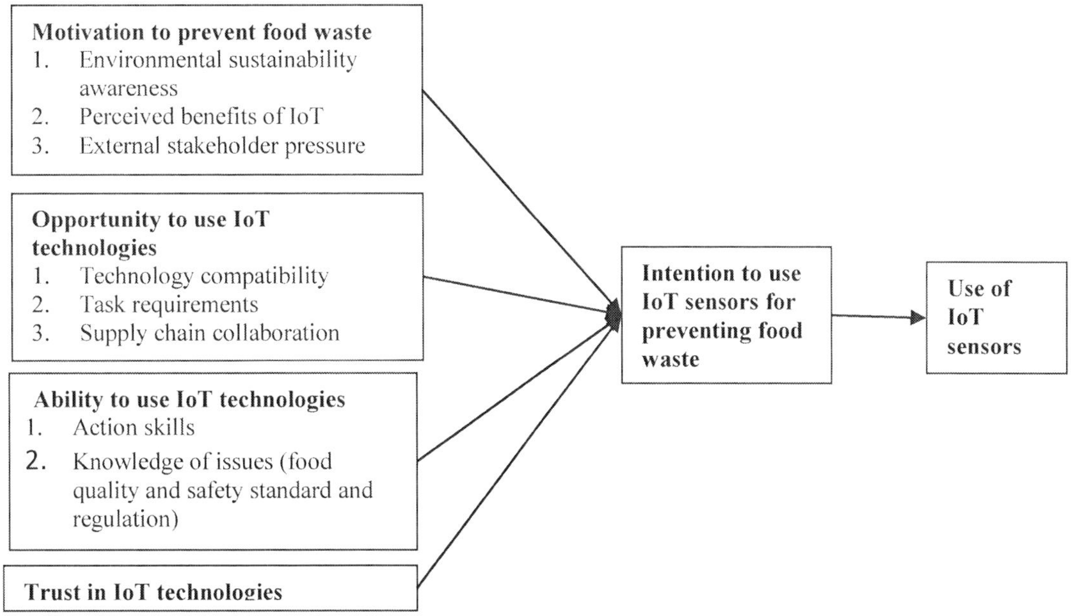

Figure 1. A research model for IoT sensors adoption for food waste prevention

Trust – IoT trust is defined by Koohang et al. (2022) as the degree to which users of IoT devices trust that the IoT service providers that provide products and applications are trustworthy, benevolent, and skillful enough in protecting the users against security/privacy threats and risks. In the context of this research, trust refers to business managers perception on the reliability, trustworthiness and data protection of using IoT sensors for the purpose of food waste prevention.

Opportunity – Opportunity refers to the extent that the company can use IoT without restrictions and the business environment and requirements favor the use of the technology. Opportunity related factors include technology compatibility, task requirements, and supply chain collaboration requirements.

Motivation – Motivational factors refer to intrinsic and extrinsic factors motivate people to take certain actions. it reflects the willingness and/or desire of managers to adopt IoT sensors in this research context. Motivation affects behavioral intention and use behavior (MacInnis & Jaworski, 1989). It can act as a driver for IoT adoption. Factors related to motivation include intrinsic factors of the manager's awareness of the environmental sustainability, perceived benefits (performance expectancy and cost-benefit), and extrinsic factor of external stakeholder pressure.

4. WORK IN PROGRESS AND EXPECTED CONTRIBUTIONS

This work-in-progress paper addresses an important research topic on improving the adoption of IoT sensors for preventing food waste. To achieve the research aim, authors have conducted extensive literature review on the state-of-the-art development of using emerging ICTs, especially IoT devices, for food waste prevention. Technology adoption theories are examined and their relevance to the context of IoT sensors adoption for food waste reduction are analyzed. Based on the literature review, a theoretical model has been proposed and reported in this paper to examine the influence of ability, trust, opportunity, motivation on IoT sensors adoption. An online questionnaire will be designed to collect data to test the proposed research model and its associated hypotheses. Measurement items for the model constructs are being adapted and developed for data collection. The survey will target food companies in the UK. The outcome of this ongoing research will make the following contributions:

- A better understanding of the state-of-the-art development and associated research challenges on the use of IoT sensors in preventing food waste.
- A theoretical framework to examine if and to what extent ability, trust, opportunity, motivation influence the behavioral intention and actual use of IoT sensors by food companies.
- Contextualization and extension of the original MOA model to ATOM model to reflect the IoT characteristics and its application context in this study and contextualization and validation of ATOM constructs and measures in the context of IoT sensors for preventing food waste.
- Practical implications for improving digital technologies adoption in food supply chains.

More importantly the results of this practice-based research will motivate all stakeholders of the food business to combat waste in every possible way through digital transformation.

ACKNOWLEDGEMENT

The authors would like to acknowledge the financial support provided by EU Cohesion Policy Interreg North West Europe Programme 2014-2020 for the REAMIT project (project number NWE831).

REFERENCES

Aamer, A. M., Al-Awlaqi, M. A., Affia, I., Arumsari, S., & Mandahawi, N. 2021. The internet of things in the food supply chain: adoption challenges. *Benchmarking: An International Journal,* Vol 28, No. 8, pp 2521-2541.

Arfi, W. B., Nasr, I. B., Kondrateva, G., & Hikkerova, L. 2021. The role of trust in intention to use the IoT in eHealth: Application of the modified UTAUT in a consumer context. *Technological Forecasting and Social Change,* Vol 167, No., pp 120688.

Ben-Daya, M., Hassini, E., & Bahroun, Z. 2019. Internet of things and supply chain management: a literature review. *International Journal of Production Research,* Vol 57, No. 15-16, pp 4719-4742.

Cao, G., Duan, Y., Edwards, J. S., & Dwivedi, Y. K. 2021. Understanding managers' attitudes and behavioral intentions towards using artificial intelligence for organizational decision-making. *Technovation,* Vol 106, No., pp 102312.

Cattaneo, A., Sánchez, M. V., Torero, M., & Vos, R. 2021. Reducing food loss and waste: Five challenges for policy and research. *Food Policy,* Vol 98, No., pp 101974.

Dong, X., Jiang, B., Zeng, H., & Kassoh, F. S. 2022. Impact of trust and knowledge in the food chain on motivation-behavior gap in green consumption. *Journal of Retailing and Consumer Services,* Vol 66, pp 102955.

FAO. 2020. The state of food security and nutrition in the world 2020. *Food and Agriculture Organization of the United Nations (FAO),* Vol, No.

Jayashankar, P., Nilakanta, S., Johnston, W. J., Gill, P., & Burres, R. 2018. IoT adoption in agriculture: the role of trust, perceived value and risk. *Journal of Business & Industrial Marketing,* Vol, No.

Koohang, A., Sargent, C. S., Nord, J. H., & Paliszkiewicz, J. 2022. Internet of Things (IoT): From awareness to continued use. *International Journal of Information Management,* Vol 62, No., pp 102442.

Li, D., Xu, X., Chen, C.-f., & Menassa, C. 2019. Understanding energy-saving behaviors in the American workplace: A unified theory of motivation, opportunity, and ability. *Energy Research & Social Science,* Vol 51, pp 198-209.

MacInnis, D. J., & Jaworski, B. J. 1989. Information processing from advertisements: Toward an integrative framework. *Journal of Marketing,* Vol 53, No. 4, pp 1-23.

MacInnis, D. J., Moorman, C., & Jaworski, B. J. 1991. Enhancing and measuring consumers' motivation, opportunity, and ability to process brand information from ads. *Journal of Marketing,* Vol 55, No. 4, pp 32-53.

Mogale, D., Kumar, S. K., & Tiwari, M. K. 2020. Green food supply chain design considering risk and post-harvest losses: A case study. *Annals of Operations Research,* Vol 295, No. 1, pp 257-284.

Nord, J. H., Koohang, A., & Paliszkiewicz, J. 2019. The Internet of Things: Review and theoretical framework. *Expert Systems with Applications,* Vol 133, No., pp 97-108.

Onwude, D. I., Chen, G., Eke-Emezie, N., Kabutey, A., Khaled, A. Y., & Sturm, B. 2020. Recent advances in reducing food losses in the supply chain of fresh agricultural produce. *Processes,* Vol 8, No. 11, pp 1431.

Sohail, M., Sun, D.-W., & Zhu, Z. 2018. Recent developments in intelligent packaging for enhancing food quality and safety. *Critical reviews in food science and nutrition,* Vol 58, No. 15, pp 2650-2662.

Teh, P.-L., & Ahmed, P. K. (2011). *MOA and TRA in social commerce: An integrated model.* Paper presented at the 2011 IEEE International Conference on Industrial Engineering and Engineering Management.

Tsang, Y., Choy, K., Wu, C.-H., Ho, G., Lam, H., & Tang, V. 2018. An intelligent model for assuring food quality in managing a multi-temperature food distribution centre. *Food Control,* Vol 90, No., pp 81-97.

van Geffen, L., van Herpen, E., Sijtsema, S., & van Trijp, H. 2020. Food waste as the consequence of competing motivations, lack of opportunities, and insufficient abilities. *Resources, Conservation & Recycling: X,* Vol 5, pp 100026.

Zhang, Y., Zhao, L., & Qian, C. 2017. Modeling of an IoT-enabled supply chain for perishable food with two-echelon supply hubs. *Industrial Management & Data Systems,* Vol 117, No. 9, pp 1890-1905.

ACTIVE-PASSIVE FRAMEWORK FOR DEVELOPING COMMUNICATION STRATEGIES TO COMBAT MISINFORMATION

Safat Siddiqui and Mary Lou Maher
University of North Carolina at Charlotte, NC, USA

ABSTRACT

Social media encourages users' participation that often leads to the spread of misinformation on social platforms. To mitigate the viral spread of misinformation, we have developed an Active-Passive (AP) framework that takes into consideration individuals' social media usage preferences when developing effective communication techniques. This framework leverages users' interaction tendencies and facilitates designing usage-focused interventions for combating the spread of misinformation. The AP framework has emerged from a review of the literature that describes users' social media interactions as a continuum from active to passive, where active users express high interaction tendencies compared to passive users. In this paper, we present the theoretical development of the AP framework and show its relevance for regulating communication and nudging strategies for platform-based interventions that combat the spread of misinformation.

KEYWORDS

Fake News, Misinformation, Combat, Mitigate, Social Media Usage

1. INTRODUCTION

The spread of misinformation, that is, unintentional distribution of false information, has adverse effects on social issues, such as elections, public health, public safety, and the loss of trust in science and media (Lewandowsky et al. 2017). Researchers from multiple disciplines (Lazer et al, 2017) identified 3 courses of immediate actions that combat misinformation: 1. make the discussion bipartisan, 2. make the truth louder, 3. introduce an interdisciplinary initiative for advancing the study of misinformation. The research findings and discussions in (Lazer et al, 2017) have not considered users' social media usage as a basis for mitigating the negative effect of misinformation. This paper introduces the AP (Active-Passive) framework to facilitate usage-focused interventions that take into account users' interaction tendencies to direct their interactions for combating misinformation.

The Active-Passive (AP) framework is developed from a review of existing literature to distinguish social media users based on their usage preferences and interaction patterns. The framework consists of 3 types of social media usage: Producing (e.g., users produce new content), Participating (e.g., users share or rate content), and Consuming (e.g., users read content). We describe social media users' preferences and tendencies along a continuum of users from active to passive:

- Active users: social media users who produce original content, share information on social platforms, and maintain virtual relationships with communities and other users (Chen et al. 2014; Gerson et al. 2017). Active users have a strong preference for interactions in all 3 types of social media usage.
- Passive users: social media users who consume content and avoid online interactions or participation with content and other users (Gerson et al. 2017). Passive users have a strong preference for interactions towards consuming information.

We have developed the AP framework as a basis for researchers to design usage-focused communication and nudging techniques for platform-based interventions that combat misinformation. Platform-based interventions apply nudging prompts to alert users, provide suggestions and recommendations to steer users' behavior in particular directions without sacrificing their freedom of choice. For instance, Facebook uses prompts to alert users when users decide to share any questionable information (Smith 2017; Su 2017); Twitter warns users about the potential harmful information on their feed (Roth et al. 2020). In this paper, we show the effectiveness of the AP framework in providing structure that enables exploring usage-focused communication and nudging techniques to more effectively mitigate the spread of misinformation on social media.

2. THE AP FRAMEWORK

The Active-Passive (AP) framework is developed by identifying the dimensions of interactions for social media users to interact with social media content and users (Figure 1). The framework uses the interaction dimensions to distinguish users based on their interaction patterns and to design prompts that can direct users' interactions to mitigate the spread of misinformation. We identified 5 dimensions of users' interactions: Content Creation, Content Transmission, Relationship Building, Relationship Maintenance, and Content Consumption. The first 4 dimensions of interactions were collected from (Chen et al. 2014), where Chen et al. (2014) studied the active users' interactions on social platforms. In addition to these 4 dimensions, we identified another dimension of interaction on social platforms from (Geeng et al. 2020), which is Content Consumption.

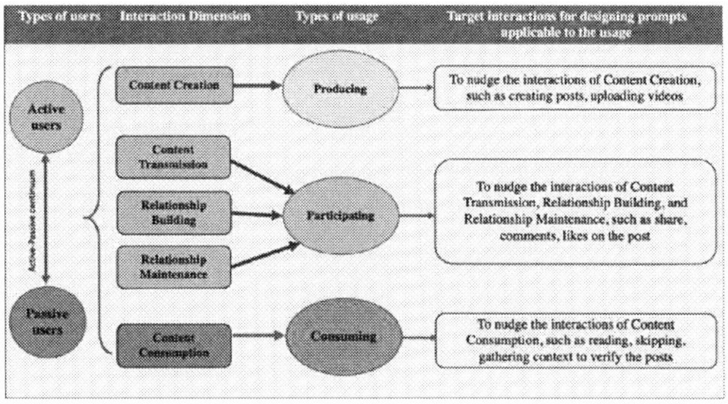

Figure 1. The AP framework addresses users' interaction tendencies to design usage-focused communication prompts

When a particular user's interactions on the platform spread across the 5 dimensions of interactions, the user indicates the tendency of being an active user. Interaction items that are used to create content, such as posting blogs, articles, photos, or videos, are in the dimension of Content Creation. Content Transmission includes interaction items such as sharing friends' posts/videos that are used to spread content on the platform. The Interaction items used to maintain online relationships, such as commenting on posts, chatting with friends through the platforms, are in Relationship Maintenance. Similarly, interaction items used to build virtual relationships, such as creating groups, sending invitations to friends and non-friend to join groups, are in the Relationship Building dimension. Finally, the content consumption includes users' interaction items associated with consuming content, such as scrolling, reading post/articles, and watching videos. In contrast to the active users, when most of the interactions of a user are involved in the dimension of Content Consumption, the user displays the tendency of being a passive user. The AP framework reduces the 5 interaction dimensions to 3 social media usage patterns (Figure 1): Producing, Participating, and Consuming (Shao 2009). We identified that the interaction items of the Content Creation dimensions described in (Chen et al. 2014) are associated with Producing usage. Content Transmission, Relationship Building, and Relationship Maintenance described in (Chen et al. 2014) are associated with Participating usage. Consuming usage is related to the content consumption dimension. These 3 kinds of usage can overlap while individuals interact with different types of content on social platforms.

The AP framework provides a basis for researchers to design usage-focused prompts to communicate to users based on individuals' social media usage preferences. The Reader-to-Leader Framework (Preece et al. 2009) highlighted the importance of various interface supports to increase participation, whereas the AP framework focuses on platform-based communication to mitigate the negative effect of misinformation. Siddiqui et al. (2021) introduced 3 principles of social media interactions to transform active and passive users' interaction tendencies to make the truth louder. This AP framework informs researchers to address Producing usage to direct users' interactions related to content creation, Participating usage to direct users' interactions associated with content transmission, relationship building, and relationship maintenance. Likewise, the AP framework enables researchers to focus on Consuming usage to develop communication strategies and nudging techniques while users are involved in consuming content. We present the implications of the AP framework for combating misinformation in the following section.

3. IMPLICATIONS OF THE AP FRAMEWORK

The AP framework shows that the usage-focused communication prompts can be applied on 3 types of social media usage to direct the interactions associated with the usage (Figure 2). Fake news is a consequence of the type of usage in the AP framework called Producing. The spread of fake news is amplified by the type of usage in the AP framework called Participating when the user shares fake content. Likewise, users are trapped into filter-bubble and echo-chamber when they are involved in the Consuming usage of the framework. In this section, we describe the implications of the AP framework in organizing the directions of the communication and nudging prompts to mitigate the negative effect of misinformation.

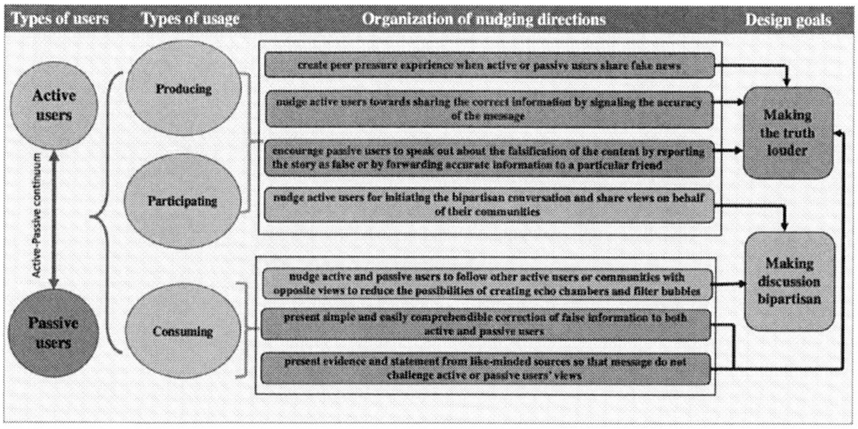

Figure 2. The implication of the AP framework in developing usage-focused communication strategies

3.1 Nudging Directions for Consuming Usages

The prompts for the Consuming usage focus on directing users' information consumption behavior so that individuals develop the habit of consuming verified information and remaining less affected when exposed to unverified information. Insincere consuming related interactions can be harmful to the users. For instance, users have a tendency to follow like-minded sources (Lazer et al. 2017), but these interactions can lead users to remain in echo chambers and be surrounded by filter bubbles. Nudging prompts that assist users in verifying information, getting additional context, and different perspectives are applied in the Consuming usage to nudge users' interactions in the content consumption dimension.

Interventions for Consuming usage can nudge users to follow the communities and users of opposite views. Instead of merely suggesting users follow others on social media, the interventions can communicate with users and inform them about the necessity of getting different perspectives on the topics, and how the interventions can help individuals to gain that perspective. The platform-based affordances that help users to verify information or get additional context are applied to the Consuming usage. The information button ('i') and the 'Related Article' section of Facebook (Smith 2017, Su 2017), are applied when users are involved in the interactions related to the content consumption. Twitter (Roth et al. 2020) warns users about harmful information when users are involved in Consuming usage. The additional resources to support users in verifying information are also applicable for this usage.

The inoculation techniques (Van der Linden et al. 2017; Cook et al. 2017) that inoculate users against misinformation can be implemented as the interventions for Consuming usage. Similarly, presenting the corrections of misinformation is also associated with the interventions of Consuming usage. As the corrections should not directly challenge individuals' worldviews (Lewandowsky et al. 2017) and people tend to accept the correction when the confirmation comes from similar ideological sources (Berinsky 2017), interventions can present the statement and evidence about the corrections from the like-minded sources and active users. Curiosity has been applied as the basis of encouraging learners to learn (Siddiqui et al. 2022) and can be the basis of platform-based interventions for Consuming usage and encourage people to be curious about learning the corrections and opposite viewpoints. Interventions of Consuming usage must limit the exposure of misinformation and should not show the misinformation in the context of correcting the message as the repeated exposure of misinformation can be harmful to the users (Greenhill et al. 2017).

3.2 Nudging Directions for Producing and Participating Usages

Fake content is created and spread on social platforms when users are involved in Producing and Participating usage. Platforms such as Facebook and Twitter block social media accounts that misuse platforms' Producing and Participating usages to spread unverified information. In general, people prefer to share accurate information (Pennycook et al. 2020; Fazio 2020) and interventions for Participating usage can highlight accurate messages to nudge the active users to contribute to the distribution of credible information. Bhuiyan et al. (2018) developed a browser extension, FeedReflect, that uses visual cues to indicate whether the information source is mainstream or non-mainstream, and nudges users towards critical thinking before consuming the information. While FeedReflect (Bhuiyan et al. 2018) focuses on nudging users' information consumption behavior, similar kinds of approaches can be developed to direct users to distribute credible information and limit their interaction with unveiled information. Siddiqui et al. (2021) developed 3 design principles to increase users' participation in spreading credible information and reduce users' participation in unverified content. Such interaction principles are applicable to the platform's Participating usage.

Insincere interactions of the Participating usage can lead to the spread of misinformation – active users due to their interaction tendency may often use the sharing functionalities in the context of unverified information. Facebook alerts individuals when the users press to share any questionable content - such nudges can remind users to be reflective about their interactions and minimizes the insincere spread of misinformation. Creating peer pressure is suitable for these usages to reduce users' tendency to post or share misinformation. As the communication bridges across cultures foster the production of more neutral and factual content (Lazer et al, 2017), interventions can nudge the interactions of the active users to get involved in the bipartisan discussions, make comments, and share their views on behalf of their communities. Communications can be personalized to the passive users by simplifying the interactions for them that require limited digital footprints, such as sharing the correct information to a particular friend or reporting the story as fake.

The nudging prompts for the active users can be applied in Producing or Participating usage to direct users' interactions in the usage for distributing credible information and limiting the spread of unverified information. Likewise, prompts for the passive users can be applied in the Consuming usage and focus on inspiring passive users to get involved in Participating usage and share credible information with their friends. Accordingly, the AP framework opens possibilities and empowers platform-based interventions to design communication prompts personalized to users' interaction tendencies that direct users' interactions towards mitigating the negative effects of fake news on social platforms.

4. CONCLUSION AND FUTURE WORK

This paper addresses users' interaction tendencies on social platforms as a basis for further research on mitigating the negative effect of misinformation. We present the Active-Passive framework that distinguishes users based on their interaction tendencies and enables designing usage-focused communication techniques and prompts to direct user interactions. The AP framework addresses social media users' active-passive tendencies and leverages the tendencies toward making the truth louder and making the discussion bipartisan. The framework opens possibilities for researchers to develop and study the effect of communication and nudging techniques personalized to individuals' social media usage preferences that can mitigate the negative effect of misinformation on social platforms.

REFERENCES

Berinsky, A.J., 2017. Rumors and health care reform: Experiments in political misinformation. British journal of political science, 47(2), pp.241-262.

Bhuiyan, M.M. et al, 2018, October. FeedReflect: A tool for nudging users to assess news credibility on twitter. In Companion of the 2018 ACM Conference on Computer Supported Cooperative Work and Social Computing (pp. 205-208).

Chen, A., Lu, Y., Chau, P.Y. and Gupta, S., 2014. Classifying, measuring, and predicting users' overall active behavior on social networking sites. Journal of Management Information Systems, 31(3), pp.213-253.

Cook, J., Lewandowsky, S. and Ecker, U.K., 2017. Neutralizing misinformation through inoculation: Exposing misleading argumentation techniques reduces their influence. PloS one, 12(5), p.e0175799.

Fazio, L., 2020. Pausing to consider why a headline is true or false can help reduce the sharing of false news. Harvard Kennedy School Misinformation Review, 1(2).

Geeng, C., Yee, S. and Roesner, F., 2020, April. Fake news on Facebook and Twitter: Investigating how people (don't) investigate. In Proceedings of the 2020 CHI conference on human factors in computing systems (pp. 1-14).

Gerson, J., Plagnol, A.C. and Corr, P.J., 2017. Passive and active Facebook use measure (PAUM): Validation and relationship to the reinforcement sensitivity theory. Personality and Individual Differences, 117, pp.81-90.

Greenhill, K.M. and Oppenheim, B., 2017. Rumor has it: The adoption of unverified information in conflict zones. International Studies Quarterly, 61(3), pp.660-676.

Lazer, D. et al, 2017. Combating fake news: An agenda for research and action.

Pennycook, G. et al, 2020. Fighting COVID-19 misinformation on social media: Experimental evidence for a scalable accuracy-nudge intervention. Psychological science, 31(7), pp.770-780.

Preece, J. and Shneiderman, B., 2009. The reader-to-leader framework: Motivating technology-mediated social participation. AIS transactions on human-computer interaction, 1(1), pp.13-32.

Roth, Y. and Pickles, N. 2020. Updating our approach to misleading information. https://blog.twitter.com/en_us/topics/company/2018/how-twitter-is-fighting-spam-and-malicious-automation

Shao, G., 2009. Understanding the appeal of user-generated media: a uses and gratification perspective. Internet research.

Smith, J. 2017 Designing Against Misinformation. (December 2017) https://medium.com/facebook-design/designing-against-misinformation-e5846b3aa1e2

Siddiqui, S. and Maher, M. 2021. Reframing the Fake News Problem: Social Media Interaction Design to Make the Truth Louder. In Proceedings of the 5th International Conference on Computer-Human Interaction Research and Applications - CHIRA, ISBN 978-989-758-538-8, pages 158-165. DOI: 10.5220/0010658200003060

Siddiqui, S. et al. 2022. Personalized Curiosity Engine (Pique): A Curiosity Inspiring Cognitive System for Student Directed Learning. In Proceedings of the 14th International Conference on Computer Supported Education - Volume 1, ISBN 978-989-758-562-3, ISSN 2184-5026, pages 17-28.

Su, S. 2017. New Test with Related Articles (2017) https://about.fb.com/news/2017/04/news-feed-fyi-new-test-with-related-articles/

Lewandowsky, S., Ecker, U.K. and Cook, J., 2017. Beyond misinformation: Understanding and coping with the "post-truth" era. Journal of applied research in memory and cognition, 6(4), pp.353-369.

Van der Linden, S. et al. 2017. Inoculating the public against misinformation about climate change. Global Challenges, 1(2), p.1600008.

PROMOTING SOCIAL ACTIVITIES IN AN ONLINE CONFERENCE DURING COVID TIMES: THE CASE OF THE EHSEMI CONFERENCE

Eliza Oliveira[1], Ana Margarida Almeida[1], Rita Oliveira[1], Nuno Ribeiro[2], Oksana Tymoshchuk[1], Rita Santos[3], Andreia Sousa[4] and Lersi Duran[1]

[1]*Department of Communication and Art / Digital Media and Interaction Research Centre, University of Aveiro*
[2]*Ipatimup, Digital Media and Interaction Research Centre, University of Aveiro*
[3]*ESTGA, Digital Media and Interaction Research Centre, University of Aveiro*
[4]*HEI-Lab/ULusófona, Digital Media and Interaction Research Centre, University of Aveiro*
Campus Universitário de Santiago, 3810-193 Aveiro Portugal

ABSTRACT

While the lockdown significantly reduced face-to-face meetings, video conferencing platforms began to be used widely to leverage different sectors, such as academia. In this scope, scientific events such as seminars and conferences were migrating to the online environment. However, online events usually do not dedicate time to group activities, decreasing the chances of building academic and professional partnerships and decreasing social interactions and networking opportunities that naturally occur in face-to-face events. This paper aims to present a series of dynamics carried out using Zoom and Mentimeter platforms within the scope of the "ehSemi" Conference to create interaction between people and encourage potential scientific networking, which happens in face-to-face conferences and is significantly lost online. Furthermore, the paper discusses the role of timing, and moderators, among other essential factors that boost the social components in the online environment. Despite, given the current pandemic situation, studying the new role of videoconferencing and interactions therein appears timely and interesting, limitations and ideas for future work are presented to improve results and provide details that were not covered in this work.

KEYWORDS

COVID-19, Social Interaction, Virtual Conference, Networking

1. INTRODUCTION

Since the lockdown caused by COVID-19 started in 2020, there has been a significant increase in using videoconferencing platforms as alternative solutions for holding scientific events such as seminars and conferences. Videoconferencing platforms such as Zoom Meetings, Cisco WebEx, Live Stream, Demio, Google Hangouts, Skype, and Microsoft Teams, enable innovative strategies to reorganize academic events by the scientific community in an online environment (Valenti et al., 2021). Among these tools, Zoom has been previously reported as the most used option for online events in several areas of knowledge due to its user-friendliness, being free of charge, and transcription functionality (Gisondi et al., 2021; Aljamaan et al., 2022; Naroo, Morgan, Shinde, & Ewbank, 2022; Lima et al., 2020). Further features such as being a multifunctional platform (with solutions for meetings, webinars, marketplace, among others) also contributed to the emergence of a new worldwide phenomenon called "Zoombombing" (Lee, 2022; Zoom, 2022).

Additionally, other interactive tools have been used to build interactive presentations, promote alternative dynamics in events, and collect feedback from the audience, such as Mentimeter, which has been commonly used with Zoom for different purposes (Lima et al., 2020; Mason et al., 2021).

Despite the value recognized to these platforms, online events are not taking full advantage of the potential of those tools to encourage the exchange of informal chats between the participants. Moreover, online events usually do not dedicate time to group activities, decreasing the chances of building academic and professional partnerships. Therefore, social interactions and networking opportunities that occur in face-to-face events tend to decrease drastically in online settings.

In this context, this paper presents social dynamics to boost social interaction and the development of scientific networking, carried out during the second edition of the ehSemi Conference, held by the Ehealth & Wellbeing group of the Digital Media and Interaction Research Centre, part of the University of Aveiro, in Portugal (DigiMedia, 2022). This event aims to be a national discussion forum for students who develop research leading to innovative digital media solutions to improve health, well-being, and quality of life, meeting experienced people in the field, and broadening their knowledge. In this sense, the reasoning behind the socially planned events is precisely to use common elements of the games, such as rules, voluntariness, and metrics, to promote social interaction between the participants and the engagement in the event.

The goal of this paper is to present a series of dynamics carried out using Zoom and Mentimeter platforms within the scope of the "ehSemi" Conference to promote social engagement between event participants. These social activities were proposed to create interaction between people and encourage potential scientific networking, which happens naturally in face-to-face conferences and is significantly lost online. The proposal of holding social dynamics in the videoconference further aimed to provide moments of leisure and fun during the event, increasing the participant's involvement in ehSemi. In this sense, the reasoning behind the socially planned events is to use common elements of the games, such as rules and fun (Mora et al., 2015), to promote social interaction between the participants and the engagement in the event.

2. METHOD

The dynamics were held on February 3, 2022, during the second edition of the ehSemi Conference, which was carried out through the Zoom platform using an institutional (Pro) account. The average number of people who participated in the event was 36, including the authors. Two moments of interactive activities were performed, one in the morning and one in the afternoon. These moments lasted 20 minutes each.

The event started at 9:30 am and ended approximately at 5:45 pm. The first social moment took place one hour after the beginning of the event, after the keynote speaker. A moderator gave instructions to the participants, asking them to pick up their smartphones to read a QR code that would appear on the screen. In addition, it was explained that the interaction would also be possible through a link generated by Mentimeter that would be shared on Zoom chat. The conference was recorded on Zoom, allowing the results of interactions and observations to be checked later.

Afterward, participants had to answer six questions presented consecutively by the moderator through Mentimeter on their smartphones or computer. The answers appeared in real-time on the Zoom to all participants as they responded. The first question, "Who am I?" aimed to characterize the participants. The answer was a single choice and included: Researcher, Undergraduate Student, Master's Student, Doctoral Student, Higher Education Professor, Health and Wellness Professional, Digital Technologies Professional, and Other. Responses appeared in cluster arrangement in each option. The second, third, and fourth questions were answered by writing words by the user. In this sense, question two was directed to identify the participant's area of research (What is my research area?), creating a word cloud in real-time while the answers were submitted. The third question was: "What do I expect from ehSemi?". The answers were presented in small boxes on the screen in this query. The fourth question, "Where am I right now?" appeared as a word cloud. Finally, the fifth and the sixth questions were directed to aspects of each person's personal life. More specifically, "How do I feel this morning?" and "Where do I like to be?" respectively. Both questions were asked in single-choice questions, and the answers appeared in real-time through bar graphs on the screen. In this sense, question five presented "excited, happy, sad, nervous, scared, tired, and sleeping" as alternative answers, while images of a sofa, countryside, city, and desert island appeared as options in the sixth query.

The game "Two truths and one lie" was performed at the second social moment, which was realized after the first parallel session of the afternoon. The proposal was to gather two participants in different breakout rooms every 3 minutes. Within that time, both participants would have to tell one lie and two truths about their academic life, while the other player would try to guess which one was the lie. All participants, except the moderator, participated in both social moments.

3. RESULTS AND DISCUSSION

Overall, the activities had high adherence by the participants since a significant number of answers were obtained in each query of the first social moment. In this sense, six participants were master's students, three were researchers, two were health and well-being professionals, ten were doctoral students, and four were high school teachers, totaling 25 responses. The second question ("Qual a minha área de investigação" - What is my research area?) had 29 responses and culminated in a word cloud that shows that "design" was the most common area among participants (Figure 1). Next, the word "knowledge" gained relevance in the third question's answer, showing that the general expectation includes generating and acquiring knowledge regarding the seminar fields. Also, the word cloud from the fourth question ("Onde estou neste momento?" - Where am I right now?) shows that the 26 participants who answered were in central ("Aveiro") and northern ("Porto") Portugal (Figure 1). Lastly, the fifth and sixth queries show that most participants, among 27 answers, were excited (15) and preferred to be in the countryside (16).

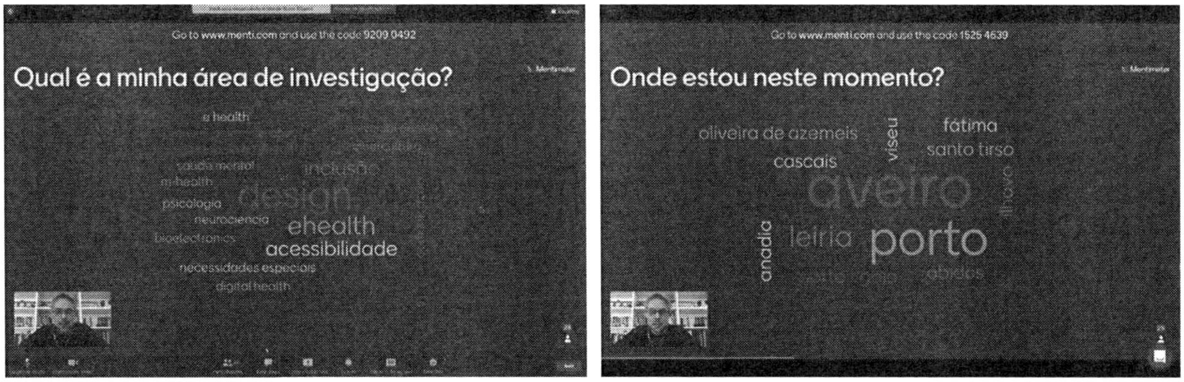

Figure 1. Word clouds generated by Mentimeter that shows that "design" was the most common area among participants (second question), and that 26 participants who answered were in central and northern Portugal (fourth question)

Regarding the second social moment, the participants requested a time increase after the first game round since not all could finish the game proposal timely. Therefore, consecutive 4-minute games were performed until reaching a total time of 20 minutes of activity.

In this context, the dynamics were a timely and feasible option to boost social interaction in the online seminar. Specifically, it was shown that the activities carried out through Mentimeter and Zoom were successful since the participation rate was significant. Therefore, it was possible to characterize and know more about the participants. The moderator successfully guided the activities, considering that people followed both social moments well, with a slight difference in the number of responses between the queries of the first activity. The importance of the moderator also became evident when it was necessary to adapt the game time. In this direction, the fact that participants asked to increase the interaction time suggests interest in the game activity and getting to know other people present in the event.

Furthermore, the Zoom platform has features that allow the immediate time adaptation of the dynamics since it was necessary to adapt the time from 3 to 4 minutes per round of interaction in the second social moment. Further, Mentimeter's interactive solutions enriched the event since the results were shown in real-time and were clear and objective. Moreover, the dynamics acted as a moment of leisure to let people know each other and promote possible networking, which commonly used to happen in coffee breaks of face-to-face events.

4. CONCLUSION

Contextual and social transformations naturally lead to changes in modes of online engagement (Jungselius & Weilenmann, 2019). This work intends to contribute to the scientific scope with a case study of social engagement in online conferences during COVID-19 pandemic, intending to present innovative ways to promote the social component in formal online events. In this sense, through the dynamics it was possible to know more about the participants and promote interaction between them. Therefore, the dynamics previously reported boosted the social strategies in the online formal ehSemi academic event. This allows hypothesizing that videoconference tools and platforms such as Zoom and Mentimeter can help to promote social interaction in online events. Further, dynamics such as these proved to be a successful choice in alternative to the socializing moments that used to take place during coffee breaks, fostering a network among the academics. The possibility to see the results in real-time made the dynamic more exciting and fun. Additionally, it made it possible to verify that people joined the dynamics in real-time. Also, it is noteworthy the importance of the moderator for the success of these social dynamics.

Given the current pandemic situation, studying the new role of videoconferencing and interactions is timely and interesting. However, some limitations of the study include the lack of information regarding other demographic details, such as gender. Also, the study provides results from 20-30 active participants of the social interactions, thus technically the number of participants is small to generalize conclusions, requiring further investigations with a larger sample. In this sense, future work comprises applying the same methods to other online conferences/events and investigating if they can also be applied to hybrid conferences. Further improvement includes enrich the demographic data of participants and seeking to know the participant's reactions and opinions about the social activities.

ACKNOWLEDGEMENT

Our thanks to FCT/MCTES for the financial support to DigiMedia - Digital Media and Interaction Research Centre (UIDP/05460/2020 + UIBD/05460/2020), through national funds.

REFERENCES

Aljamaan, F., Alkhattabi, F., Al-Eyadhy, A., Alhaboob, A., Alharbi, N. S., Alherbish, A., ... Temsah, M. H. (2022). Faculty Members' Perspective on Virtual Interviews for Medical Residency Matching during the COVID-19 Crisis: A National Survey. *Healthcare (Switzerland)*, Vol. 10, No. 1. https://doi.org/10.3390/healthcare10010016

DigiMedia. (2022). *Research Groups*. https://digimedia.web.ua.pt/organization#research-groups

Gisondi, M. A., Chambers, D., La, T. M., Ryan, A., Shankar, A., Xue, A., & Barber, R. A. (2021). A Qualitative Thematic Analysis of 'INFODEMIC: A Stanford Conference on Social Media, Ethics, and COVID-19 Misinformation' (Preprint). *Journal of Medical Internet Research*, Vol. 24, No. 2, pp. 1–16. https://doi.org/10.2196/35707

Jungselius, B., & Weilenmann, A. (2019). Same Same But Different: Changes in Social Media Practices Over Time. In *Proceedings of the 10th International Conference on Social Media and Society* (pp. 184–193). New York, NY, USA: Association for Computing Machinery. https://doi.org/10.1145/3328529.3328559

Lee, C. S. (2022). Analyzing Zoombombing as a new communication tool of cyberhate in the COVID-19 era. *Online Information Review*, Vol. 46, No. 1, pp. 147–163. https://doi.org/https://doi.org/10.1108/OIR-05-2020-0203

Lima, K. R., Neves, B. H. S. Das, Ramires, C. C., Soares, M. D. S., Martini, V. A., Lopes, L. F., & Mello-Carpes, P. B. (2020). Student assessment of online tools to foster engagement during the COVID-19 quarantine. *Advances in Physiology Education*, Vol. 44, No. 4, pp. 679–683. https://doi.org/10.1152/advan.00131.2020

Mason, S., Ling, J., Mosoiu, D., Arantzamendi, M., Tserkezoglou, A. J., Predoiu, O., & Payne, S. (2021). Undertaking Research Using Online Nominal Group Technique: Lessons from an International Study (RESPACC). *Journal of Palliative Medicine*, Vol. 24, No. 12, pp. 1867–1871. https://doi.org/10.1089/jpm.2021.0216

Mora, A., Riera, D., Gonzalez, C., Arnedo-Moreno, J. (2015). A Literature Review of Gamification Design Frameworks, *2015 7th International Conference on Games and Virtual Worlds for Serious Applications (VS-Games)*. Scovde, Sweden, pp. 1-8. doi: 10.1109/VS-GAMES.2015.7295760.

Naroo, S. A., Morgan, P. B., Shinde, L., & Ewbank, A. (2022). The impact of COVID-19 on global contact lens education. *Journal of Optometry*, Vol. 15, No. 1, pp. 60–68. https://doi.org/10.1016/j.optom.2020.11.002

Valenti, A., Fortuna, G., Barillari, C., Cannone, E., Boccuni, V., & Iavicoli, S. (2021). The future of scientific conferences in the era of the covid-19 pandemic: Critical analysis and future perspectives. *Industrial Health*, Vol. 59, No. 5, pp. 334–339. https://doi.org/10.2486/indhealth.2021-0102

Zoom: Video Conferencing, Cloud Phone, Webinars, Chat.. (2022). Retrieved from https://explore.zoom.us/docs/pt-pt/communications-platform.html

SOCIAL MEDIA PRESENCE OF PUBLIC ADMINISTRATION AS A TOOL TO EDUCATE TAXPAYERS

Tereza Zichová
Faculty of Informatics and Statistics, Prague University of Economics and Business, Czech Republic

ABSTRACT

Today's social networks affect the functioning of societies around the world. They influence our thinking and reasoning on fundamental issues of life, as was clearly demonstrated during the COVID-19 pandemic and the war in Ukraine. National governments and public institutions often use social media to communicate very important information to citizens. The public administration of the Czech Republic uses many information networks to communicate with citizens. Although it tries to publish information regularly, it does not make full use of its potential to educate citizens about their obligations to the state. At the same time, it rarely targets this information to the specific situations of citizens, bar exceptions, such as the newly developed Czech government web portal portal.gov.cz. Educating citizens on tax issues is key to building a tax culture and promoting tax morale. This paper aims to analyse Facebook posts in order to typify the posts and propose recommendations to increase taxpayer education. For the analysis, 110 posts were selected on the most followed social media account of the Czech tax administration on Facebook. The posts were categorized based on a customized framework for classifying social media posts. The study yields several significant findings. The Czech Financial Administration uses Facebook almost exclusively to provide information on compliance deadlines, while there is still a large gap in the education sector. The results can be beneficial not only for public administrations of different states but also for other government institutions using the power of the online world.

KEYWORDS

E-government, Public Administration, Content Analysis, Social Networks, Taxpayers

1. INTRODUCTION

Social media represent an interactive platform that supports user-generated content and real-time interactions according to a person's preferences (Kaplan & Haenlein, 2010). They have developed rapidly, promoting the formation of an intricate ecology of social networks (Yank, 2021). The unique characteristics of social media and the benefits they promise have seen them grow exponentially (Lin & Lu, 2011). Social media represent an important communication channel for government institutions. By connecting directly with citizens, they can convey information quickly and in a personalized manner, more effectively than offline communication in many cities. The results of Olinski and Szamrowski (2021) suggest that despite the undeniable benefits of social media, public organizations are only using a small part of this potential.

Social networks can be used by public institutions not only to disseminate information but also to build relationships and increase learning (Wonhyuk & Melisa, 2021). Many researchers confirm a positive relationship between tax education and taxpayers' attitudes toward tax morale motivation to pay taxes (Indriyarti & Christian, 2020; Triandani & Apollo, 2020). One of the tools to raise awareness for the exercise of tax citizenship may be an e-tax-learning platform for citizens (Zichová, 2021). Elements of tax awareness can also be integrated into social networks.

The Czech Ministry of Finance is the financial arm of the executive branch. It is subordinate to the government. The Financial Administration is subordinate to the Ministry of Finance. It is a system of administrative bodies of the Czech Republic intended for tax administration. It consists of the following authorities with nationwide coverage: General Finance Directorate, Appellate Finance Directorate, and regional Tax Offices. The Financial Administration of the Czech Republic uses its website as the main online means of informing citizens. In addition, it manages four social networks: Twitter, YouTube, Facebook and

Instagram. Facebook has the largest number of fans and followers. According to the number of citizens at the beginning of 2021, based on the results of the census of population, houses and apartments, up to 0.12% of citizens follow the Facebook page.

Acknowledging the important role of citizens' tax education and of the social networks used by public administration, the paper focuses on the following research questions. RQ1: What are the most common typologies of Financial Administration Facebook posts? RQ2: How could the Financial Administration change its activities on social networks to educate taxpayers?

2. RESEARCH METHODS

The qualitative research method consisted of a content analysis of Facebook posts by the Czech Financial Administration and their categorization according to the Tafesse & Wien (2016) framework for categorizing social media posts. This framework was chosen because it draws on several theoretical approaches related to post categorization (Kim et al., 2015; Taecharungroj, 2016; Jahn & Kunz, 2012). Other typologies for posts by government institutions have not yet been explored and placed in the context of studies. The framework of Tafesse & Wien (2016) was adapted to categorize posts related to government activities. The final version of the framework contains a total of 11 identified categories of contributions by the state administration. As a data sample, 110 publicly published posts on the main Facebook page during the period of 1 January – 31 March 2022 were selected. The financial administration's Facebook page was visited from a personal Facebook profile that has the page listed as a favourite. Therefore, the page was displayed in the same way as it is displayed to its fans and followers.

For all posts, the text and emoticon content of the caption is monitored, in some cases supplemented by information in an attached photo or graphic. For this reason, posts were analysed in detail individually according to the date of publication. They were transcribed and analysed manually in Excel, which offers a sufficient array of analytical tools. Manual transcription and analysis of the posts were chosen because they helped to better capture all of the above aspects of the posts, not just the written text. The analytical programs do not always take into account the detailed meaning of photos and text in graphics.

Posts that contained multiple topics were included in all of the relevant categories. Reactions, comments, and the visual style of posts were not addressed in the examination, which was limited to textual content and associated emoticons only. The indicative target audience, the group of citizens for whom the post makes the most sense, is also continuously recorded.

3. RESULTS

In order to fit the posts into the Tafesse & Wien (2016) framework, it was first necessary to adapt the typologies to match state institutions. Table 1 describes the modified framework for categorizing social media posts. Also included are the modified definitions and topic areas of these typologies.

Table 1. Framework for categorizing social media posts. Source: Tafesse & Wien (2016), Framework for categorizing social media posts of public administration. Modified and compiled by the author

Categories of brand posts (Tafesse & Wien, 2016)	Modification of categories for public administration posts (author)	Definition and common message themes of modified categories
Emotional brand posts	Emotional posts	Posts that evoke citizens' emotions. They typically employ emotion-laden language, inspiring stories and jokes to arouse fun, enthusiasm and wonder. Common themes: emotionally expressed posts, storytelling, jokes.
Functional brand posts	Organizational posts	Posts that highlight the organizational attributes of public administration services. Common themes: opening hours, deadlines.

Educational brand posts	Educational posts	Posts that educate citizens. They help consumers acquire new skills or knowledge. Common themes: tips, instructions, blog posts, external articles.
Brand resonance	Resonance with public administration	Posts that highlight the main symbols and associations of public administration identity. Common themes: public administration image (e.g., logo, motto, slogan, characteristics), photos, and institution history.
Experiential brand posts	Public administration merits	Posts that evoke consumers' sensory and behavioural responses. They often associate the institution with pleasurable experiences. Common themes: sensory stimulation (e.g., sight, hearing, taste, smell), physical stimulation (e.g., physical actions, performances).
Current event	Current event	Posts that comment on themes that describe events like holidays, anniversaries, cultural events, and the weather or season. Common themes: holidays and other special days and celebrations.
Personal brand posts	Personal public administration posts	These posts centre around citizens' relationships, preferences, and experiences. Common themes: friends, family, personal preferences, stories, plans.
Employee brand posts	Employee public administration posts	Posts that present employees' perspectives on a range of issues. Common themes: employees' work, technical expertise, personal interests, hobbies.
Brand community	Public administration community	Posts that promote and reinforce the institution's online community. They also encourage participation from current members. Common themes: encouraging fans to become members, acknowledging fans (e.g. mentioning their name, tagging them), using user-generated content.
Customer relationship	Citizens relationship	Posts that solicit information and feedback about citizens' needs, expectations, and experiences. They seek to strengthen the impact of citizens' relationships on social media channels. Common themes: customer feedback, testimony, reviews, services.
Cause-related brand posts	Cause-related posts	Posts that highlight socially responsive programs. They promote valuable social issues and initiatives and encourage customers and fans to support them. Common themes: financial or material collection for a non-profit organisation, for refugees.
Sales promotion	N/A	N/A

Table 2 describes the typological distribution of social media posts published during the period of 1 January 2022 – 31 March 2021 on the Financial Administration's Facebook page. 92% of the posts were of an informative character. They mainly discussed deadlines for fulfilling obligations. Most of them also contained a link to more information. 18% of posts included educational information, but organizational information mostly prevailed. On the other hand, the posts did not contain any emotions, jokes or community-building aspects.

Table 2. Distribution of categories for public administration posts. Source: author

Categories for public administration posts	Frequency
Emotional posts	0%
Organizational posts	92%
Educational posts	18%
Resonance with public administration	3%
Public administration merits	12%
Current event	12%
Personal public administration posts	0%
Employee public administration posts	6%
Public administration community	0%
Citizens relationship	0%
Cause-related posts	6%

Given the importance of taxpayer education, it would be very useful to take advantage of the educational potential of the online environment and try to publish more articles to educate interested parties. At the same time, the language of the posts is very careful and formal. The texts of the contributions are intended for different groups of citizens without clear differentiation, complicating orientation. This area could also benefit from more pages for different life situations or a clear segmentation of the content, that is, the differentiation of topics for different target groups.

4. DISCUSSION

Czech tax administration authorities maintain a very formal and reserved presence on Facebook, which is probably expected of such an institution. On the other hand, this approach may evoke excessive rigidity and discourage more potential fans and followers, who could benefit from informational support. It is also important to note that many contributions only serve a certain group of citizens and are not relevant to others. Meanwhile, targeted content could produce higher engagement rates (Jukić, T. & Merlak, M., 2017; Warren et al., 2014). In order to more specifically direct the communication of a state institution on Facebook, it would be good to conduct additional research directly among the various citizens who use this social network and find out their perceptions. Despite the theoretical basis, citizens have many cultural specificities and practices that may differ significantly.

The research method used has some limitations, mainly due to the potential subjective perception of the text by the researcher. However, the high level of precision of the framework categorization meant that none of the assessed posts suffered from ambiguous classification.

5. CONCLUSION

The paper looks at the potential benefits of the government's use of social networks. The importance of tax education has been confirmed in several studies and has become a motivation for finding ways to educate citizens in different forms and channels. The research offers a framework modification of categories for public administration social network posts. It then classifies posts from recent months into different typologies. The results show that communication by the Czech Financial Administration is mainly informative (94% of the posts studied) and highly formal. Therefore, it is not making full use of its educational potential. At the same time, better targeting of the text of the posts – by segmenting the content for specific target groups – could be very useful. A solution could also be to set up more Facebook pages based only on citizens' life situations.

As part of future research, it would be very interesting to conduct additional sentiment analyses of posts and analyses of comments on posts to determine follower reactions. Facebook interactions could also be compared with other social media sites in use, such as Instagram and Twitter. Content analysis, the visualisation of posts and infographics used, as well as video analysis, could also provide insightful conclusions. In the future, it would also be useful to focus on comparing the Financial Administration's communication with other institutions, both Czech and international, in order to assess its effectiveness and create a general methodological recommendation for communication with citizens.

REFERENCES

Kaplan, A. M., & Haenlein, M. (2010). Users of the world, unite! The challenges and opportunities of Social Media. *Business Horizons*, 53, 59–68. http://dx.doi.org/10.1016/j.bushor.2009.09.003

Lin, K., & Lu, H. (2011). Why people use social networking sites: An empirical study integrating network externalities and motivation theory. *Computers in Human Behavior*, 27, 1152–1161. http://dx.doi.org/10.1016/j.chb.2010.12.009

Wondwesen, T. & Anders, W. (2017). A framework for categorizing social media posts, *Cogent Business & Management*, 4:1, 1284390, https://doi.org/10.1080/23311975.2017.1284390

Yang, C. (2021). Research in the Instagram Context: Approaches and Methods, *The Journal of Social Sciences Research*, 7(1), 15–21, 2021, https://doi.org/10.32861/jssr.71.15.21

Olinski M. & Szamrowski P (2021). Facebook as an engagement tool: How are public benefit organizations building relationships with their public?, *PLoS ONE* ,16(9): e0256880. https://doi.org/10.1371/journal.pone.0256880

Triandani, M. & Apollo, A. (2020). Effect The Understanding Of Taxation, Tax Sanctions And Taxpayer Awareness of Taxpayer Compliance: Research On Taxpayers Of Individual Entrepreneurs In Tangerang Region, *Dinasti International Journal of Digital Business Management* (DIJDBM), 2(1), 87–93. https://doi.org/10.31933/dijdbm.v2i1.638

Zichová, T. (2021) 'ICT in Public Education: E-Tax-Learning for Generation Y and Generation Z', in *International Conference on Efficiency and Responsibility in Education, proceedings of the International Scientific Conference ERIE*, Faculty of Economics and Management, Czech University of Life Sciences Prague, Prague, 175–183.

Indriyarti, E.K. & Christian, M. (2020) 'The Impact of Internal and External Factors on Taxpayer Compliance', *Journal of Business & Applied Management*, vol. 13, no. 1, pp. 33–48.

Maier M.A. & Allen, M. (2017). Content Analysis: Advantages and Disadvantages, *The sage encyclopedia of communication research methods*, 1–4, 240–242. https://doi.org/10.4135/9781483381411

Taecharungroj, V. (2016). Starbucks' marketing communications strategy on Twitter. *Journal of Marketing Communications*, 1–19. http://dx.doi.org/10.1080/13527266.2016.1138139

Kim, D., Spiller, L., & Hettche, M. (2015). Analyzing media types and content orientations in Facebook for global brands. *Journal of Research in Interactive Marketing*, 9, 4–30. http://dx.doi.org/10.1108/JRIM-05-2014-0023

Jahn, B., & Kunz, W. (2012). How to transform consumers into fans of your brand. *Journal of Service Management*, 23, 344–361. http://dx.doi.org/10.1108/09564231211248444

Cho, W. & Melisa, W.D. (2021). Citizen Coproduction and Social Media Communication: Delivering a Municipal Government's Urban Services through Digital Participation, *Administrative Sciences*, 11, 59. https://doi.org/10.3390/admsci11020059

Jukić, T. & Merlak, M. (2017). The Use of Social Networking Sites in Public Administration: The Case of Slovenia, *The Electronic Journal of e-Government*, 15(1), 2–18.

Warren, A. M., Sulaiman, A. & Jaafar, N. I. (2014). Social media effects on fostering online civic engagement and building citizen trust and trust in institutions. *Government Information Quarterly*, 31(2), 291-301.

INVESTIGATING USE AND IMPACT OF SOCIAL MEDIA ON STUDENT ACADEMIC PERFORMANCE: CASE OF A UNIVERSITY IN SOUTH AFRICA

Ruth Wario
Department of Computer Science and Informatics, University of the Free State, South Africa
Private Bag X13, Kestell Road, QwaQwa, South Africa

ABSTRACT

In today's world, social media has become an integral part of our social life. Social media is seen as a communication and interacting platform that could be utilized to enhance our connectivity, research, and learning. In recent years, its usage has increased dramatically among the youth and young adults, particularly students being the primary users of social media. With excessive use and a high number of students spending time online, raises the question whether excessive use of social media can affect academic performance. This research therefore investigates the use of social media networking sites and related impact on student academic performance. The study further explores which social media network is the most popular amongst South African university students. A questionnaire survey method was administered to undergraduates at South African University during 2019 academic year. One hundred students participated in the study and the data was analyzed using the Statistical Package for Social Sciences (SPSS) software, version 27. The findings show the negative impact of social media use on student academic performance and also underlines the need to control and manage social media use in academic settings

KEYWORDS

Social Media Usage, Learning, Academic Performance

1. INTRODUCTION

The proliferation of social media in the present age has revolutionized our way of communicating and learning to the extent that it has become our preferred medium of everyday communication and learning. Social media is also seen as a learning tool that could be utilized to enhance student engagement and improve learning and performance. It offers multiple opportunities to both students and institutions to improve teaching and learning methods. Through these networks, students can communicate, get in touch, access information, research, and collaborate. Additionally, institutions can communicate and share important information such as campus news as well as learning resources to students who are connected to the relevant networks and sources. Because of its pivotal role in aiding our communication, the use of social network sites has increased globally and continues to increase.

There were 3.48 billion social media users globally in 2019, growing by 288 million (9 percent) since 2018 (Digital trend stats, 2019). Similar trends were also seen in South Africa. According to South African Business tech reports (2019), about 54 % of the South Africa population has direct access to the internet representing over 31 million people online. South Africa is one of the largest consumers of social media with more than 40% of the population active on social media. The report added that WhatsApp, YouTube, Facebook, Instagram and Twitter are the most popular social media sites with young audiences leading the use. Social media marketing platform Global State of Digital (2019) report found that the typical South African internet user spends a third more time online than Americans and almost double that of Germans. The report further states that a South African user spends 8 hours 23 minutes on the internet per day with a third of that time spent on social media, compared to 7 hours 2 minutes in Singapore and 6 hours 38 minutes in the USA. However, studies concerning South Africa learners' use and impact of social media networking sites on their academic performance has not been done, given the excessive use of social media, especially by college-aged individuals.A previous study on social media activity focused on general use within the South African population as well as the different types of platforms used (Budree et al, 2019; Dlamini & Johnston, 2018; Ogbonnaya & Mji, 2014). Ogbonnaya and Mji (2014) examined the use of social media among students in South Africa. A survey of 200 students from two South African universities showed that almost all the students (99%) adopted social media platforms to connect with friends and relatives and as well as for academic purposes. Recent research indicated the impact social media had on the culture and lifestyle of people, particularly the youth (Nagle, 2018; Jacob, 2015). The prevailing problems affecting the youth with regard to social media are addiction, time consumption, cyberbullying, social isolation, monophobia, poor academic performance and introversion, (Qiaolei et al, 2018; Apuke, 2017; Kumar, et al, 2018; Primack et al. 2017). As reported in South African City Press (2019), the World Federation for Mental Health finds that social media can increase depression and self- harm in young people if excessively used. The report further highlighted the addictive nature of social media such that people spend much of their time on the platform, consequently affecting their studies and behaviour. This research therefore investigates the use of social media networking sites and its impact on student academic performance. The study further explores which social media network is the most popular amongst South African university students. A questionnaire survey method was administered to undergraduate students at a South African university during 2019 academic year. One hundred students participated in the study and the data was analyzed using the Statistical Package for Social Sciences (SPSS) software, version 27. The findings are discussed, and a conclusion drawn.

2. LITERATURE REVIEW

With the proliferation of social media and its excessive use in our institutions, there are questions about its impact on academic performance. There are mixed results regarding use and impacts of social media in an institution of learning. Some studies reported a significant negative relationship between social media and academic performance (Habes, et al., 2018; Owusu-Acheaw & Larson, 2015; Maya, 2015; Baker & Cochran, 2012). Others reported significant positive relationships between social media and academic performance (Lampe, et al., 2015; Sarwar, et al., 2019). Maya (2015) found that spending excessive time on social networking sites has a negative impact on academic performance. According to other studies, this negative impact mainly occurs when social media sites are used solely for social networking, making new friends and chatting, which consequently diminishes student academic performance as they spend more time doing non-academic activities (Bellur, Nowaka & Hullb, 2015; Wood et al., 2012).

Social media is also seen as a distractor in students' ability to concentrate, especially when they study or work on projects and assignments. While studying, students keep checking their social media account for updates, messages and notifications. This is also observed during lecture and classroom teaching, where students pay less attention to the lecture as they are busy chatting with friends or reading unnecessary, non-academic related material. As a result, they miss important information related to academic activities. It seems that students who use social media spend less time studying, with adverse effects on their academic outcomes. Social media has hampered students writing skills in such a way that short forms of words or phrases are always used (Obi et al., 2012). This type of writing negatively affects students' exams, assignments, projects, and ultimately their grades. However, other studies have found no relationship between the use of social media networking sites and academic performance (Lampe, et al., 2015; Sarwar, et al., 2019; Smith, et al., 2017; Park, et al., 2018). Reports show that responding to or posting tweets of an academic nature does not affect learning (Jeffrey et al., 2015).Moreover, some researchers suggest that social networking sites offer added value in educational settings, support collaboration, facilitate discussion and assimilation of knowledge during teaching practices, educational methodologies and theories (Macià & García, 2016; Ricoy & Feliz, 2016) thus, creating the conditions necessary for developing new methodologies (Putnik et al., 2016). The main benefits that social media offer in educational settings stem from their value as a tool for information exchange and sharing (Asterhan & Bouton, 2017) and as a means of socialisation and communication (Balakrishnan & Lay, 2016; Macià & García, 2016). Social media platforms come with many educational materials, which help students broaden their scope of knowledge as well as develop various good skills and talents (Dahlstrom, 2012). The popularity of social media has infiltrated institutions of learning, and is seen as a supporting tool, which aids teaching and learning (Moran & Tinti-Kane, 2012). Teachers are embracing social media sites for effective discussions and dealing with students in matters relating to academics, which improves learning benefits through better within and outside class interactions.

3. RESEARCH METHODOLOGY

The target population in this study was undergraduate students at a South African University during 2019 academic year. A pilot study was conducted to test the validity of the questions. The questions ranged from the type of social media used, its purposes and the time spent on social media sites. One hundred students participated in the study with a return rate of 100%. The data was analyzed using SPSS software, version 27.

4. RESULT AND DISCUSSION

Most the respondents (68%) were female and 32% were male. All respondents (100%) used some form/type of social media and for more than 5 years. The results showed that the most dominant social media tools used by respondents were WhatsApp (99%), Facebook (93%), Instagram (84%) and YouTube (65%).

As indicated in figure 1, the primary reasons for social media site use was for socialising and making new friends (73%) compared to 27% who indicated their reasons for use was study and collaborating with fellow students. Approximately a third (32%) of respondents indicated the spent between 6 to 8 hours on social media on a daily basis, 31 % spent 3 to 5 hours, 28% spent more than 8 hours on social media, and 9% spent 1 to 2 hours (figure 2). Because of the long hours spent on social media platforms, the majority of respondents (64%) indicated social media had a negative impact on their academic performance in contrast to 4% of respondents who disagreed (figure 3). This finding corroborates the arguments made by other researchers who suggest that students who spend much time on social media platforms for chatting and socialising are likely to perform poorly in their academics (Owusu-Acheaw, & Larson, 2015; Asemah, & Okpanachi, 2013). The researchers further report that as time spent on social media platforms increase, academic performance of students deteriorate, because they have less time to study. The majority (73%) of the respondents indicated use of social media tools for non-academic activities, and 27 % used such platforms for academic related activities. Activities included study and collaboration with fellow students. O'keeffe & Clake-pearson (2011) reported that social media benefited students by connecting them to one another for class assignments and projects. This finding indicates the important role social media platforms play in supporting student learning once it is adopted and integrated into classroom instruction. Indeed, one cannot dispute the fact that social media platforms contribute to students' academic life when used judiciously.

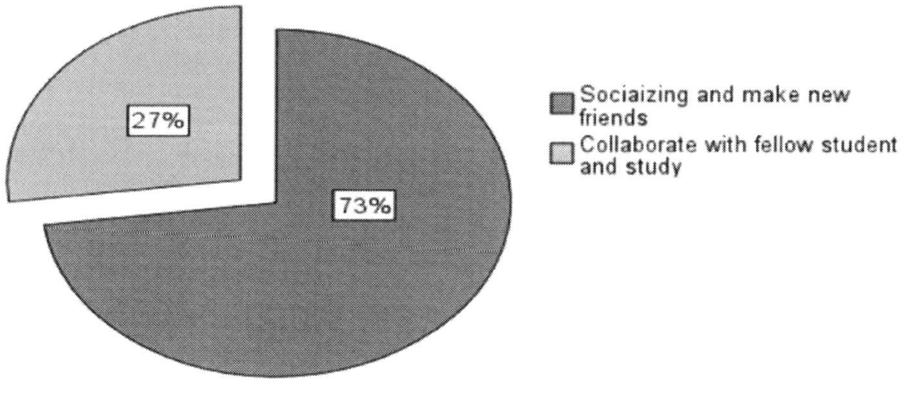

Figure 1. Reasons for using social media

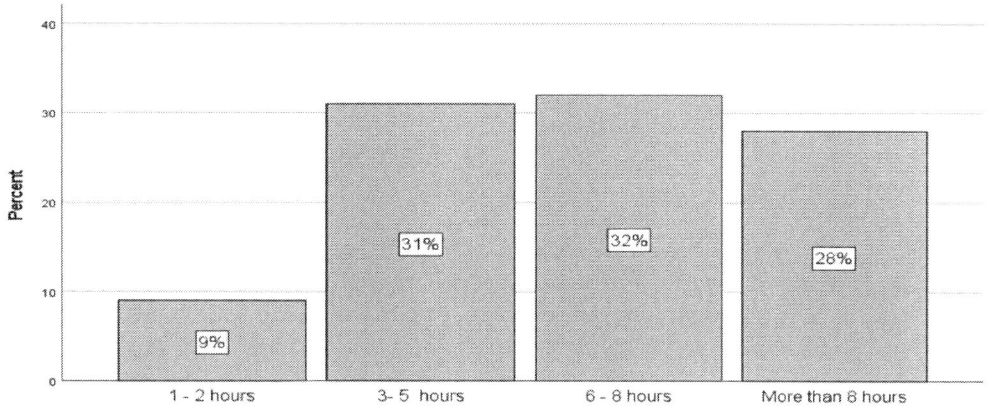

Figure 2. Hours spent daily on social media site

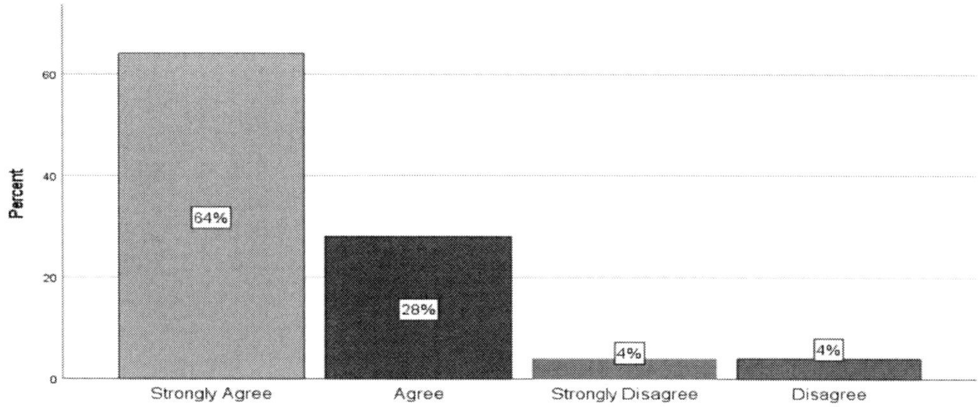

Figure 3. Negative effect of social media on student sudy

The important role played by social media in student life is evident. As shown in Figure 4, when respondents were asked on whether use of social media had any positive effect on their lives. Most of the respondents (68%) agreed that social media indeed positively impacted their lives. It is evident that social media networks play an important role in our day-to-day life, hence should be embraced and integrated.

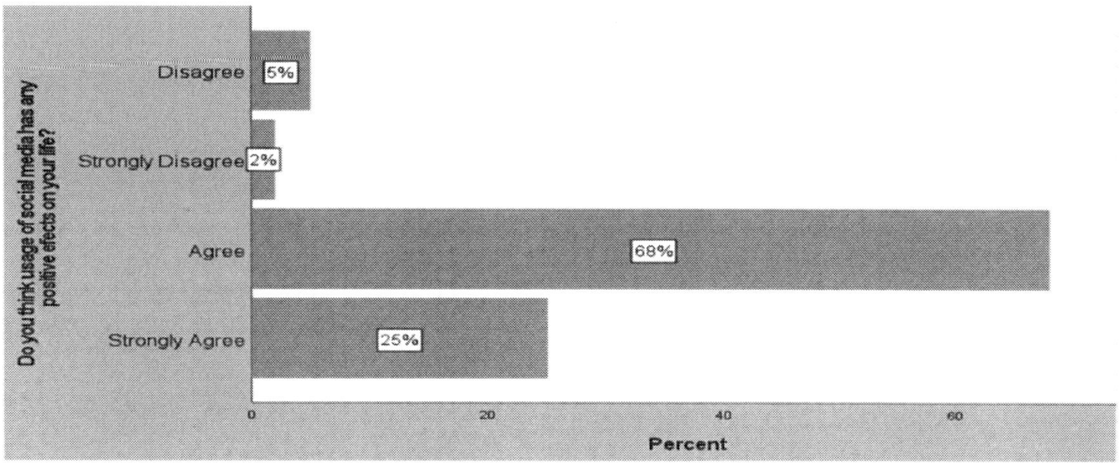

Figure 4. Do you think social media has any positive effect of your life

5. CONCLUSION

Even though social media play a vital role in student life because of its popularity, its benefits have not been fully achieved within institutions of higher learning. This is evident from the findings where the majority of the respondents spend much of their time daily (between three to more than 8 hours) on social media, engaging in non- academic activities such as socialising and making friends. Excessive use of social media on non-academic activities, if unchecked, can affect student academic performance and consequently student drop-out. Studies has proven the benefits that comes with social media use in an institution of learning and this benefit can only be achieved if social media use is geared towards education purposes. Some of these benefits include information and idea sharing, group discussions, and student collaboration and engagement. Therefore, it is imperative that the institutions of higher learning authorities take interventional steps to help students by informing them of the negative consequences of excessive social media use. Uncontrolled use of social media reduces study time, causes fatigue and sleep disruption, which have negative impacts on student concentration level in class, consequently affecting academic performance

6. RECOMMENDATION

Recommendations in the light of the findings include:

- The institution of higher learning needs to adopt a new strategy in integrating social media tools into student teaching and learning. This can be achieved through channelling students' assignments, projects and discussions on social media tools to help inculcate the habit of using these tools for academic purposes.
- There is need for an awareness campaign among the students to promote social media networks as a tool not only for communication, entertainment and making friends but also for learning. This awareness campaign should include the negative consequences of excessive social media use to avoid obsession with these sites.
- The institution of higher learning authorities should restrict access to certain social media sites that may distract students' attention during class/school hours.
- Students should be encouraged to use social media network sites judiciously to ensure that they do not influence their academic performance negatively.

REFERENCES

Amedie, J. (2015). The Impact of social media on Society. Advanced Writing: Pop Culture Intersections. http://scholarcommons.scu.edu/engl_176/2

Apuke O. (2017). The Influence of social media on Academic Performance of Taraba State University Undergraduate Students, *Journal of Communication and Media Technologies* 7(4).

Asemah, S., Okpanachi, R. (2013). Influence of social media on the academic performance of the undergraduate students of Kogi State University, Anyigba, Nigeria, *Research on Humanities and Social Sciences*, 3(12), 90-96

Asterhan, C., & Bouton, E. (2017). Teenage peer-to-peer knowledge sharing through social network sites in secondary schools. *Computers & Education, 110,* 16-34. doi: 10.1016/j.compedu.2017.03.007

Baker, P. & Cochran, D. (2012). Effect of Online Social Networking on Student Academic Performance. *Computers in Human Behavior,* 28(6),2117-2127. http://dx.doi.org/10.1016/j.chb.2012.06.016

Bellur, K. Nowaka, & Hullb, S. (2015). Make it our time: In class multitaskers have lower academic performance, *Computers in Human Behavior,* 53(1), 63-70. doi: 10.1016/j.chb.2015.06.027

Budree, A., Fietkiewicz, K. & Lins, E. (2019). Investigating usage of social media platforms in South Africa, *The African Journal of Information Systems*: 11(4); 314-336.

Dahlstrom, E. (2012). *ECAR* study of undergraduate students and information technology. (Research Report). Louisville, CO: EDUCAUSE Center for Applied Research.

Digital trend stats (2019). Digital trends 2019: Every single stat you need to know about the internet. Retrieved March 2019 https://thenextweb.com/contributors/2019/01/30/digital-trends-2019-every-single-stat-you-need-to-know-about-the-internet/

Dlamini; N. & Johnston, K. (2018). The use of social media by South African organisations *Journal of Advances in Management Research* 15(5). doi: 10.1108/JAMR-05-2017-0063

Global State of Digital (2019). South African internet users spend much more time online than Americans and Europeans. Retrieved March 2019 from https://www.businessinsider.co.za/south-africa-one-of-the-worlds-top-internet-users-hootsuite-report-2019-2

Habes, M., Alghizzawi, M., Khalaf, R., Salloum, A. & Mazuri, G. (2018). The Relationship between social media and Academic Performance: Facebook Perspective. *International Journal of Information Technology and Language Studies*. 2(1), 12-18.

Jeffrey H., Kuznekoff, M. & Scott T. (2015). Mobile Phones in the Classroom: Examining the Effects of Texting, Twitter, and Message Content on Student Learning, *Communication Education*, 64:3, 344-365, doi: 10.1080/03634523.2015.1038727

Lampe, C., Wohn, D., Vitak, J., Ellison, N., & Wash, R. (2015). Student use of Facebook for organizing collaborative classroom activities. *Computer Supported Collaborative Learning, 6*, 329–347.

Macià, M., & García, I. (2016). Informal online communities and networks as a source of teacher professional development: A review. *Teaching and Teacher Education, 55,* 291-307. doi: 10.1016/j.tate.2016.01.021

Maya, k., (2015). Achievement scripts, media influences on Blacks students' academic performance, self-perceptions and carrier interests. *Journal of Black psychology*, 42(3),.195-220. doi: 10.1177/0095798414566510.

Moran, M., Seaman, J. & Tinti-Kane, H. (2012). *Blogs, wikis, podcasts and Facebook: how today's higher education faculty use social media, 2012*. (Babson Survey Research Group). Boston, MA: Pearson Learning Solutions.

Nagle, J. (2018). Twitter, cyber-violence, and the need for a critical social media literacy in teacher education: A review of the literature. *Teaching and Teacher Education, 76*, 86-94.

Obi, C., Bulus, L., Adamu, M., & Sala'at, A. (2012). The need for safety consciousness among Youths on social Networking Sites. *Journal of Applied Science and Management (JASM),* 14(1).

Ogbonnaya, U. & Mji, A. (2014) Use of social media by university students in South Africa, *EDULEARN.* https://library.iated.org/view/OGBONNAYA2014USE

O'Keeffe G, Clarke-Pearson K. (2011). Council on Communications and Media. The impact of social media on children, adolescents, and families. Pediatrics.;127(4):800-4. doi: 10.1542/peds.2011-0054. Epub 2011 Mar 28. PMID: 21444588

Owusu-Acheaw, M., & Larson, A. (2015). Use of social media and its impact on academic performance of tertiary institution students: A study of students of Koforidua Polytechnic, Ghana. *Journal of Education and Practice*, *6*(6), 94-101.

Park, E., Song, H., & Hong, A. (2018). The use of social networking services for classroom engagement? The effects of Facebook usage and the moderating role of user motivation. *Active Learning in Higher Education.* https://doi.org/10.1177/1469787418809227

Primack, B., Shensa, A., Sidani, J. (2017). Social Media Use and Perceived Social Isolation Among Young Adults in the U.S. *American Journal of Preventive Medicine 53(1).* Doi: https://doi.org/10.1016/j.amepre.2017.01.010

Putnik, G., Costa, E., Alves, C., Castro, H., Varela, L., & Shahl, V. (2016). Analysing the correlation between social network analysis measures and performance of students in social network-based engineering education. *International Journal of Technology and Design Education*, *26*(3), 413-437. doi: 10.1007/s10798-015 9318-z

Qiaolei, J., Xiuqi, T. & Ran, T. (2018). Examining factors influencing internet addiction and adolescent risk behaviors among ecessive internet users. Health communication, 33, 1424-1444.

Sarwar, B., Zulfiqar, S., Aziz, S., & Ejaz, C. (2019). Usage of social media tools for collaborative learning: The effect on learning success with the moderating role of cyberbullying. *Journal of Educational Computing Research, 57*, 246–279.

South Africa City press (2019). Social media can increase depression and self-harm in young people. Retrieved January 2019 https://www.news24.com/citypress/voices/social-media-can-increase-depression-and-self-harm-in-young-people-20181027

South African Business tech reports. (2019). Biggest social media and chat platforms in 2019. Retrieved April 2019 from https://businesstech.co.za/news/internet/296752/these-are-the-biggest-social-media-and-chat-platforms-in-2019/

Smith, E. (2017). Social media in undergraduate learning: Categories and characteristics. *International Journal of Educational Technology in Higher Education, 14*(1), 1–24.

Wood, E., Zivcakova, L., Gentile, P., Archer, K., De Pasquale, D., & Nosko, A. (2012). Examining the impact of off-task multi-tasking with technology on real-time classroom learning. *Computers & Education,* 58(1), 365–374.

MINDSET*PLUS*: THE 'MANAGEMENT AND INFORMATION DECISION SUPPORT EPILEPSY TOOL' TO PROMOTE ASSESSMENT, GOAL-BASED SKILLS TRAINING, AND SERVICE LINKAGE FOR PEOPLE WITH EPILEPSY

Ross Shegog[1], Refugio Sepulveda[2], Katarzyna Czerniak[3], Rosalia Guerrero[4], Alejandra Garcia-Quintana[5], Robert Addy[6], Kimberly Martin[7], Latasha Jackson[8] and David Labiner[9]

[1]*PhD; UTHealth School of Public Health, Houston, Texas, 7000 Fannin St., Suite 2668, Houston, Texas, 77030, USA*
[2]*PhD; Dept. of Neurology, University of Arizona, 1501 N. Campbell, Tucson AZ 85724, USA*
[3]*MLA, MPH; UTHealth School of Public Health, Houston, Texas, 7000 Fannin St., Suite 2672-2, Houston TX 77030, USA*
[4]*MPH; UTHealth School of Public Health, Houston, Texas, 1200 Pressler St., RAS E905, Houston TX 77030, USA*
[5]*DDS; UTHealth School of Public Health, Houston, Texas, 7000 Fannin St., Suite 2672-1, Houston TX 77030, USA*
[6]*PhD; UTHealth School of Public Health, Houston, Texas, 7000 Fannin St., Suite 2659, Houston TX 77030, USA*
[7]*LVN; Epilepsy Foundation Central & South Texas, 8601 Village Dr., Suite 220, San Antonio, TX 78217, USA*
[8]*BS; Epilepsy Foundation of Texas, 2401 Fountain View Drive, Houston, Texas 77057, USA*
[9]*MD; Head of Neurology Department, University of Arizona, 1501 N Campbell Ave, Tucson, AZ 85724-5023, USA*

ABSTRACT

Introduction: People with epilepsy can adhere to epilepsy self-management behaviors to improve seizure control, medication adherence, and lifestyle factors that contribute to seizures. Responsive online interventions can assist patients and providers to assess self-management, set treatment goals, and decide on education and social service programs. The Management Information & Decision Support Epilepsy Tool (MINDSET) is a bilingual online program designed to improve patient-provider communication to enhance epilepsy self-management. MINDSET may have utility for community health workers when assisting patients to improve their self-management. **Purpose**: To enhance MINDSET to include recommendations for education and social service programs ('MINDSETPlus') and to establish an implementation framework to facilitate use of MINDSETPlus by community health workers in community-based neurology clinics. **Methods**: An expert advisory group, comprising stakeholders from the Epilepsy Foundations in Texas and the Universities of Texas and Arizona provided formative review and consensus on MINDSET enhancements. Implementation theory and expert consensus informed a phased implementation framework. **Results**: MINDSETPlus enables patients to assess their self-management, select behavioral goals (for seizure, medication and lifestyle management), receive recommendations for further training tailored on current self-management and/or co-morbidities (depression and memory), and cue their community health worker to priority social determinants. A phased framework was derived for onboarding, training and implementing the MINDSETPlus-mediated intervention in neurology clinics. **Conclusion**: MINDSETPlus provides decision support for community health workers that may improve fidelity and metrics for quality improvement and assist to navigate, assess, reinforce, educate, and link epilepsy patients to community programs and services. Feasibility and efficacy testing of the intervention is in progress.

KEYWORDS

Epilepsy, eHealth, Self-Management, Community Health Workers, Chronic Disease, Decision Support

1. INTRODUCTION

Epilepsy is among the most common neurological conditions affecting approximately 3.4 million people in the United States (IOM, 2012). People with epilepsy can adhere to self-management behaviors to improve seizure control, medication adherence, and lifestyle factors that contribute to seizures (Helmers et al., 2017). The Center for Disease Control's Managing Epilepsy Well Network (MEWN) 2.0 initiative is designed to increase the evidence base for existing epilepsy self-management programs (Helmers et al., 2017; Sajatovic et al., 2021). These programs include PACES (self-management skills training), UPLIFT (depression management), HOBSCOTCH (memory management) and MINDSET (described below) (Fraser et al., 2015; Thompson et al., 2015; Caller et al., 2016). Community health workers are certified public health workers who can link patients to needed health and educational programs, and social services in the community. They share lived experiences with the population served (Crespo et al., 2020) and are well positioned to use eHealth decision support to assess and assist patients in meeting their self-management goals. Responsive, mobile, web-based interventions for patient self-management assessment, action plan development, and linkage to evidence-based programs and social services may enhance the role of community health works in assisting patients in the neurology clinics. An eHealth-mediated protocol may also benefit health care practice by standardizing care, increasing practice fidelity, and providing ongoing quality assurance.

The Management Information & Decision Support Epilepsy Tool (MINDSET) is bilingual tablet-based online decision support to improve patient-provider communication to enhance self-management of epilepsy patients through tailored behavioral goal selection (Fig. 1). At their clinic visit patients input data on seizure, medication, and lifestyle management and then decide on self-management goals for 3 behaviors identified as needing improved adherence (Fig. 2). The patient and provider can then print and review a dynamically constructed tailored action plan. MINDSET has demonstrated acceptability and feasibility for use by epilepsy patients and health care providers during regular clinic visits, and effectiveness in significantly increasing epilepsy self-management behaviors (Shegog, 2020). However, the original MINDSET program was outdated, requiring upgrade, and was not designed for community health workers to assess patient eligibility for recommended programs and services. The purpose of this study was to 1) update the MINDSET architecture and function (MINDSET*Plus*) and 2) establish a framework for community health workers to use MINDSETPlus to navigate, assess, and educate patients in community-based neurology clinics and to link them to needed programs and services.

Figure 1. MINDSET splash screen and use

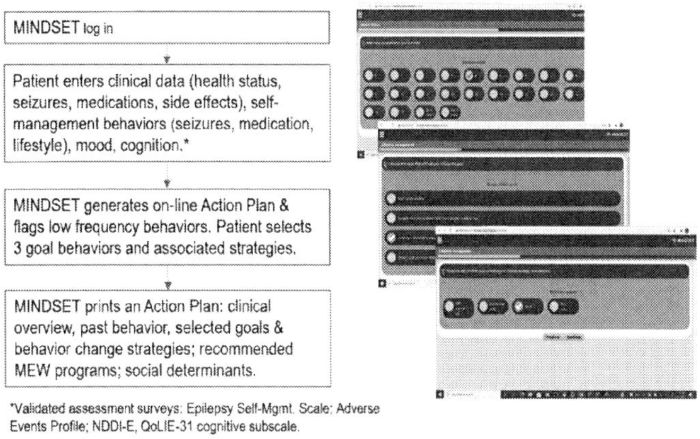

Figure 2. MINDSET flow, data entry, and example screens

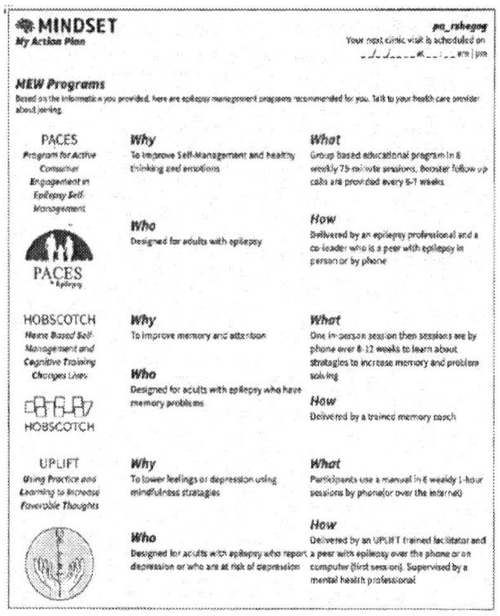

Figure 3. Sample Action Plan page

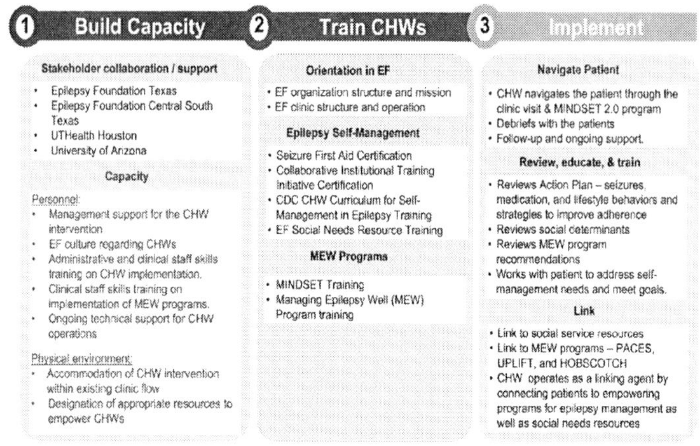

Figure 4. Three phase implementation

2. METHODS

An expert advisory group comprised stakeholders from the Epilepsy Foundation (EF) Texas (admin., neurologist, clinic coordinator, program managers), and Central South Texas (admin./educator), UTHealth (community health worker certification specialist, behavioral scientists, and analyst), and the University of Arizona (neurologist, behavioral scientist). The expert advisory group provided formative review and consensus on MINDSET enhancements. A phased development protocol was used that comprised: (1) content analysis to identify needed enhancements to specifications and functions including the back-end database; (2) literature review to determine the decision criteria for inclusion of epilepsy patients into Managing Epilepsy Well self-management programs and needed social support services; (3) expert review of proposed decision algorithms and design features; (4) user manual and design document development and expert review; (5) software upgrade to a responsive cross-platform Internet-accessible website; and (6) alpha, usability and feasibility evaluations. Implementation theory (Interactive Systems Framework & RE-AIM Model) and expert consensus informed the identification of facilitators and barriers for community health worker onboarding, training, and implementation (Wandersman et al., 2008; Glasgow et al., 1999). Weekly expert advisory group meetings provided consensus on a stepped process model to onboard, train, and integrate community health workers with MINDSET*Plus* decision support into neurology clinics.

3. RESULTS

3.1 *MINDSETPlus*

MINDSET*Plus* incorporates validated assessment surveys to assess epilepsy patients. These include the Epilepsy Self-management (ESM) scale (DiOrio et al., 2004), Neurological Disorders Depression Inventory in Epilepsy (NDDI-E)(Gilliam et al., 2006; Friedman et al, 2009), QoLIE-31 cognitive subscale (Cramer et al., 1998), and Health Leads social determinant inventory (Health Leads, 2017). MINDSETPlus provides decision support to the community health worker to determine eligibility of a patient to enroll in evidence-based epilepsy self-management programs (PACES, UPLIFT, HOBSCOTCH) by using scale and item scores. The ESM indicates eligibility for PACES, a general epilepsy self-management program suitable for all patients with epilepsy and particularly those challenged in self-management (Fraser et al., 2015). The NDDI-E provides indicates eligibility for UPLIFT, a training program focused on managing depression (Thompson et al., 2015) and the QoLIE-31 cognitive scale indicates eligibility for HOBSCOTCH, a training program to enhance memory (Caller et al.,2016). The Healthy Leads inventory indicates patient priorities on social determinants of health (Health Leads, 2017). MINDSET was upgraded from Adobe AIR technology to contemporary web-based standards (HTML, CSS, and Javascript on a Cordova framework) for cross-platform access from smart phone (iOS, Android), tablet, and Windows desktop devices (Chrome and Safari browsers). MINDSETPlus provides a printable tailored Action Plan of patient selected ESM behavioral goals (in seizure, medication and lifestyle domains). Plan enhancements include cues for PACES, UPLIFT, and/or HOBSCOTCH use and social determinants possibly requiring social services (Figure 3).

3.2 Community Health Worker Implementation Framework

A 3-phase framework describes the steps for implementing the community health worker (CHW) intervention with MINDSETPlus decision support in neurology clinics, providing a blueprint to address core elements of capacity building, training, and implementation (Fig. 4). The framework is designed to address identified individual and organizational level facilitators (n=8) and barriers (n=8) of implementation. These included: (1) individual level factors for community health workers (knowledge and skills in assessment, intervention, and support linkage) for epilepsy self-management, co-morbidities (depression, cognitive deficit), and social determinants; (2) individual level factors for clinic coordinators (knowledge and skills in managing community health worker's in these roles); (3) organizational capacity factors for neurology clinics (i.e. mission, resource, management support, personnel, clinic flow and function, availability of self-management education programs (PACES, UPLIFT, HOBSCOTCH), social support programs, and mental health counselling); and (4) motivational factors related to the intervention (i.e. relative advantage, compatibility, complexity, trialability, and observability). The framework requires collaboration and resources from state credentialed community health worker training programs (UTHealth), the epilepsy community health worker training program (CDC, Dartmouth), social service resources (EF), self-management programs (CDC MEWN), and digital decision support (UTHealth, Univ. Arizona).

4. CONCLUSION

This study is significant in using eHealth to assist community health workers in neurology clinic settings. MINDSETPlus provides decision support by compiling a tailored action plan that confirms patients' epilepsy self-management behaviors and goals. It assists community health workers to navigate, assess, reinforce, educate, and link epilepsy patients to community educational programs and services appropriate to the patient's needs. This is supportive of the CDC MEWN 2.0 national initiative to encourage epilepsy patients' use of evidence-based self-management programs. Identification of individual, organizational, and intervention level implementation facilitators and barriers enables advanced planning for implementing eHealth decision-support in neurology clinics. Providing community health workers with eHealth decision support in this setting is innovative. It can enhance collection of quality improvement metrics on service fidelity and provide professional assessment and feedback. Future work is indicated to study the human factors and usability parameters of MINDSET*Plus* with community health workers and the feasibility and efficacy of this combination in the neurology clinic setting.

ACKNOWLEDGEMENT

This work was funded by the Center for Disease Control (CDC) Special Interest Projects SIP19-003; SIP20-006.

REFERENCES

Caller TA et al, 2016. A cognitive behavioral intervention (HOBSCOTCH) improves quality of life and attention in epilepsy. *Epilepsy and Behavior*, 57(pt A):111–7.

Cramer JA et al, 1998, Development and Cross-Cultural Translations of a 3 1 –Item Quality of Life in Epilepsy Inventory. *Epilepsia*; 39(1):81-88.

Crespo R et al, 2020. An emerging model for community health worker-based chronic care management for patients with high health care costs in rural Appalachia. Preventing Chronic Disease. *Public Health Research, Practice, and Policy*. 2020, 12, E13.

DiIorio C et al, 2004. Project EASE: a study to test a psychosocial model of epilepsy medication management. *Epilepsy and Behavior*, 5:926–36.

Fraser RT et al, 2015. PACES in epilepsy: results of a self-management randomized controlled trial. *Epilepsia*, 56: 1264–74.

Friedman DE et al, 2009, Identifying depression in epilepsy in a busy clinical setting is enhanced with systematic screening. *Seizure*, 18:429–33.

Gilliam FG et al, 2006. Rapid detection of major depression in epilepsy: a multi-centre study. *Lancet Neurol*, 5:399–405.

Glasgow REM et al, 1999, Evaluating the Public Health Impact of Health Promotion Interventions: The RE-AIM Framework, *AJPH*, 89:1322-1327.

Health Leads Inc. *Social Needs Screening Tool*, © 2018. Available at https://healthleadsusa.org/resources/the-health-leads-screening-toolkit/ Accessed May 19th, 2022.

Helmers SL et al, 2017. Self-management in epilepsy—Why and how you should incorporate self-management into your practice. *Epilepsy and Behavior*, 68:220–4.

IOM (2012). Institute of Medicine (US) Committee on the Public Health Dimensions of the Epilepsies. *Epilepsy Across the Spectrum: Promoting Health and Understanding*. Washington (DC): National Academies Press.

Sajatovic M et al, 2021. The Managing Epilepsy Well (MEW) Network database: Lessons learned in refining and implementing an integrated data tool in service of a national U.S. research collaborative. *Epilepsy and Behavior*. 115, Feb., 107650.

Shegog R et al, 2020. MINDSET: Clinic-based decision support demonstrates longitudinal efficacy for increased epilepsy self-management adherence among Spanish speaking patients. *Epilepsy and Behavior*. 113.

Thompson NJ et al, 2015. Expanding the efficacy of project UPLIFT: distance delivery of mindfulness based depression prevention to people with epilepsy. *J Consult Clin Psychol*, 83(2):304–13.

Wandersman A et al, 2008, Bridging the Gap Between Prevention Research and Practice: The Interactive Systems Framework for Dissemination and Implementation. *Am. J. Community Pychol*, 41:171-181. DOI 10.1007/s10464-008-9174-z

REFLECTION PAPERS

FEMINIST THEMATIC DISCOURSE ANALYSIS IN CS

Alice Ashcroft
Lancaster University
School of Computing and Communications, Lancaster University, Lancaster, LA1 4WA, UK

ABSTRACT

How conversation takes place is a well-researched area in the field of Linguistics, particularly when it comes to how Feminist Methodologies are applied to this. What remains to be seen, however, is the application of these techniques in the field of Computer Science. Although some research has started to emerge in this area, this paper argues that there needs to be an examination of the subtleties in conversation, through a union of three methodological practices: thematic analysis, feminist methodologies, and discourse analysis. This reflective paper summarises a consideration on discourse and conversation analysis, the existing cross over with feminist research, and presents areas of further research within the field of Computer Science.

KEYWORDS

Conversation, Discourse, Feminism, Methodologies, Design

1. INTRODUCTION

When considering the analysis of conversation in design, there are a number of methodologies outlined in the field of linguistics, many of which have already been adapted to be in line with theories surrounding Feminist Methodologies (Sprague, 2016) and therefore may prove valuable. Conversation Analysis (CA) and Discourse Analysis (DA) both offer differing advantages, and pose contrasting limitations, but both do allow for the analysis and understanding of how conversation takes place (Wooffitt, 2005). How these can be brought into alignment with Feminist Methodologies, as stated, is an area that has been researched over many years within linguistics, with Boden (1994) and Holmes (1986) looking at this more historically, and scholars such as Stokoe and Weatherall (2002) leading the narrative more recently. What remains to be seen, however, is how these can be applied thematically to the field of Computer Science (CS) and the sub-fields within this.

This paper will outline the existing areas of research and understanding, and how these could overlap when it comes to methodologies within the field of CS, not just linguistics. The effect a field of research has on a methodology is well understood (Wisniewski, et al., 2018), as context is always significant in research, and perhaps even more so when Feminist Methodologies are applied. As this applies to CS, and Human Computer Interaction (HCI) within this, it is important to consider that Feminist Methodologies and "Feminism seems well positioned to support HCI's increasing awareness and accountability for its own social and cultural consequences" (Bardzell & Bardzell, 2011). When it comes to Feminist Epistemology, however, "there has been debate between feminists about whether there can be feminist epistemology" (Barbour, 2018), but if epistemology refers to the theory of knowledge and understanding, and the subjects of the research 'have gender', then the argument from feminists that "gender and individual identity are significant in the process of becoming a subject and a knower" (Flax, 1993) (Barbour, 2018), then surely they must be relevant in social research. How this applied in CS and HCI, however must be considered, they are areas of research where often logic and structure are seen as important values, and therefore this may clash with the more general understanding of how knowledge is formed. As stated by Hancox-Li & Kumar (2021), "feminist epistemology has long taken a critical stance towards fully formalized systems, instead emphasizing the interactive nature of knowledge creation and the importance of exploring multiple possible meanings". This could be argued to be even more important to consider where gendered language is the topic of research, as this research quickly became, as the misunderstanding of language, or its interpretation is key to understanding the effect of gendered language in CS.

2. CONVERSATION AND DISCOURSE ANALYSIS

CA and DA offer similar approaches in the "qualitative analysis of the functional and sense-making properties of language" (Wooffitt, 2005). The similarities in these methodologies cover 'talk' as a topic for analysis, and the way in which this is done in both approaches could be argued to be quite similar when it comes to their break down of conversation. The main differences between CA and DA lie in substantive and methodological issues (Wooffitt, 2005). These differences lend themselves to discussions surrounding Feminist Methodologies by each method, allowing a different level of detail to be applied to conversation depending on the aim of the analysis. It could be suggested that it should be the aim of both practices to allow for Feminist Methodologies to be applied, but with the combination of Thematic Analysis, this may allow a broader approach to be applied when it comes to CS.

2.1 Thematic Conversation Analysis

Thematic Conversation Analysis (TCA) is the process of applying Thematic Analysis methodologies onto Conversation (Ashcroft, 2020). Whilst CA relies heavily upon the coding of conversation and specific attention being given to intonation, TCA is more concerned with how things are said through phrasing, as opposed to the overlaps, and more detailed structure (Ashcroft, 2020) than CA is traditionally interested in (Wooffitt, 2005).

TCA does, however, bear quite a few resemblances to DA, in that it focuses on the interactions as opposed to the more detailed parts of the conversation, allowing for themes to then be extracted from this. How this applies to feminist theories is already well understood by the linguistics community, but the impact this has on CS and the products which are continuously designed, used, and built by the CS community has yet to be uncovered in much depth. There is an understanding of the need for variety when it comes to gender to be present and involved in the process, but seemingly very little practical research has been done when it comes to the direct impact gendered language has on CS. Furthermore, this could be supported with a systematic study or literature review to fully understand the potential areas of CS this may impact. However, to understand how this can be applied further, the principles of Feminist Conversation Analysis should be comprehended, and then considered with regards to CS itself.

2.2 Feminist Conversation Analysis

The fundamental principle of Feminist Methodologies lies in the assumption that any prior research or literature may be built upon a patriarchal bias (Sprague, 2016). The removal of this could be argued to be simply good research practice, and any researcher should strive to have no bias in their work, yet since bias is often unconscious, a conscious effort should therefore be made in order to overcome this. Furthermore, Bardzell and Bardzell (2011) clearly outline the differences between "Gender and Computing" as a field of research, and "Feminist HCI methodologies" as the application of feminist methodologies to the field of HCI. This important distinction must be considered throughout this reflective paper, as although the field of this researcher is Gender and CS, and within this HCI, the aim of this reflection is to understand and explore Feminist CA in context. Therefore, when Feminist Methodologies are applied to CA, work by Stokoe and Weatherall outlines how an understanding of gender, and the way in which boys and girls are taught to speak, leads onto how men and women do speak. This understanding is paramount when carrying out any observation, recording, and analysis of conversation (Stokoe & Weatherall, 2002). Throughout their work, they pose that many classic CA traits are not immune to being affected by gender (Stokoe & Weatherall, 2002). Furthermore, the researchers make comment to the widely argued discussion that "gender difference research is counterproductive for feminism because it reifies the gender dualism and perpetuates stereotypes". However, they also state that although gender is something they "have" and not something they "do" (Stokoe & Weatherall, 2002), it is important to consider the affect that this, and any other trait a person has, will have on conversation that takes place. This includes not only how words are spoken, but how it they are received. What shall be discussed in Section 3 of this paper, is the impact this may have on the field of CS, when conversation takes place.

3. REFLECTION

Applying a Thematic Analysis to discourse, with the understanding of the principles of Feminist Conversation Analysis, could be argued to be a vital part of understanding any process with Computer Science. Conversation is seemingly one of the main ways in which decisions are made and interactions take place, even if these are done digitally (Brooke, 2021). Although efforts are continuously made to implore the significance and importance of interdisciplinary research, there seem to be only a few areas of CS where the intersection of gender, language and CS have begun to be uncovered (Ashcroft, 2020). The importance of this cannot be understated; only by looking for existing areas of research, their suggested methodologies, and applying these to CS, can we uncover if there are any issues caused by gender, and only then action be taken, to overcome these issues.

Traits of conversation, such as turn-taking and overlap, consistently referenced throughout a range of CA literature, must be analysed in industry practices within CS, and other fields within CS (e.g. CS Education), to uncover their significance. Failing to do so would be a waste of an opportunity provided by the field of linguistics.

4. FURTHER RESEARCH REQUIRED AND CONCLUSION

If all members of the CS community are not only recruited into the sector without bias, but then listened to and respected within the community, the advantages of this cannot only be seen in the bottom line of an organization (Hunt, et al., 2018), but lead to higher employee retention (Holtzblatt & Marsden, 2018), an increased sense of belonging (Widdicks, et al., 2021), and the creation of services which contain fewer amounts of bias, and therefore advantage the customers more significantly (Criado Perez, 2019). Therefore, further research in this area is essential to uncover any potential changes which should be implemented in the sector to ensure that when women, or any under-represented group for that matter, are in the room, they are heard, listened to, and their views are given the same heed as any other.

ACKNOWLEDGEMENT

With thanks to my PhD supervisors Dr Mark Rouncefield and Dr Lynne Blair for their support and guidance, to Rebecca Hall for her helpful suggestions, and to the reviewers of this reflective paper for their comments and critique.

REFERENCES

Ashcroft, A., 2020. *Gender Differences in Innovation Design: A Thematic Conversation Analysis.* s.l., 32nd Australian Conference on Human-Computer Interaction.

Barbour, K., 2018. Embodied Ways of Knowing: Revisiting Feminist Epistemology. In: L. Mansfield, J. Caudwell, B. Wheaton & B. Watson, eds. *The Palgrave Handbook of Feminism and Sport, Leisure and Physical Education.* London: Palgrave Macmillan UK, pp. 209-226.

Bardzell, S. & Bardzell, J., 2011. Towards a feminist HCI methodology: social science, feminism, and HCI. *Proceedings of the SIGCHI conference on human factors in computing systems* , pp. 675-684.

Boden, D., 1994. *The Business of Talk: Organizations in Action.* s.l.:Cambridge: Polity Press.

Brooke, S. J., 2021. Trouble in programmer's paradise: gender-biases in sharing and recognising technical knowledge on Stack Overflow. *Information, Communication & Society,* 24(14), pp. 2091-2112.

Criado Perez, C., 2019. *Invisible Women: Exposing Data Bias in a World Designed for Men.* 10th Edition ed. s.l.:Vintage.

Flax, J., 1993. Disputed Subjects: Essays on Psychoanalysis, Politics, and Philosophy.

Hancox-Li, L. & Kumar, I. E., 2021. Epistemic Values in Feature Importance Methods: Lessons from Feminist Epistemology. In: *Proceedings of the 2021 ACM Conference on Fairness, Accountability, and Transparency.* s.l.:Association for Computing Machinery, p. 817–826.

Holmes, J., 1986. Functions of You Know in Women's and Men's Speech Language in Society. *Language in Society.*

Holtzblatt, K. & Marsden, N., 2018. *Conference on Human Factors in Computing Systems - Proceedings.* s.l.:s.n.

Hunt, V., Yee, L., Prince, S. & Dixon-Fyle, S., 2018. *McKinsey & Company: McKinsey & Company Home People & Organizational Performance.* [Online] Available at: https://www.mckinsey.com/business-functions/people-and-organizational-performance/our-insights/delivering-through-diversity [Accessed 13 January 2022].

Sprague, J., 2016. *Feminist methodologies for critical researchers: Bridging difference.* 2nd Edition ed. s.l.:Walnut Creek.

Stokoe, E. & Weatherall, A., 2002. Gender, language, conversation analysis and feminism. *Discourse and Society,* 13(6), pp. 707-713.

Widdicks, K., Ashcroft, A., Winter, E. & Blair, L., 2021. *Women's Sense of Belonging in Computer Science Education: The Need for a Collective Response.* s.l., United Kingdom and Ireland Computing Education Research conference.

Wisniewski, P. J. et al., 2018. *Intersectionality as a Lens to Promote Equity and Inclusivity within SIGCHI.* s.l., s.n.

Wooffitt, R., 2005. *Conversation Analysis and Discourse Analysis.* s.l.:SAGE Publications.

AN INCENTIVE MODEL FOR PATIENT ADHERENCE TO A HEALTH APP

Cândida Sofia Machado and Cláudia Cardoso
Polytechnic Institute of Cávado and Ave – School of Management
Barcelos – Portugal

ABSTRACT

Electronic health records show potential gains both in economic and health terms, especially when such records are as complete as possible, and patient centred. Its use can be enhanced when integrated into an app, easily accessible and used by patients. However, both patient centred electronic records and health apps are not yet widely used by patients, nor integrated into health systems. The question we propose to discuss is whether health app developers' business models have the right incentives for patients to add to these apps, feeding them with data; or whether patients should be treated as data providers and be remunerated for it.

KEYWORDS

Health Apps, Electronic Health Records, Multi-Sided Markets

1. INTRODUCTION

We have good news and bad news about the future of health systems and their ability to offer adequate health care. The bad news is that demand for health services is increasing rapidly, mainly due to an ageing population; the good news is that technological innovation has presented more and better solutions to improve these same health systems. As a good example, we have information technology tools that can help solve many problems of organization, interoperability, and rapid responses of health services.

However, we observe a slow adoption of the tools available and a cluster of problems that could be mitigated if these tools were used massively.

With this reflection paper, we propose to discuss how the incorrect definition of business models, adopted for the electronic personal health records (through an app), is being a barrier on their dissemination. To this end, we begin by presenting a review of the obstacles to patient adherence to it; then, review the main concepts associated with multi-sided markets; and finally, we propose a new approach for patient reward when using a health app.

2. OBSTACLES TO PATIENT ADHERENCE TO ELECTRONIC PERSONAL HEALTH RECORDS

Electronic personal health records (EPHR) are widely used by health providers and public health services. However, patient's participation and involvement remain limited. In most cases, records are fed and used by health professionals. However, mainly, the objective was for these records to be patient-centred, encompassing both information produced in the context of health care delivery or information produced by the patient, and accompanying the patient both throughout life and in the different health services.

The increasingly widespread use of all types of apps opens the possibility of using this kind of platform for patients to control their EPHR. However, the adoption and use of health apps has also faced several obstacles. Therefore, it is important to account for the factors that explain the adherence or resistance to EPHR (usable through an app). The literature presents factors related to the technology, the characteristics of patients or the characteristics of the health system or health services.

Lui et al. (2013) showed that the physician-patient relationship and perceived usefulness for patients had significant effects on their behavioural intentions to use Internet-based personal health record systems. Also, they showed that the perceived ease of use affected the patients' behavioural intentions indirectly, through the perceived usefulness. For mobile health (m-health) perceived usefulness, perceived ease of use, perceived reliability, and perceived security and privacy are identified as variables that can influence the attitude towards adopting the technology (Shareef et al., 2014).

The involvement of physicians helps to overcome the resistance to share information. In fact, odds of providing consent are significantly higher for the patients whom a primary care physician has been involved in their medical care, and, as the number of different physicians involved in the care of the patient increases, the odd of providing consent slightly increases (Yaraghi et al., 2015).

Along with the characteristics of technology, patient's characteristics can also influence the persistent use of this same technology. Woods et al. (2017), studying patients who recently completed identity proofing to use the Veterans Affairs patient portal, concluded that portal usage was associated with digital inclusion, access to broadband Internet and digital skills. Tavares and Oliveira (2016) tested an information technology acceptance model for electronic health records portals. They concluded that habit is the only factor that significantly explains both intention and use.

Esmaeilzadeh and Sambasivan (2017) reviewed literature on patient's support for health information exchange, essential to guarantee the success of apps for EPHR. They summarize the main explicative factors as: perceived pros and cons; the type of health information to be shared and identity of recipients; patient characteristics, patient participation level and preferences regarding consent and features. Similar factors were presented on what influences patients to engage and interact with their clinical data online (Crameri et al., 2020). Demographic characteristics, patient's perceptions and patient's empowerment are key areas to understand patient's engagement and interaction with their clinical data online. However, the authors alleged that patient's participation on EPHR is still low and the reasons to explain that remain unknown.

The analysis of patients' adherence factors to technology usually starts from the premise that the use of EPHR is positive for health outcomes and the patient has an advantage in the use of this technology and in sharing health information. However, one of the main factors shown is the patient's perception on the usefulness, reliability and/or safety of the tools used. Therefore, the gain (if it exists) is not certain for the patients. Therefore, if persistent use of EPHR involves time-consuming data entry and risk of sharing sensitive data, the benefits must clearly overcome the costs. The hypothesis of paying for patient's use of the records had not been subject of analysis.

3. MULTI-SIDED MARKETS IN E-HEALTH

The transformation in healthcare services may take place by using e-health models which have the potential for enabling better access to healthcare services while generating efficiency gains, revenue increases, cost-savings, and increased service quality in healthcare (Mettler and Eurich, 2012).

E-health models present characteristics of two-sided markets or, more generically, multi-sided markets (Kuziemsky and Vimarlund, 2018). In multi-sided markets a platform facilitates the interaction between two or more groups of participants who need each other to generate value from the interaction (Schmalensee and Evans, 2007). Therefore, value creation for one participant depends on the presence of other participants, on one hand, and, on the other hand, the value created is higher the greater the number of users in the platform. Moreover, the theory on multi-sided markets focuses on the fact that participants do not internalize the welfare impact of their platform use on other participants (Rochet and Tirole, 2006). Platform participation thus generates strong externalities for its members.

These network effects and network externalities are essential for the sustainability of this market structure and for healthcare markets based on platforms. Hence, participants' adhesion is vital for the success of e-health business models. As stated by Jia et al. (2019: 6) "the more users a network has, the more valuable that network will be to each user". Economic value is created by the interactions or transactions between users, measured by the platform.

Besides network effects and network externalities, another key feature in multi-sided markets is the price structure. In effect, in multi-sided e-health markets the price structure is crucial for participation and influences interactions between platform participants (Vimarlund and Mettler, 2017). The price structure refers to the decomposition or total price level allocation between the buyer and the seller.

Like the price level, the price structure also influences profits and economic efficiency (Rochet and Tirole, 2006). Moreover, economic efficiency may be improved by charging more to one side of the market and less to the other side, that is, if the platform can cross-subsidize between different groups of users that take part in the transaction (Rochet and Tirole, 2003).

Overall, in markets characterized by strong network effects, like multi-sided markets, users' adherence and participation is vital for value creation and value capture (Song et al., 2018). E-health platforms should be able to attract different groups of users to be successful and that is why platforms devote particular attention to its business model.

4. THE INCENTIVE MODEL FOR HEALTH DATA PROVIDERS

In most e-health models, personal health information or EPHR are the foundation for the provision of these healthcare services. As a subsegment of e-health, m-health platforms collect health data and can share this information among healthcare providers, researchers, and patients, among others. But we are often dealing with multiple platforms (publicly or privately owned) from different healthcare providers which poses major concerns about the compatibility and the interoperability of these platforms. Moreover, literature shows that health care providers do not have an incentive to implement interoperable EPHR' systems (Ozdemir et al., 2009; Stephanie and Sharma, 2018).

And this is where the patient can play a central role. If we move the ownership of patient electronic health records to the patient - to whom the information ultimately belongs to – via cloud data sharing systems, the patient will become responsible for gathering, organizing, and sharing his/her personal health information on a platform. By doing so, we would obtain a patient centred system with integrated personal health records (Parente, 2021).

Patient centred m-health platforms can collect and combine information on patients' health but also on lifestyle and/or well-being. This information could be relevant not only to healthcare service providers to assess patient's risk and monitorization of medical treatment, but this information could also be important to explore how the patient-specific risk map can be used to evaluate health costs throughout the patient's life. The latter can be of importance for insurance companies (for insurance cots estimation), for the pharmaceutical industry, for researchers, for public policy planners and for advertisers (nutrition, well-being, etc.). For instance, for pharmaceutical companies, patient health records combined with big data analytics can provide valuable information about how medicines perform in the real world, and it can be a step forward from traditional and expensive clinical trials.

We need to have in mind that these are sensitive data, that belong to the patient. The ultimate decision on providing and sharing personal health records should be of the patient. In addition, issues such as trust, security and transparency of m-health platforms are vital.

The success of m-health platforms relies heavily on the quantity and quality of the data supplied by healthcare services but, mainly, by patients who own it. In patient centred platforms, patient participation is crucial for success and incentives are key to participation and engagement (Philipson, 2001). Thus, it is essential to define the proper economic incentives for patients to gather, organize and share personal health information.

With this in mind, and since personal health information belongs to the patient, it should be up to the patient to use it, make it available or sell it to whoever he/she chooses to. Health data is valuable for firms and researchers. If gains are generated from its use, the owner of the data should get his/her share of those gains.

Therefore, to overcome the obstacles of patients' adherence to health apps mentioned in section 2, we propose a business model where patients (data owners/providers) are paid for their personal health data whenever the data is used by firms or researchers. Patients could be paid through direct payments (e.g., pharmaceutical companies), rewards (e.g., reduced insurance premium from insurance companies) or in-app rewards (e.g., vouchers, access to certain services). A patient centred platform with a business model that offers this type of incentives for data owners/providers would allow for a sustainable e-health ecosystem.

5. CONCLUSION

The discussion of this article is a starting point for a new approach to business models for EPHR apps. We try to show that the obstacles to patients' adherence to these apps could be overcome with the appropriate business model. In multi-sided markets the correct sharing of earnings is essential for the adding of all parties and, in this case, this sharing must include patients, as these provide an essential element to the network: data.

ACKNOWLEDGEMENT

This work was funded by the project "NORTE-01-0145-FEDER-000045", supported by Northern Portugal Regional Operational Programme (Norte2020), under the Portugal 2020 Partnership Agreement, through the European Regional Development Fund (FEDER).

REFERENCES

Abd-alrazaq, A. A. et al, 2019. Factors that affect the use of electronic personal health records among patients: a systematic review. *International journal of medical informatics*, 126, pp 164-175.

Crameri, K.-A. et al., 2020. Personal electronic healthcare records: What influences consumers to engage with their clinical data online? A literature review. *Health Information Management Journal*, 51(1), pp 3-12.

Esmaeilzadeh, P., and Sambasivan, M., 2017. Patients' support for health information exchange: a literature review and classification of key factors. *BMC Medical Informatics and Decision Making*, 17 (33), pp 1-21.

Jia, X. et al., 2019. An analysis of multi-sided platform research over the past three decades: framework and discussion. *MIT Sloan Working Paper 5891-19*. Cambridge, MA: MIT Sloan School of Management.

Kuziemsky, C., and Vimarlund, V., 2018. Multi-sided markets for transforming healthcare service delivery. *Studies in Health Technology and Informatics*, 247, pp 626-630.

Liu, C. et al., 2013. Patients' acceptance towards a web-based personal health record system: an empirical study in Taiwan. *International Journal of Environmental Research and Public Health*, 10 (10), pp 5191-5208.

Mettler, T., and Eurich, M., 2012. A "design-pattern"-based approach for analysing e-health business models. *Health Policy and Technology*, 1(2), pp 77-85.

Ozdemir, Z.D. et al., 2009. Adoption of electronic and personal health records: an economic analysis. *AMCIS 2009 Proceedings*.755.

Parente, S.T., 2021. Health information technology and financing's next frontier: the potential of medical banking. *Business Economics,* 44(19, pp 41-50.

Philipson, T., 2001. Data markets, missing data, and incentive pay. *Econometrica*, 69(4), pp 1091-1111.

Rochet, J-C., and Tirole, J., 2003. Platform competition in two-sided markets. *Journal of the European Economic Association*, 1(4), pp 990-1029.

Rochet, J-C., and Tirole, J., 2006. Two-sided markets: a progress report. *RAND Journal of Economics*, 37(3), pp 645-667.

Shareef, M.A. et al., 2014. Predicting mobile health adoption behaviour: A demand side perspective. *Journal of Customer Behaviour*, 13 (3), pp 187-205.

Song, P. et al., 2018. The ecosystem of software platform: a study of asymmetric cross-side network effects and platform governance. *MIS Quarterly*, 42(1), pp 121-142.

Stephanie, L. and Sharma, R., 2018. Modelling digital and value flows in e-health: a game-theoretic analysis. *CONF-IRM 2018 Proceedings*. 27.

Tavares, J., and Oliveira, T., 2016. Electronic Health Record Patient Portal Adoption by Health Care Consumers: An Acceptance Model and Survey. *Journal of Medical Internet Research*, 18 (3), e5069.

Vimarlund, V., and Mettler, T., 2017. Introduction to the ecosystem for two-sided markets, barriers and facilitators. In *E-Health Two-Sided Markets, Implementation and Business Models*, pp 3-14. Elsevier Press.

Woods, S.S. et al., 2017. The Association of Patient Factors, Digital Access, and Online Behavior on Sustained Patient Portal Use: A Prospective Cohort of Enrolled Users. *Journal of Medical Internet Research*, 19 (10), e7895.

Yaraghi, N. et al., 2015. Drivers of information disclosure on health information exchange platforms: insights from an exploratory empirical study. *Journal of the American Medical Informatics Association*, 22 (6), pp 1183-1186.